Heritage Building Information Modelling

Building Information Modelling (BIM) is being debated, tested and implemented wherever you look across the built environment sector. This book is about Heritage Building Information Modelling (HBIM), which necessarily differs from the commonplace applications of BIM to new construction.

Where BIM is being used, the focus is still very much on design and construction. However, its use as an operational and management tool for existing buildings, particularly heritage buildings, is lagging behind.

The first of its kind, this book aims to clearly define the scope for HBIM and present cutting-edge research findings alongside international case studies, before outlining challenges for the future of HBIM research and practice.

After an extensive introduction to HBIM, the core themes of the book are arranged into four parts:

- Restoration philosophies in practice
- Data capture and visualisation for maintenance and repair
- Building performance
- Stakeholder engagement

This book will be a key reference for built environment practitioners, researchers, academics and students engaged in BIM, HBIM, building energy modelling, building surveying, facilities management and heritage conservation more widely.

Yusuf Arayici is a Professor of Civil Engineering and the Dean of the Faculty of Engineering at Hasan Kalyoncu University, Gaziantep, Turkey, and his research spans from BIM and process modelling to sustainable urban regeneration and heritage modelling through research and enterprise projects.

John Counsell is an academic and Head of Postgraduate Studies at Cardiff School of Art and Design, Cardiff Metropolitan University, Cardiff, UK. He has expertise in conservation, Heritage-BIM (HBIM), BIM, GIS, architectural technologies, building pathologies, RFID and dynamic monitoring of building performance using sensors.

Lamine Mahdjoubi is Professor of ICT at the Department of Architecture and the Built Environment, UWE, Bristol, UK, where he leads the interdisciplinary Building Information Modelling Research Group.

Gehan Nagy is a Researcher in the Faculty of Engineering, the British University, Cairo, Egypt.

Soheir Hawas is Professor of Architecture at the Department of Architecture and Urban Planning, Cairo University, Egypt.

Khaled Dewidar is Professor of History and Theories of Architecture at the British University, Cairo, Egypt, and he is the Associate Dean for Teaching and Learning at the Faculty of Engineering.

Heritage Building Information Modelling

Edited by Yusuf Arayici, John Counsell,
Lamine Mahdjoubi, Gehan Nagy, Soheir Hawas
and Khaled Dewidar

LONDON AND NEW YORK

First published 2017
by Routledge
2 Park Square, Milton Park, Abingdon, Oxon OX14 4RN

and by Routledge
711 Third Avenue, New York, NY 10017

Routledge is an imprint of the Taylor & Francis Group, an Informa business

© 2017 selection and editorial matter, Yusuf Arayici, John Counsell, Lamine Mahdjoubi, Gehan Nagy, Soheir Hawas and Khaled Dewidar; individual chapters, the contributors

The right of Yusuf Arayici, John Counsell, Lamine Mahdjoubi, Gehan Nagy, Soheir Hawas and Khaled Dewidar to be identified as the authors of the editorial material, and of the authors for their individual chapters, has been asserted in accordance with sections 77 and 78 of the Copyright, Designs and Patents Act 1988.

All rights reserved. No part of this book may be reprinted or reproduced or utilised in any form or by any electronic, mechanical, or other means, now known or hereafter invented, including photocopying and recording, or in any information storage or retrieval system, without permission in writing from the publishers.

Trademark notice: Product or corporate names may be trademarks or registered trademarks, and are used only for identification and explanation without intent to infringe.

British Library Cataloguing-in-Publication Data
A catalogue record for this book is available from the British Library

Library of Congress Cataloging-in-Publication Data
A catalog record for this book has been requested

ISBN: 978-1-138-64568-4 (hbk)
ISBN: 978-1-315-62801-1 (ebk)

Typeset in Bembo
by Apex CoVantage, LLC

Printed in the United Kingdom
by Henry Ling Limited

Contents

	List of contributors	vii
1	Introduction JOHN COUNSELL AND YUSUF ARAYICI	1
2	It's BIM – but not as we know it! JOHN EDWARDS	6
3	What are the goals of HBIM? JOHN COUNSELL AND TIM TAYLOR	15
4	Heritage and time: mapping what is not there DAVID LITTLEFIELD	32
5	From history to heritage: Viollet-le-Duc's concept on historic preservation KHALED DEWIDAR	45
6	Integrating value map with Building Information Modeling approach for documenting historic buildings in Egypt SOHEIR HAWAS AND MOHAMED MARZOUK	50
7	Capturing heritage data with 3D laser scanners ANTONY PIDDUCK	61
8	Evaluation of historic masonry: towards greater objectivity and efficiency ENRIQUE VALERO, FRÉDÉRIC BOSCHÉ, ALAN FORSTER, LYN WILSON AND ALICK LESLIE	75
9	HBIM applications in Egyptian heritage sites YASMINE SABRY HEGAZI	102

10	Planning of sustainable bridges using building information modeling MOHAMED MARZOUK AND MOHAMED HISHAM	114
11	Jeddah Heritage Building Information Modelling (JHBIM) AHMAD BAIK AND JAN BOEHM	133
12	Algorithmic approaches to BIM modelling from reality EBENHAESER JOUBERT AND YUSUF ARAYICI	154
13	HBIM and environmental simulation: possibilities and challenges HUSAM BAKR KHALIL	190
14	Green BIM in heritage building: integrating Building Energy Models (BEM) with Building Information Modeling (BIM) for sustainable retrofit of heritage buildings LAILA KHODEIR	203
15	HBIM, a case study perspective for building performance YUSUF ARAYICI	218
16	From LiDAR data towards HBIM for energy analysis LUCÍA DÍAZ-VILARIÑO, PAWEL BOGUSLAWSKI, MIGUEL AZENHA, LAMINE MAHDJOUBI, PAULO B. LOURENÇO AND PEDRO ARIAS	224
17	Participatory sensing for community engagement with HBIM JOHN COUNSELL AND GEHAN NAGY	242
18	Development of OntEIR framework to support heritage clients SHADAN DWAIRI AND LAMINE MAHDJOUBI	257
19	Conclusion YUSUF ARAYICI, JOHN COUNSELL AND LAMINE MAHDJOUBI	269
	Index	275

List of contributors

Editors' biographies

Yusuf Arayici is a professor in Hasan Kalyoncu University in Gaziantep, Turkey, and his research spans from BIM and process modelling to sustainable urban regeneration and Heritage Building Information Modelling through research and enterprise projects. He built up his academic career mainly in the School of the Built Environment at the University of Salford, UK. He undertook research in user requirements capture, process modelling and BIM. He graduated 11 PhD students and authored over 70 publications, including journal and conference papers, articles and monographs, book chapters and two books, titled *Requirements Engineering for Computer Integrated Environments in Construction* and *Building Information Modelling*.

John Counsell is an academic and head of postgraduate studies in Cardiff School of Art and Design of Cardiff Metropolitan University. He has expertise in Conservation, Heritage-BIM (HBIM), BIM, GIS, Architectural Technologies, Building Pathologies, RFID and dynamic monitoring of building performance using sensors. John Counsell is an MSc Architect RIBA ACIAT and originally a conservation architect and one of the early adopters of 3D digital modelling and simulation of buildings from 1985. He has significant experience of CAD and BIM in architectural and conservation practice and 20 years of experience of research into BIM/GIS and web-based collaborative tools, leading to over 40 publications.

Khaled Dewidar is professor of history and theories of architecture in the British University of Cairo, and he is currently associate dean for teaching and learning of the Faculty of Engineering. He was the head of the Architecture Department from 2009–2014. Prior to 2009, Prof. Dewidar was a professor at Ain Shams University, where he led research teams documenting heritage and analysing historic buildings. He has supervised 12 PhD degrees as well as 49 masters' theses in this domain – architecture, heritage documentation, ICT in architectural practice, history and theories of architecture, architectural design, building techniques and building construction, green architecture and sustainable development.

Soheir Hawas is a professor of architecture in the Department of Architecture and Urban Planning of Cairo University. She was the secretary of the Council of the Department of Architecture from 1998–2005. She is the founder and former chair of the Research Studies & Politics Department at the National Organization for Urban Harmony (NOUH), Ministry of Culture, Egypt. Prof. Hawas is a leading heritage researcher and has supervised some 65 PhD and masters' theses.

Lamine Mahdjoubi is professor of ICT in the Built Environment, UWE, Bristol. He leads the interdisciplinary Building Information Modelling Research Group, at the Department of Architecture and the Built Environment. His research spans the social, health and environmental aspects of the built environment. He has extensive experience in developing novel computer-based simulation techniques and designing decision-support tools to examine user-built environment interaction. He has presented to various conferences and seminars on BIM and safety and risk management. He is co-organising the current UK workshop series for practitioners on "Exploring the Impact of BIM on Health and Safety". He is a member of the board of International Making Cities Livable (USA), visiting professor at Robert Gordon University and chief external examiner at Heriot-Watt University.

Gehan Nagy has worked in several research projects and taught several architectural courses in a number of Egyptian and UK schools of architecture during her doctoral and postdoctoral studies. She then worked in a number of private higher education institutions in Egypt, where she taught many architectural courses. In 2012 she helped in developing the curriculum and establishing the Architectural Department at one of the higher educational intuitions. Dr Nagy is an experienced researcher into the impacts of the built environment on people (including on health and human bio energy), ecosystems and passive technologies of local heritage.

Chapter authors' biographies

Pedro Arias is a professor at the University of Vigo. He got his PhD from the University of Vigo, where he developed photogrammetric methods for the documentation of historic agro-industrial constructions. His current research covers a wide range of applications of geotechnologies to civil and building engineering, such as monitoring and intelligent management of infrastructures and energy efficiency of historic buildings.

Miguel Azenha is an assistant professor at the Department of Civil Engineering of the University of Minho in Guimarães, Portugal. He is a full member of the Institute in Sustainability and Innovation in Structural Engineering. He is experienced in three main fields: (i) multi-physics simulation of the overall behaviour of construction materials, based on binders such as cement or lime; (ii) development and application of new experimental techniques for material characterisation of cement-based materials, namely in terms of thermal and mechanical properties; (iii) Building Information Modelling, particularly in concern to a direct contact between the academic and industrial environment for technology transfer.

Ahmad Baik is an architect and a lecturer at King Abdul-Aziz University (KAU) and researcher at University College London (UCL) who specialises in Islamic Architecture and Historic Buildings "Documentation and Management", since 2003. Baik's PhD thesis is about using Building Information Modelling (BIM) to document and manage the historic buildings in Saudi Arabia. Baik has published numerous papers about the subject of Jeddah Historical Building Information Modelling (JHBIM).

Jan Boehm's research focuses on photogrammetry, image understanding and robotics. With his background in computer science he wants to bridge the remaining gap between photogrammetry and computer vision. The latter provides key components to increased productivity in the geomatic processing pipeline. In past projects he already successfully leveraged the productivity in terrestrial laser scanning by introducing automation to georeferencing by direct georeferencing, automated registration using intensity features and automated modelling strategies.

Pawel Boguslawski is a researcher in the Faculty of Environment and Technology at the University of the West of England (UWE), United Kingdom. His research is focused on 3D spatial modelling, data structures and emergency response in the built environment. He has been leading several research projects and supervising postdoc, PhD students and research assistants at Universiti Teknologi Malaysia (UTM), where he was working as senior lecturer in the Department of Geo-information. He got his PhD from the University of Glamorgan in 2011 in the field of GIS, where he worked on data structures for 3D building modelling.

Frédéric Bosché is an international graduate and researcher, having studied and then worked in five different countries over the last 15 years. Since 2011, Frédéric has been a lecturer within the School of Energy, Geoscience, Infrastructure and Society at Heriot-Watt University, Edinburgh (UK). Frédéric has been involved in several research projects in the fields of sensing, data processing, visualisation and analytics in the AEC sector (and beyond). His current and recent fields of interest include automated construction dimensional control using 3D imaging technologies; immersive mixed reality systems; and wearable technologies for health and safety.

Lucía Díaz-Vilariño is a postdoctoral researcher in geomatics, surveying and remote sensing at the University of Vigo. She has been a visiting researcher at several research and academic institutions: the Faculty of Geo-information Science and Earth Observation, Enschede (Netherlands, 2013), the Institute Für Photogrammetry, Stuttgart (Germany, 2014), the Department of Civil Engineering of the University of Minho, Guimarães (Portugal, 2015) and the Department of Architecture and the Built Environment of the University of the West of England, Bristol (UK, 2015).

Shadan Dwairi is currently a PhD student in the University of the West of England, Architecture Department; she has a master's degree in architecture from Jordan University of Science and Technology. Dwairi has nine years of experience as an architect, three of them as senior architect and the head of tender department, in which she managed to land numerous projects in conservation, rehabilitation and restoration; she also worked as a lecturer in the Department of Architecture.

John Edwards was, until April 2014, assistant director of Cadw, the Welsh government's historic environment service. On behalf of Welsh ministers, John was responsible for the management of 129 historic sites in state care, including ICOMOS World Heritage Sites at five locations. Whilst at Cadw John initiated and led the Welsh government's first live energy efficiency research project for traditional buildings, which included HBIM. Prior to Cadw John worked for English Heritage. John is now a director of Edwards Hart, a multidisciplined consultancy practice, where he leads on building conservation, sustainability, energy efficiency, standard setting and training.

Alan Forster is associate professor (senior lecturer) in building conservation, low carbon design and construction technology at Heriot-Watt University, Edinburgh. He is programme leader for MSc in building conservation (technology & management) and has 10 years' industrial experience in building surveying, general construction and building conservation practice. Professionally, Alan is a Fellow of the Chartered Institute of Building (FCIOB); an elected member of the Royal Institution of Chartered Surveyors (RICS) Building Surveying Professional Group (Scotland); a member of the Editorial Advisory Group for the *RICS Conservation Journal* and a member of the Editorial Advisory Board for *Structural Survey: Journal of Building Pathology & Refurbishment*.

Yasmine Sabry Hegazi is a university professor in the Architecture Department of the Faculty of Engineering at Zagazig University, and PhD holder of architecture with a specialisation in world heritage sites management and monitoring. She has 14 years' experience in architectural conservation including on-site supervision, projects management and planning. She is skilled and experienced in producing technical assessments and confident with legal regulations concerning cultural heritage sites. Her PhD was about world heritage site management and monitoring.

Mohamed Hisham is a civil engineer. He received his BSc and MSc in civil engineering from Cairo University in 2007 and 2011, respectively. His expertise has been in the fields of structural engineering, construction management (planning, scheduling and cost control) and execution of different project types such as residential buildings, administrative buildings and industrial projects. His experience covers different phases of projects, including design, planning and construction. His research interests include Building Information Modelling (BIM) applications, cost estimation models, construction equipment management, bridge management and optimisation of construction processes.

Ebenhaeser Joubert is a construction process engineer with 20 years of experience in a varied range of construction and engineering applications. He received a BSc from the University of Pretoria in 2001 and an MSc from the University of Salford in 2015. His interests include coding, linear algebra and robotics. Joubert is currently studying automated vectorising methods for laser scanned data. This growing field of study is closely linked to machine learning and robotics, hence Joubert's continuing interest and research in its algorithmic development.

Husam Bakr Khalil is a professor in the Faculty of Engineering and Architectural Engineering Department in the British University in Egypt. He is the deputy director of the Center of Sustainability & Future Studies (CSFS). Dr Khalil has 30 years of teaching and research experience. He taught at a number of academic institutes in Egypt and abroad, including Helwan University and King Abdul Aziz University. He is currently a professor at British University in Egypt (BUE) and teaches a number of courses including architectural design, computer applications in architecture, 3D Max, environmental simulation and BIM.

Laila Khodeir is an associate professor of architecture who graduated from Ain Shams University in 2002. Her master's degree in 2005 focused on "Social Sustainability", namely "The Impact of Local Communities on Ecolodge Design Criteria", whereas her PhD in 2010 focused on a newly emerging discipline in Egypt at that time, which is "Facility Management". Since then, the research work of Dr. Khodeir has focused generally on facility management, construction project management and the application of management as a broad term in teaching architecture. In her work, she has focused mainly on the importance, impact and value added from applying efficient management practices throughout the whole building life cycle, while shedding light on the local practices in Egypt. In general, the recent research work of Dr. Khodeir tackled four major themes: efficient decision making, construction project management practices, facility and maintenance management of both heritage and new buildings and the application of the principles of management in teaching architecture.

Alick Leslie is a Chartered Geologist with over 20 years' experience of the analysis of building materials. He joined Historic Scotland (now Historic Environment Scotland) in 2013 as Conservation Science Manager, with a small team of scientists working to provide analysis and interpretation of HES's Properties in Care. In addition to analysis at PICs the team is involved in collaborative research with various research initiatives in the UK and are involved in co-supervision of 12 PhD projects. The team are also involved in several European projects.

David Littlefield is a senior lecturer in the Department of Architecture and the Built Environment at the University of the West of England, Bristol, UK, where he teaches advanced cultural studies and interior architecture. David has authored, or made major contributions to, more than a dozen books on architecture and cities, including *Architectural Voices: Listening to Old Buildings* (Wiley 2007), *Space Craft: Developments in Architectural Computing* (RIBA 2008) and *London (Re)generation* (Wiley 2012). He curated the V&A exhibition "Unseen Hands: 100 Years of Structural Engineering" (2008) and co-organised the international conference "Transgression" on behalf of the Architectural Humanities Research Association (2013). David has worked as an artist in residence at the Roman baths, in Bath, England. He is chair for Research & Enterprise for the Interior Educators network and co-runs the research network Estranged Space.

Paulo B. Lourenço is a full professor and head of the Structural Group at University of Minho. His research domains include non-destructive testing, advanced experimental and numerical analysis, innovative strengthening techniques and earthquake engineering, with a focus on masonry and timber. He worked as consultant in more than 75 monuments in Portugal and abroad. He supervised more than 40 PhD and 50 MSc students. He is editor of the International Journal of Architectural Heritage, coordinator of the European MSc in Structural Analysis of Monuments and Historical Constructions and Convener of the revision of Eurocode 6.

Mohamed Marzouk is a professor of construction engineering and management at the Department of Structural Engineering, Faculty of Engineering, Cairo University. He received his BSc and MSc in Civil Engineering from Cairo University in 1995 and 1997, respectively. He received his PhD from Concordia University in 2002. His research interest includes computer simulation, optimisation of construction processes, Building Information Modelling (BIM), fuzzy logic and its applications in construction, risk analysis, sustainability and decision analysis.

Anthony Pidduck is a Senior Lecturer working within Property Management and Development in Nottingham Trent University, focused on Building Surveying in Design, Building Services and Sustainability. Antony's special interest is in Heritage and Conservation including retrofit and recording. Antony started out in building services with a specialism in lighting design before returning to NTU to complete an MA in The Designed Environment. Antony has worked as a Building Surveyor and with an architectural practice leading refurbishment projects and lighting design. Antony works on sustainability in buildings through the IsBET group. His key area of interest is the use of BIM in Heritage and Conservation recording and interpretation.

Tim Taylor works at Atkins Global. His research interests are focused on two principal areas: building physics and building forensics. After completing an undergraduate degree in engineering at Cambridge University, he subsequently worked with the Building Research Establishment and Loughborough University on the development of a certification scheme for the responsible sourcing of construction products. The scheme aimed to address ethical and environmental standards in material supply chains. He completed his doctorate at Cardiff Metropolitan University, working in collaboration with a social housing developer based in Swansea (Coastal Housing Group). Through the research project, he investigated how non-destructive thermal imaging surveys could be used to assess the thermal performance of new housing at different stages of the building construction process.

Enrique Valero is a research associate at Heriot-Watt University. After having studied and worked in the Universidad de Castilla-La Mancha and Universidad Nacional de Educacion a Distancia (Spain), he has secured research visits in prestigious research departments at Carnegie Mellon University (USA) and Katholieke Universiteit Leuven (Belgium), working on the acquisition and analysis of 3D data in the construction field. In 2014, he joined the School of Energy, Geoscience, Infrastructure and Society at Heriot-Watt University in Edinburgh (UK). His research interests include 3D data analysis, object modelling and representation, laser scanning, RFID technologies and health and safety in construction.

Lyn Wilson is the Digital Documentation Manager at Historic Environment Scotland, and also manager of the Centre for Digital Documentation and Visualisation partnership with The Glasgow School of Art, which delivered the Scottish Ten Project (www.scottishten.org). Her primary areas of interest lie in the scientific application of 3D digital documentation within the historic environment, emerging technologies in heritage science and heritage-BIM applications. She is a passionate advocate of applied science and technology within the heritage sector. A heritage scientist with 20 years' experience, Lyn gained her B.Sc. (Hons) in Archaeology from the University of Glasgow in 1997. She received an MA (1999) and PhD (2004) from the University of Bradford, specializing in Computational Archaeological Science.

1 Introduction

John Counsell and Yusuf Arayici

1.1 Introduction

This book is about Heritage Building Information Modelling (HBIM). This is a term that has only begun to be used in the latter part of the last decade, since Building Information Modelling (BIM) superseded 3D digital modelling and computer aided design (CAD) as the term generally used to describe the use of information and communication technology (ICT) for the design, construction, and procurement of the modern built environment. It is sometimes also defined as Historic Building Information Modelling, a somewhat narrower term (Murphy et al 2009). One of the earliest definitions of the principles and purpose of Heritage Information Modelling stemmed from the Getty Conservation Institute's Recording, Documentation, & Information Management (RecorDIM) Initiative (2003–2007) (Eppich & Chabbi 2007).

This book was inspired by a workshop held in March 2015 in Luxor, Egypt, entitled "Trends in Heritage Focused Building Information Modelling and Collaboration for Sustainability", supported by the British Council's Newton Fund and the Egyptian government's Science and Technological Development Fund. The workshop was attended by a number of Egyptian and UK researchers in the field of HBIM. The book provides a critical analysis of current HBIM development and research. Building Information Modelling is still variously defined and described, and so it is that current approaches that are claimed to be HBIM display equal or greater diversity. This introductory chapter first discusses the origins of HBIM, then outlines the format of the book and gives a context to that format. It will be of use to all those who care for our built heritage and seek better tools with which to record its value, clarify its significance, and conserve it in the long term.

In preparing for the workshop on HBIM in March 2015 in Luxor, Egypt, the authors claimed a need for "analysis of the long-term potential for sustainable redeployment and reuse of Heritage". They argued that "increasing resource scarcities require improved analytical tools for conserving existing buildings in general, and mitigating climate change; and that Heritage buildings form a particular challenge, due to the need to conserve their historic and aesthetic worth, addressing environmental social and economic sustainability." These pillars of sustainability are not fixed; perceptions of quality of life change over time, with resulting dissatisfaction with heritage structures and a view that they have become unfit for their purpose. The authors also claimed that the "Heritage Buildings that are most relevant and at risk were not so much unoccupied national or internationally important monuments, but the cultural backdrop of occupied architecture that: has historic and aesthetic value; demands particular maintenance and refurbishment skills and analysis; deploys expensive materials that may be in increasing shortage; and generally cannot affordably be replaced with better performing new constructions."

1.2 HBIM in context

A UK government (2012) publication on BIM states that it "is the first truly global digital construction technology and will soon be deployed in every country in the world". A Cabinet Office publication (2012) further defined BIM as "the process of generating and managing information about a built asset over its whole life". Nevertheless, there is a strong focus on new construction (and at best major refurbishment) in these definitions. In predicting the future goal of level 3 of BIM, the UK government (2015) refined their definition as "a collaborative way of working, underpinned by the digital technologies which unlock more efficient methods of designing, delivering, and maintaining physical built assets. BIM embeds key product and asset data in a 3D computer model that can be used for effective management of information throughout an assets lifecycle – from earliest concept through to operation." This aspirational view of BIM focuses on the widest participative use by all possible stakeholders, throughout the lifecycle of the built asset. Yet an NBS report stressed, "Where BIM is being used, the focus is still very much on design and construction, with use as an operational and management tool for buildings lagging behind" (NBS 2013).

By contrast with 'new construction BIM' where the assets do not yet exist, a key question for HBIM is "Why model when the structure can be experienced in reality?" It may therefore be more appropriate to consider the need for a detailed 3D reference system to be "used for effective management of information about the heritage asset". This could then also support information overlaid via augmented reality tools. Yet much of relevance in understanding and valuing heritage is not fully visible or accessible, whether through previous destruction, reconstruction, or re-concealment, for which modelling may be considered appropriate. Uses change over time for much of our built heritage, and the repurposing that then takes place may now also justify modelling to fully test and explore the impacts and the sustainability of the resources required before implementation. Chapters in this book address each of these premises.

1.3 Progress to date

It seems appropriate, in a book on heritage and BIM, to briefly outline the antecedents of the 3D digital information modelling of heritage. Ivan Sutherland (1963) is often attributed with the initiation of CAD in 1963, although his 'Sketchpad' programme was 2D rather than 3D. One of the authors first saw 'fly-rounds' of 3D digital building models, fully lit and colour rendered by both Intergraph and Calcomp, around 1980. However, the emphasis for design and construction remained focused on 2D digital replications of drawing boards, since at the time these 3D modelling approaches were not commercially viable, taking some three weeks to rewrite the program to change the location of a window. The potential commercial impact of 3D digital models of heritage arrived in the mid 1980s. For example, the 1986 JP Morgan Bank Headquarters in London, designed by BDP, used 3D hidden-line-based depictions of the listed City of London Boys' School on the Thames frontage to examine the impact of the new construction on its context, and a similar approach to examine the new building's impact on the city of London Heights, devised to protect key views of St Paul's Cathedral. These could both be explored interactively, but in wireframe graphics only. By the mid-1990s the first major implementations of virtual reality were taking place, such as the 1995 interactive full-colour VR reconstruction of the UNESCO World Heritage Site Lascaux Caves, closed to the public due to damage from earlier visitor numbers (Britton 1998).

Worthing and Counsell (1999) described "issues arising from computer-based recording of heritage sites" focused upon the 1996 modelling of the Tower of London in its central London context. These visualisations were used by Historic Royal Palaces and the Tower Environs Partnership in order to serve technical studies to underpin proposed enhancement of the surroundings of the

Tower of London. However, the 1995 study that gave impetus for that modelling (outlined in the paper) was that Historic Royal Palaces wanted visualisable output from a '3D Geographic Information System (3D GIS)' that merged their Oracle Text-based maintenance management system with their newly commissioned highly detailed digital survey drawings, commissioned following the disastrous Hampton Court fire of 1986.

It may therefore be considered that practical and affordable HBIM was only realisable from the mid-1990s. Ogleby (1995) described the process as "much of the input for these systems is produced using CAD, and as a result of the advances mentioned previously much of the input to the CAD models comes from photogrammetry." Ogleby (op. cit.) also affirmed that "work like that of Cooper et al. . . . on the Tomb of Christ in Jerusalem, for example, would not have been possible even 10 years ago". The Cyrax time-of-flight laser scanner was patented in 1998 and initiated a new era in high-speed 3D data capture.

In the two decades that followed, there were many research projects focused on digital heritage and digitising heritage. This chapter will review and to an extent categorise the varying approaches that have formed the genesis of HBIM.

Some of these emergent approaches used GIS, including 3D data that could be used to reconstruct heritage visualisations. UNESCO (2002) held one of its 30th anniversary World Heritage conferences at the newly opened library in Alexandria, Egypt, focused on "Heritage Management Mapping, GIS and Multimedia", at which there were a range of digital heritage projects described, including 3D GIS applications for both built and natural heritage (Counsell 2005). UNESCO (2003) then launched its "Charter for the Preservation of Digital Heritage", claiming that "where resources are 'born digital', there is no other format but the digital original" that should be preserved. These digital materials include "texts, databases, still and moving images, audio, graphics, software, and web pages, among a wide and growing range of formats. They are frequently ephemeral, and require purposeful production, maintenance and management to be retained" (ibid.). A later UNESCO (2012) conference went beyond this, using the title of "The Memory of the World in the Digital Age: Digitization and Preservation". This supports an argument for an HBIM deployable 3D reference system for all cultural heritage sites in order to maintain, manage, and view these diverse materials in context, with geospatial analytical tools to filter information or augment reality. In a similar vein, recent calls for wider stakeholder engagement in levels 2 and 3 BIM in the UK call for it to be described as "Digitising the built environment" or "Digital Built Britain" "and not BIM"! (Chawla 2015).

1.4 This book

The main thrust of this book is to identify how emerging and forecast technologies may be used to support better restoration philosophies; enhance data capture and management; widen public engagement; improve energy performance; and plan for and enhance resilience in the face of climate change and natural hazards. The book is comprised of four heritage themes following setting the scene for HBIM, including restoration philosophies in practice; data capture and visualisation for maintenance and repair; stakeholder engagement; and building performance.

In Setting the Scene for HBIM, Chapter 2 takes the viewpoint of a heritage practitioner in defining how HBIM can enhance management activities, while Chapter 3 discusses the goals of HBIM to determine progress to date and progress still required. Within the section on Restoration Philosophies in Practice, Chapters 4, 5, 6, and 7 explore restoration methods and philosophies in the context of HBIM. The chapters in Data Capture and Visualisation for Maintenance and Repair provide insights into data capture and modelling approaches using laser scanning, photogrammetry, and other means for HBIM modelling, visualisation, and heritage maintenance. This then leads into

Building Performance. Chapters 14, 15, and 16 elaborate on the performance modelling of heritage buildings to inform planning to fully test and explore the impacts and outcomes and determine the sustainability of the resources required before implementation. In the final section on Stakeholder Engagement, in Chapters 17 and 18 there is discussion of how the public, local communities, and other stakeholders can be more inclusively engaged in collaborative heritage valuation and management, and how HBIM can facilitate this.

1.5 Conclusion

It is argued that the focus of HBIM should be primarily on realising the value and significance, and supporting the long-term sustainable conservation, of cultural heritage assets in the built environment for all stakeholders. There is little research literature that addresses such broad-scale user engagement in the specifications of or development of HBIM. There are a few published prototypes for Heritage-focused BIM that can be claimed to contrast with these definitions of 'new construction BIM', but there is little widespread use. Equally, there is little research literature that addresses the significantly differing requirements of Heritage-focused BIM from new construction BIM. It is argued, therefore, that this is an area worthy of study.

References

Britton, B.J. (1998). 'LASCAUX Virtual Reality Project'. Online at: www.hamiltonarts.net/lascaux.html.
Cabinet Office (2012). 'Government Construction Trial Projects July 2012'. Online at: www.cabinetoffice.gov.uk, www.gov.uk/government/uploads/system/uploads/attachment_data/file/62628/Trial-Projects-July-2012.pdf.
Chawla, R. (2015). 'Let's Talk about Digitising the Built Environment'. Online at: www.bimplus.co.uk/people/lets-talk-about-digitising-built-environment/.
Counsell, J. (2005). 'An approach to adding value while recording historic gardens and landscapes' Chapter 19 (pp. 175–185) in 'Digital Applications for Cultural and Heritage Institutions' eds. Hemsley J.R., Cappelini V., Stanke G. Ashgate, Aldershot, England. ISBN: 0–7546–3359 4.
Eppich, Rand, and Chabbi, Amel (eds.) (2007). 'Recording, Documentation and Information Management for the Conservation of Heritage Places: Illustrated Examples'. Los Angeles, CA: Getty Conservation Institute. Online at: http://hdl.handle.net/10020/gci_pubs/recordim_vol2.
Murphy, M., McGovern, E., & Pavia, S. (2009). 'Historic building information modelling (HBIM)'. *Structural Survey Journal*, 27(4), 311–327.
NBS (2013). 'NBS International BIM Report 2013'. Online at: www.thenbs.com/~/media/files/pdf/nbs-international-bim-report_2013.pdf.
Ogleby, C.L. (June 1995). 'Advances in the digital recording of cultural monuments'. *ISPRS Journal of Photogrammetry and Remote Sensing*, 50(3), 8–19.
Sutherland, I. (1963). 'Sketchpad, a Man-Machine Graphical Communication System'. MIT. Online at: http://images.designworldonline.com.s3.amazonaws.com/CADhistory/Sketchpad_A_Man-Machine_Graphical_Communication_System_Jan63.pdf.
UK Government (2012). 'Industrial strategy: government and industry in partnership: building information modelling'. HM Government, URN 12/1327 published by Department for Business Innovation & Skills. Online at: www.bis.gov.uk, www.gov.uk/government/uploads/system/uploads/attachment_data/file/34710/12–1327-building-information-modelling.pdf.
UK Government (2015). 'Digital built Britain level 3 building information modelling – strategic plan'. URN BIS/15/155 published by Department for Business Innovation & Skills. Online at: www.bis.gov.uk, www.gov.uk/government/uploads/system/uploads/attachment_data/file/410096/bis-15–155-digital-built-britain-level-3-strategy.pdf.
UNESCO (2002). 'Heritage Management Mapping, GIS and Multimedia'. whc.unesco.org/document/9202 & Markaz al-Qawmī li-Tawthīq al-Turāth al-Ḥaḍārī wa-al-Ṭabī'ī 'Heritage Management Mapping: GIS and

Multimedia, Alexandria, Egypt, October 21–23, 2002'. Egyptian Center for Documentation of Cultural and Natural Heritage.

UNESCO (2003). 'Concept of Digital Heritage'. Online at: www.unesco.org/new/en/communication-and-information/access-to-knowledge/preservation-of-documentary-heritage/digital-heritage/concept-of-digital-heritage/.

UNESCO (2012). 'The Memory of the World in the Digital Age: Digitization and Preservation'. Online at: www.unesco.org/new/en/communication-and-information/events/calendar-of-events/events-websites/the-memory-of-the-world-in-the-digital-age-digitization-and-preservation/.

Worthing, D., & Counsell, J. (1999). 'Issues arising from computer-based recording of heritage sites'. *Structural Survey Journal*, 17(4), 200–210.

2 It's BIM – but not as we know it!

John Edwards

Most of the focus for Building Information Modelling (BIM) is on new construction, with BIM protocols being developed to make the construction process more efficient. A BIM protocol for existing buildings might end up as being very similar to new construction, and this chapter considers that this might not be the most appropriate way of approaching BIM for existing buildings. The public debate on BIM for existing buildings is often confusing and on occasions lacks a clear vision on end objectives. Heritage BIM (HBIM) would exclude the greater proportion of existing buildings, and only if there is a groundswell of support for using BIM on existing buildings will it become commonplace. It is suggested that a pure focus on HBIM will put it into a niche and delay its take-up for the majority of existing buildings.

This chapter will suggest that the first question should be: What do we want to achieve? This could be a very accurate digital model with the most detailed information on significance, technical data, thermal data, building services and so on. On the other hand, it could be that a most accurate digital model is not required, and something that describes locations with maintenance and use issues will suffice. However, the latter can hardly be described as an HBIM.

In order to help develop a structure for an end user focused BIM for existing buildings, it is proposed that it should be developed in two broad areas. Existing Building Information Modelling (EBIM) contains the basics required to maintain and operate a building which includes data on building fabric and building services, and an HBIM is an additional layer. The HBIM would also contain historical and heritage data, significance values, conservation policies and perhaps a much more enhanced form of digitisation, but this depends on what the overall objective is.

EBIM and HBIM are ultimately about managing and maintaining buildings properly and efficiently, and whether it is an EBIM or an HBIM depends on whether the buildings have any heritage value or not. The nearest we come to describing this in accordance with current thinking is to call it BIM for FM (facilities management), and descriptions for EBIM and HBIM above would naturally sit within FM. Ideally, E/HBIM should also be suitable for energy and sustainability along with refurbishment and retrofit, but this depends on the extent and quality of the data within the E/HBIM. Most importantly, the E/HBIM should permit the continuous enhancement of data input.

2.1 Introduction

From a UK perspective, best practice in the management and conservation of historic and traditional buildings is contained in the latest British Standard BS 7913, published on 31 December 2013 (1). This details the two essential ingredients of technicalities and significance, along with appropriate processes which include surveys, management and supervision. BS 7913 in essence details what should be done and how it should be done as well as detailing the competencies required by those involved. Building Information Modelling (BIM) or Building Information Modelling for heritage buildings (HBIM) will not necessarily deliver the best practice described in BS 7913. The goal of HBIM, however, should be to put BS 7913 into practice. From a non-UK perspective, BS 7913 travels well, but there will be other home-grown standards for application in other countries and there is also international guidance such as that contained in ICOMOS (International Council on Monuments and Sites) charters.

Whilst ensuring that best practice is followed and deployed, one must question whether the current concept of BIM will ever achieve this. Most of the focus for BIM is on new construction, with BIM protocols being developed to make the construction process more efficient. In the UK there will be a requirement for centrally publically funded projects to be developed using BIM sometime during 2016, which will drive up the number of projects adopting BIM. Here the emphasis is on projects to level 2 BIM. Let's be clear about what this means. These are the BIM maturity levels as described in the Strategy Paper for the UK Government Construction Client Group in 2011 (2):

- Level 0: Unmanaged CAD, probably 2D, with paper (or electronic paper) as the most likely data exchange mechanism.
- Level 1: Managed CAD in 2D or 3D format using BS 1192: 2007 with a collaboration tool providing a common data environment and possibly some standard data structures and formats. Commercial data managed by stand-alone finance and cost management packages with no integration.
- Level 2: Managed 3D environment held in separate 'BIM(M)' tools with attached data. Commercial data managed by an ERP (enterprise resource planning) system. Integration on the basis of proprietary interfaces or bespoke middleware could be regarded as 'pBIM' proprietary. The approach may utilise 4D programme data and 5D cost elements.
- Level 3: Fully open process and data integration enabled by IFC/IFD (Industry Foundation Classes/International Framework for Dictionaries). Managed by a collaborative model server. Could be regarded as iBIM or integrated BIM(M) potentially employing concurrent engineering processes.

Working up to level 2 means all parties working collaboratively with their own 3D CAD models, but not necessarily working on a single, shared model. The collaboration comes in the form of how the information is exchanged between different parties – and is the crucial aspect of this level. Design information is shared through a common file format, which enables any organisation to be able to combine that data with their own in order to make a federated BIM model and to carry out interrogative checks on it. Hence any CAD software that each party used must be capable of exporting to one of the common file formats such as IFC (Industry Foundation Class) (3) or COBie (Construction Operations Building Information Exchange) (4). Level 3 means all parties working on one model.

There is believed to be relatively little activity in the UK beyond level 3 – level 4 involving timing and programming with level 5 also including cost information and that completes the 'construction phase'. Level 6 is for Facilities Management and therefore using BIM in the everyday life of a building and taking BIM beyond the construction phase.

The emphasis of BIM, whilst being described as taking a project from cradle to grave, is in fact about projects. Most of the wonderful examples of BIM used for existing buildings are because of project work. Part of Edinburgh Castle and the Engine Shed conservation centre development in Stirling, both by Historic Scotland, are two fine examples.

It is perfectly clear that when BIM reaches level 6, the benefits of BIM are many times greater than just bringing BIM up to level 5. There are relatively few new-build construction projects using BIM up to level 6. This means that buildings being constructed today or in the near future will not reap the benefits that BIM level 6 will bring to the management of the building. Despite this, the heritage sector has been flirting with the idea of BIM for heritage projects and seems to be automatically following the current BIM protocols and levels designed for the construction of new buildings. Work by the Council on Training in Architectural Conservation (COTAC) within their BIM 4 Heritage Group has highlighted the problems faced by adopting the current BIM model (5).

Projects involving work to a heritage building will need a fairly accurate 3D model – probably derived from laser scanning, but there are alternatives, which may not be as accurate. As a 'project' it may seem very sensible to adopt BIM protocols at level 2 or 3. However, in comparative terms, there are not that many 'heritage projects' in the construction industry, so there will not be much HBIM taking place. Indeed, if one considers the amount of new construction taking place compared to work on existing infrastructure, then one can see that new construction is probably just over half the UK construction industry output; and of new construction, only a proportion of it, the larger projects and those publically funded, will see BIM used in the next few years. BIM will therefore be of active interest to a minority within the construction industry. This isn't particularly good for an industry that may greatly benefit from the more effective and efficient ways of working that BIM should deliver.

The answer may be to have some focus on BIM for existing buildings now in terms of their management, maintenance, use and improvement before waiting for BIM level 6 to take hold as part of developing 'projects'. Note the emphasis on existing buildings and not heritage buildings. Heritage buildings for most people are those which receive legal protection, such as those in the UK described as listed buildings and scheduled monuments, and only make up about 2% of the UK building stock.

2.2 Road to HBIM (Heritage BIM) via EBIM (Existing BIM)

The public debate on BIM for existing buildings is often confusing and on occasion lacks a clear vision on end objectives. Heritage BIM would exclude the greater proportion of existing buildings, and only if there is a groundswell of support for using BIM on existing buildings will it become commonplace. It is suggested that a pure focus on HBIM will put it into a niche and delay its take-up for the majority of existing buildings.

However, one must ask whether BIM for existing buildings (we could call this EBIM) is worthwhile, and the first thing to explore is why we would need it. Where projects that involve major improvements and refurbishment are concerned, there may be a strong case to justify the costs involved in laser scanning and production of a 3D model. However, the majority of buildings will not go through such major improvements and refurbishment, and one must ask whether there is any merit in producing an EBIM for these buildings.

Most buildings will not even have CAD drawings or perhaps any drawings at all, and a model of some sort would have to be produced. The quality and type of model should depend upon the reason for the EBIM. If it will be used for assisting in the development of a building then laser scanning to produce the 3D model may be essential, but if not then perhaps 3D outline drawings rendered with images where appropriate may suffice. Alternatively, there is a low-cost method of producing a 3D model with Autodesk 123 Catch photo survey software (6) and freely available Trimble SketchUp software (7).

Before considering the above, the case would need to be made for producing an EBIM. Some of the benefits are clearly understood from looking at the benefits of BIM for projects: all information in one place, the location of building services, information on components and materials. This can be built upon with information on maintenance, servicing, spare parts, cleaning and much more. One can see that there are many benefits, but there are also costs, such as the financial cost of producing an EBIM.

Again, the form of the model will be very important. If laser scanning is undertaken, then one must consider whether an accurate 3D model is really required. For many historic buildings that don't have vertical walls, horizontal ceilings and flat surfaces, the time involved in developing such a model may be excessive and too costly for some. Cadw, the Welsh government's historic environment service, set up a pilot project involving the production of an HBIM at Heritage Cottage.

Heritage Cottage is a small terraced house located in Cwmdare, South Wales. It was built in 1854 and is almost in its original state. It is the site of Cadw's first live energy efficiency research project, also including the development of a 3D model in Revit through laser scanning and the development of an HBIM. Even though this very modest cottage would seem to be fairly regular in shape and size, the laser scanning revealed many irregularities, resulting in a lot of work to mimic an exact replica of the building in the 3D model. It is, of course, not necessary to have this degree of accuracy purely for an HBIM, but from an archive recording perspective, an accurate model would be necessary (8).

Some organisations see the merit in BIM for the management and operation of buildings – for example, the Government Service Administration (GSA) of the US government. This organisation already manages its property in a professional manner with all relevant data in place within their 'computerized maintenance and management system (CMMS)' (9) They see the advantage in the relevant data produced from a BIM developed for construction being automatically input into the CMMS. They also see the BIM for facilities management being linked to the CMMS, as well as other systems, for visual coordination of facility assets with O&M (operations and management) data.

However, there are already technological solutions to optimise the management and operation of buildings. CAFM (computer-aided facilities management) software can help plan the utilisation of space, plan preventative maintenance, organise reactive maintenance, standardise services, and streamline processes and align this to the service requirements of building occupiers and against budgets (9) – in which case the added benefits of EBIM have to be sought. As stated above, the EBIM enables visualisation, and the GSA cites many benefits. Here we focus on some of these (9):

> It provides GSA Maintenance Workers with the most accurate information on field conditions and maintenance information before leaving the office, which saves time. It can provide the most accurate equipment inventory and mobile access to the BIM/EBIM and other linked and integrated data in the field provides access to all documentation without making trips back to the office.

There are also advantages for GSA building operators, and this includes reducing risk and uncertainty of performing work orders by identifying building components that are not otherwise easily identified. EBIM also maintains links to equipment histories, facilitating proper equipment condition assessments. Building performance is optimised by comparing actual to predicted energy performance.

As for the occupiers of buildings, it increases their satisfaction due to quicker resolutions to unscheduled work orders and also reduces unscheduled work orders and improves communication between tenants and building maintenance workers.

The GSA considers that these are real benefits brought about by what is, in the context of this discussion, EBIM. In summary, EBIM should enable the following when it comes to the management and operation of a building:

- cleaning methods, materials and programming
- planning and managing use

- maintenance methods, materials, components and programming
- building services locations and stop valves and switches
- energy use
- work specifications

Many organisations, however, have not progressed to the stage where the GSA is, and have not advanced to the stage where they have CAFM. Indeed, the author's research into management practices of a range of UK heritage and non-heritage organisations suggests (10) that relatively few have fully co-ordinated computerised property management systems and some do not have any computerised systems other than tracking through the compilation of information on spreadsheets. EBIM/HBIM developed in an appropriate form would no doubt aid improved property management, but one must question whether this is the most cost-effective means of achieving this. Such organisations could already be operating using CAFM, but do not.

At Heritage Cottage, Cadw also envisages information on the energy insulation qualities of the building fabric such as walls and windows to be input into the HBIM. This is only possible because tests and analysis have been undertaken to provide real information, rather than theoretical information (11). This means that, should proposals be developed for wall insulation or secondary glazing to be installed, the HBIM can be used to assess the effects on the building fabric and on the use of energy. Providing the information is available to be input into an EBIM/HBIM, then there should be no reason why this shouldn't be possible in all existing buildings.

This 'green' element takes the normal BIM or EBIM/HBIM a stage further and could be called GBIM, but it shouldn't be seen this way as an add-on, as green issues should be seen as normal practices. A BIM of a new building in design should be developed with environmental and energy performance simulated with different weather conditions and building occupancy to create an optimum building that is sustainable, energy efficient and provides both comfort and healthy conditions for the building's occupants. For sustainability purposes this should ideally be passively ventilated wherever practically possible. This would require the most accurate information input into the model and thermal information to include the thermal performance of the building envelope, the performance of plant and equipment plus a proper understanding of how spaces will be occupied and the implications of this.

This should also be the case with existing buildings, where simulated modelling is of equal importance. However, if this is to be undertaken properly, then just relying on the laser scan survey would be a mistake. A condition survey is necessary to enable information on repair works to be developed. It is well known that buildings don't perform in the way they were intended, and this affects all types of buildings of different ages. Where commercial office buildings are concerned, the London Better Buildings Partnership found that most modern office buildings of a similar age generally have fairly equal performance (12). A study by Leeds Metropolitan University established that newly built UK housing generally performs about 10% less efficiently than predicted (13), whilst many studies into traditional housing (built pre-1919) indicate that they perform better than expected (14).

This all points towards the necessity to undertake tests and analysis to properly understand the performance of a building and to therefore properly predict performance when simulating the effects of changes or improvements with an EBIM. At Heritage Cottage for example, Cadw undertook in situ U-value tests, air leakage tests and thermal co-heating tests. This means that the HBIM can contain real information which is as accurate as possible. It also means improvements can be simulated, including predicted effects on energy consumption.

In summary, the HBIM at Heritage Cottage, based on a 3D model developed from laser scanning, included:

- building fabric and components make-up and materials, including finishes and where possible their source/supplier
- thermal performance and use of energy
- building service locations and routes
- planned building maintenance
- process for reactive maintenance
- specification for repair and maintenance works, including data sheets and videos
- competencies required

All of the above could be part of an HBIM for any type of building, so why call it an HBIM? These are all issues pertinent to the maintenance, management and operation of any type of building of any age, and therefore when containing these types of ingredients, it is not an HBIM but an EBIM. It is only when one adds in the additional layer of cultural heritage information that it becomes an HBIM. At Heritage Cottage the cultural elements include:

- cultural significance of elements and groups of elements
- issues coming out of the 'Statement of Significance'/historic significance analysis
- data for authentic replication (e.g. mortars)
- particular processes, methods, etc. for authenticity
- recording requirements
- sensitivity issues concerning use–protection measures
- decision making/process requirements (e.g. Heritage Impact Assessments)

If Heritage Cottage had a conservation management plan, like many historic buildings have, then the conservation policies it contains would also be included.

It should be conceivable to produce a base model for EBIM, HBIM and GBIM. They could all have a standard set of ingredients, some of which would be essential and others essential but only if available and which could be included if and when they do become available. Such BIMs should include details of what is necessary to manage and operate buildings properly, and this should be derived from acknowledged best practice guidance. For example, where HBIM is concerned in the UK, this would be the latest British Standard for the conservation of historic buildings, BS 7913: 2013: Guide to the Conservation of Historic Buildings, and this is looked at in greater detail later.

Many may argue that existing buildings do not require EBIM, HBIM or GBIM and as buildings are adequately cared for without such a tool to manage and co-ordinate information. The author would argue that many buildings are not managed and maintained properly. This may be due to a lack of knowledge and expertise or a lack of resources and more commonly to being reliant upon third parties who do not have the most appropriate expertise.

Applying EBIM, HBIM or GBIM will not automatically save buildings from disrepair and neglect, but it would, providing it's the right type of BIM developed with appropriate expertise and competency, mean using the most appropriate knowledge and resources to ensure that what is done is done properly.

There are many examples where inappropriate work is undertaken to traditional and historic buildings. Using cement in renders and repointing mortars, replacing suspended timber floors with concrete, replacing original timber windows with modern uPVC and installing chemical damp-proof courses when they are not required are just a few that affect traditional buildings and to a lesser degree legally protected historic buildings. All such 'technical' information that leads to the correct decisions being taken could be contained in an EBIM, and information on the merits or otherwise of replacing components such as windows would be contained in a GBIM. The historic significance of the window and the historic inappropriateness of cement could be contained in the HBIM. These are all, in effect, three BIM layers in a comprehensive BIM for a heritage building and called HBIM.

Such a comprehensive concept would in effect provide 'guidance' by being included within the HBIM. It would be a checklist of all the things that should be considered and could also include things like instruction sheets on how, for example, mortar joints should be repointed, and supported by video and references to other sources of information.

As indicated at the beginning of this chapter, from a UK perspective, BS 7913: 2013 is acknowledged to be the most authoritative UK guide on the conservation and management of historic and traditional buildings, and any HBIM should set out to ensure that BS 7913 is put into practice. It provides the ideal framework on which to base the standard application of HBIM applied to traditional and historic buildings. It would encompass all the issues that one would consider at the base-level EBIM, the 'significance' issues that produce the additional layer for it to be called HBIM as well as those green issues that would be contained in GBIM.

2.3 Conclusion remarks

Here we are principally looking at HBIM from a building management, maintenance and improvement perspective, but it should also be ensured that when a development contains an historic or traditional building, that the attributes of BS 7913 should be included.

Below we describe how the HBIM can be combined with BS 7913 in an HBIM through the application of different activities:

> Meaning of conservation: There is merit in having HBIM developed for this reason alone. If HBIM is developed with common ingredients following BS 7913, then those that deal with the building will have to use the HBIM and therefore be guided by it. Where those that are dealing with a traditional building are genuine experts in historic building conservation, they may argue that this does not provide a value. However, the vast majority that deal with historic buildings are not experts. In the UK conservation accreditation is the mark of an individual with genuine expertise and there are only 793 to serve 570,000 legally protected listed buildings and monuments.

(15)

Understanding significance: HBIM should contain 'significance values'. Reference to section 4 of BS 7913 could be made to illustrate how significance should be assessed.

Managing significance: This is about using significance values in decision making. Section 5.2 of BS 7913 describes heritage management principles and provides some critical 'advice'. For example: The impact of any proposed changed should be justified; justification shall be proportionate; material for repair matching existing but note performance; the principle of minimum intervention (i.e. retention of as much fabric as possible of a building when repair or other intervention is required) is important; need to record interventions; work to be implemented and managed by competent persons.

Section 5.5 of BS 7913 is about using conservation policies within conservation management plans which should be input into the HBIM in terms of every applicable aspect and/or the relevant policy programmes when considering certain activities. For example in redecorating, a policy could be available to indicate the approach or could go further and detail the specification. Section 5.9 of BS 7913 describes the reactive approach to managing significance with heritage impact assessments (HIAs). Here the HBIM can provide the process and the pro forma.

Surveys and inspections: In section 6.2 BS 7913 explains the types of inspections and surveys, when they should be undertaken and often who by as well as the outputs required. This is both useful in developing projects, which could be the reason for which the HBIM is being developed, or in the management and operation of the building. In section 6.3 it focuses on building pathology and building performance, which again is important when developing projects but also in the management of buildings.

Energy efficiency and sustainability: BS 7913 takes a holistic approach to energy efficiency and sustainability, which again is useful when developing a project and in the management of buildings.

Building maintenance and repair: In section 7.2 of BS 7913 it says that significance should be taken into account when developing the maintenance strategy. It also says that historic buildings deserve the best and most appropriate methods and works to be deployed and that sufficient time should be allocated to ensure quality maintenance. In section 7.3 it states: "A maintenance plan should have a cyclical programme. Normally this includes routine daily, weekly and twice yearly activities to keep the fabric clear, and services with other more detailed works on a longer term cycle of one year to five years."

Managing buildings: In section 6.1 of BS 7913 it states: "When managing historic buildings, significance should be taken into account at every stage from the business strategy of the organization that owns or occupies the historic building to physical work activities."

Project management: In section 8.1 of BS 7913 it states: "Where work is to be undertaken to an historic building, project management should be integrated with heritage management." This emphasises that understanding and managing significance is essential and needs to be incorporated into the HBIM. In the same section it also states: "There should be a process in place to ensure all people working on the project are aware of the significance of the historic building." In 5.7 it states: "Major projects require a wide range of skills and resources for planning and implementation. A planning policy framework should be developed at the beginning of the project with clear statements regarding the heritage outputs of the scheme."

Managing quality and appropriateness of work: In section 8.2 of BS 7913 it makes the point that project supervision is a key element in the management of any type of work and that only work which has been undertaken properly should be accepted. It also states that the role of the project supervisor should be defined at the outset and that without robust project supervision from a competent person there is a higher risk of defective work. A standard HBIM needs to build this into management processes for projects and for the management of buildings.

Consent applications: BS 7913 doesn't state what the various consent processes are, but it contains helpful statements and tools in order to make well-considered applications and also to help reject some requirements under building regulations that are potentially damaging to an historic building. A standard HBIM needs to incorporate this.

There are alternatives to following BS 7913. The International Council on Monuments and Sites (ICOMOS) Education and Training Guidelines take one through some of the most essential areas of activity and competencies and can guide one through the areas to be covered in developing a framework for an HBIM to follow best practices (16).

BIM is ultimately about creating the correct and most appropriate information in a satisfactory format for it to be shared by all parties, brought together and managed. If the heritage sector is to reap some very obvious benefits of BIM then it needs to have some independent vision that is realistic, that where appropriate looks for benefits outside the heritage sector and also looks for the means by which BIM can improve the way we deal with historic buildings.

In order to get broader appeal and relevance, the concept of EBIM needs to be promoted as something that could benefit over 20 million UK buildings rather than just focusing on HBIM with what most people understand to be the heritage sector with less than 600,000 buildings.

A standardised HBIM will not just bring about consistency but could be the most fail-safe method of developing projects for historic buildings and in their subsequent management.

As will be noted from the above, however, many aspects of BS 7913 lay down good practice for the care and management of all existing buildings, and in particular the six million traditional buildings in the UK.

References

1. BS 7913: 2013: Guide to the Conservation of Historic Buildings, British Standards Institution, 2013.
2. UK BIM Working Party Strategy Paper, 2011.
3. https://en.wikipedia.org/wiki/Industry_Foundation_Classes.
4. https://en.wikipedia.org/wiki/COBie.
5. BIM 4 Conservation, Council on Training in Architectural Conservation, April 2014.
6. http://www.123dapp.com/howto/catch.
7. http://www.sketchup.com/.
8. http://archive.cyark.org.
9. http://www.gsa.gov/portal/content/122555.
10. National Trust Buildings Review Benchmarking, 2016, Edwards Hart Ltd (not published).
11. RICS Building Conservation Journal, January 2015.
12. http://www.betterbuildingspartnership.co.uk/resources?topic=21.
13. http://www.goodhomes.org.uk/downloads/members/003-performance-gap-release-final.pdf.
14. http://ihbconline.co.uk/newsachive/?p=2565.
15. CIOB Conservation Accreditation Research Report, 2016, Edwards Hart Ltd (not published).
16. http://www.icomos.org/en/charters-and-texts/179-articles-en-francais/ressources/charters-and-standards/187-guidelines-for-education-and-training-in-the-conservation-of-monuments-ensembles-and-sites.

3 What are the goals of HBIM?

John Counsell and Tim Taylor

3.1 Introduction

This chapter outlines some of the challenges in defining HBIM, with the limited recent progress that has been made, while clearly outlining challenges yet to come. Some of the gaps between 'parametric parts library based' commercial BIM and HBIM are discussed. It outlines geospatial information and decision making, and the strengthening relationships between GIS and BIM in the context of heritage buildings, cultural landscapes and virtual and intangible heritage. The chapter starts by indicating the 'demand' for HBIM; proceeds to outline the 'cutting edge' of current practice and challenges still to be met; and ends by analysing some supporting case studies.

3.2 An 'information resource for future generations'

There are few if any current published case studies of sophisticated, well-defined HBIMs. The Sydney Opera House is one of a few emergent case studies. Chris Linning (2014), the Sydney Opera House manager for Building Information, has recently been presenting his view of the Sydney Opera House (SOH) goal of ensuring the effective sustainable conservation and management of the heritage complex for a projected further lifespan of 250 to 300 years. He titles his presentation 'BIM – An Information Resource for Future Generations'. This seems an excellent overarching goal for HBIM – 'an information resource for future generations'. With Schevers et al. (2007), Linning defined the need for an evolutionary approach to establishing a fully formed 'Digital Facility Information Management System'. "Introducing a full scale Facility Management System is hardly feasible or desirable. A more evolving approach is necessary where the digital facility model evolves from a relatively simple information system to a more integrated and knowledge intensive system." As with the goals of the more conventional new construction focused BIMs, they also see the need for the resulting information model to act as a 'single source of truth'. "Ideally, the digital facility model should be the integrated data source for all information systems at SOH. This means that when one information system processes or changes some data, all other systems are aware of that change, eliminating information redundancy." They further point out that "an integrated information model opens up the way for more automated intelligence in the model incorporating rules and best practices."

From that definition of scope and ambition, the SOH surveyed available software and systems (via several partial trial iterations of BIMs from various consultant organisations, together with a recent point cloud survey by the 'Scottish Ten' [2013]). They concluded (McTaggart 2015) that "there was no single BIM solution that could be applied to both development and construction projects as well as ongoing facilities management of the Opera House." SOH has since tendered for a bespoke solution, and consultants are now working on the phased delivery of modules within a framework (Knutt 2015).

3.3 What should be the focus of HBIM?

Building Information Modelling is still variously defined and described, so it is not surprising that HBIM approaches show equal or greater diversity. Kemp (2014) provides a useful summary of the value that lies in BIM as a 'process':

- It converges information production with sound engineering judgement and design
- It provides wider, faster access to comprehensible and integrated information
- It fosters instinctive but rigorous collaboration and better decision making
- It harnesses innovative technologies and harvests intelligence from big data
- It enables reflective, adaptive thinking to incorporate whole life and integrated systems approach within the wider geographic context.

(Kemp 2014)

The American Institute of Architects defined a broad goal for BIM within Integrated Project Delivery (IPD), a mechanism for involving all key participants for optimal results. They stated that IPD "is a collaborative alliance of people, systems, business structures and practices into a process that harnesses the talents and insights of all participants to optimize project results, increase value to the owner, reduce waste, and maximize efficiency through all phases of design, fabrication, and construction" (AIA 2007).

That definition is particularly helpful as a benchmark against which to analyse the goal of HBIM, if all key participants are focused via HBIM on the social, economic, environmental and cultural sustainability of built heritage during its lifetime management and maintenance.

Maxwell (2014) defined the current state of the art in HBIM as 'in the shadows'!

> With conservation, repair and maintenance (CRM) currently amounting to some 42% of all industry activity, the current emphasis on new-build orientated BIM risks leaving related developments in the CRM sector in the shadows. At the heart of an HBIM / CRM approach is a fundamental requirement to establish value, significance and accurately surveyed data of the asset that is anticipated being worked upon. With little progress having been made by BIM (in its currently common accepted sense in accommodating the more difficult world of dealing with long-established existing buildings of many architectural periods, styles and structural compositions), the emergence of a meaningful Historic Building Information Modelling (HBIM) approach is virtually non-existent. In pursuing this comparatively untouched avenue, a detailed and fundamental understanding of existing structural conditions, material degradations, and performance-in-use circumstances is essential for each structure being incorporated into the approach.

3.3.1 Support for primary interpretation

Primary interpretation is defined by ICOMOS (1990) as the expert recording process via which the historic structure is analysed and comprehended. ICOMOS (op. cit.) defined robust mechanisms for 'on-site' expert recording, with the explicit focus of establishing value, significance and analysing materials, structure and pathology. The ICOMOS guidelines (op. cit.) state the "record of a building should be seen as cumulative with each stage adding both to the comprehensiveness of the record and the comprehension of the building that the record makes possible." They conclude that "recording should therefore so far as possible not only illustrate and describe a building but also demonstrate significance." This was summarised (Worthing 1999) as "the recording of an historic building is not a 'one-off' event but a continuous process that is a prerequisite of many management activities. It is important to recognise that recording is not value free. It is part of the process of understanding and interpreting the building – and in therefore attributing value to the structure." Molyneux (1991) stated similarly, "The purpose of analytical recording is not merely to ascertain the initial form of the building but rather how the use of the structure has responded to and reflects social, cultural and economic change through time." Recording should therefore, so far as possible, not only illustrate and describe a building but also demonstrate significance. At any level above the most basic, a heritage record is thus a mixture of description and analysis. There are now multiple media available for data acquisition and entry. More overt and explicit classification and codification of facts is necessary when using computers, but also clarity when evidence is uncertain or capable of multiple interpretations (such as concealed parts of the fabric, pathology not fully visible on the surface, or detectable via non-destructive tests). The chosen media and descriptors will enhance or limit the later use of that data.

3.3.2 Cultural landscapes and intangible heritage as well as heritage structures

Fai and Graham (2011) widen the scope of Heritage BIM beyond the 'expert eye' defined by Maxwell. They similarly identify the limited progress currently towards HBIM. "Despite the widespread adoption of building information modelling (BIM) for the design and lifecycle management of new buildings, very little research has been undertaken to explore the value of BIM in the management of heritage buildings and cultural landscapes." They particularly identify cultural landscapes as well as buildings and structures. The HBIM requirements of Cultural Landscapes are likely to be broader than those of infrastructure in general. However, even BIM for Infrastructure remains poorly developed. "Buildings are much better served than the infrastructure side at present in terms of BIM implementation. . . . Infrastructure is more challenging – more ambiguous, no set boundaries and consistently moving and changing – how do current BIM technologies tackle that?" (Schofield 2015).

3.3.3 Intangible heritage

Fai et al. (op. cit.) describe the need to record performance data, and what they define as qualitative assets, such as historic photographs, oral history, or music, for example. "To that end, we are investigating the construction of BIMs that incorporate both quantitative assets intelligent objects, performance data and qualitative assets historic photographs, oral histories, music." Yet even for performance data Niskanen et al. (2014) state more generally that "the benefits deriving from the integration of static BIM data to dynamic facility monitoring data are not extensively exploited or understood."

UNESCO (2001) gave a broad description of such 'qualitative assets'. They defined 'intangible cultural heritage' as "among others, language, literature, music, drama, dance, mime, games, hunting, fishing and agricultural practices, religious ceremonies, traditional skills in weaving, building and carving, cuisine, extrajudicial methods of dispute resolution, traditional medicine and traditional knowledge applied to plants and their medical, biological and agricultural properties."

UNESCO (2003) then identified in particular the need to record and store (and georeference), for future generations, the emerging palimpsest and marginalia of digital and 'Web 2.0' recordings of heritage sites, such as photographs, tweets, and other 'comments'. "Digital materials include texts, databases, still and moving images, audio, graphics, software and web pages, among a wide and growing range of formats. They are frequently ephemeral, and require purposeful production, maintenance and management to be retained."

These are all potential ingredients in the establishment of ongoing value and significance, and HBIM should enable them to be scrutinised, particularly in establishing how the cultural landscape or heritage artefact is a 'response to and reflection of social, cultural and economic change through time'.

3.3.4 Support for 'secondary interpretation' and 'bottom-up readings'

In a 2006 UK research council workshop on "Preserving Our Past", (EPSRC 2006), the cross-disciplinary experts present confirmed continuing need for research to

> move away from the concept of one-way push of information to (two-way) interactive participation and inclusion. They stress the need to embrace social inclusion, interpretation, storytelling, authenticity, and interactive design. They further enquired: is there space for multiple readings of heritage (alternative interpretations); is there opportunity to present these alternatives; How do we challenge top-down interpretations; Can we encourage bottom-up readings; Can we get away from socio-economic and political readings; Can we accommodate personal and cultural readings; and how can we make heritage more immediate – tactile / auditory / experienced?

3.3.5 An evolutionary fusion towards HBIM

Counsell (2002) argued that BIM processes alone were not enough for the full range of necessary functions in recording and using heritage data, and that GIS processes are also needed. Leading practitioners (Kemp 2015a) now identify that geospatial information (via GIS) is as critical as BIM information (via parts libraries and BIM software), referring to that integrated goal as Geo-BIM. The current state of the art appears to be that GIS is seen as complementary to BIM, but in a separate silo at a different scale, so implicitly 'less detailed' than BIM. "The next big challenge leading to 2020 will be to Geo Enable BIM" (AGI 2015), i.e. to achieve a synthesis where the silos and perceived issues of accurate detail no longer exist. There is still a tendency in this approach to focusing on the building or structure. For heritage such a focus needs to be broader, embracing cultural landscapes and the landscape curtilage of heritage buildings in a more seamless manner. This is more akin to the seamless scaling that should in future occur between BIM building units and Smart City, but the cultural landscape contains many particularly challenging ephemeral elements.

3.3.6 Who are the key participants?

In identifying the future scope for Geo-BIM, Plume (2015) states that we "may have a stewardship responsibility toward a place, or we may be required to design or plan a place, and of course, in our everyday life, we are users of places", going on to ask "what if we were to collect, store and make available in an explicit fashion all that we know about how places are valued, both collectively and personally?" and "If we are able to find a way to collect and capture our understanding of place in a structured way, would that provide a useful conceptual framework for an integrated digitally enabled environment?" This appears to be even more essential for HBIM. The ICOMOS Burra Charter, for example, states that those "with associations with the place as well as those involved in its management should be provided with opportunities to contribute to and participate in identifying and understanding the cultural significance of the place. Where appropriate they should also have opportunities to participate in its conservation and management" (Australia ICOMOS 2013). Silberman (2015) similarly refers to ongoing paradigm shifts as "process, not product; collaboration, not passive instruction; memory community, not heritage audience". There is little research literature that evidences such wide user engagement in the development of HBIM. A notable exception that does engage a broad range of 'users' is identified in the work of Fassi et al. (2015). A consequence of wider and more ad hoc user engagement will be that ease of use of HBIM, particularly by occasional users, will become increasingly critical, probably far more so than for conventional 'new construction' BIM.

3.4 A clear framework for HBIM

ICT systems tend to fail if they are too prescriptively narrow to support the full range of practice, because then users have to develop and deploy approaches beyond that ICT system's scope. Hence it is important that a framework is defined that is sufficiently embracing, even if not all modules are initially implemented. That framework does not appear to yet exist for HBIM. Plume (2015) proposes a BIM-focused "information framework to support the digital enablement of the built and natural environments", a fusion between emergent BuildingSMART and Open Geospatial Consortium approaches. The AGI Foresight Report 2020 (2015) identifies the need for a 'geography of everything':

> Identifying and exploiting the linking of multiple technologies and policies – based strictly on an outcome based approach. This will include BIM, indoor navigation, asset management, internet of things, smart cities, artificial intelligence and big data, integration of data science and analytics, creative visualisation and analytics. Location Intelligence can and should become critical ('the geography of everything') and provide the foundation on which business decisions are made.

Plume goes on to describe

> two technologies that form the foundation of these ideas: the digital models that are used to represent aspects of the physical world; and the Internet technologies that collectively capture, hold, find, interpret and deliver the information [...incorporating] the Internet to transport the information; the semantic Web to enable smart ways to find and retrieve information; geolocation technologies to enable searching based on geographic context; and RFIDs with sensors to facilitate the Internet of Things to realise a sensate environment.

This framework would need to support the full range from cultural heritage landscape to specific structure, including:

- accurately surveyed data with: recording of value, and significance; analysis of materials, structure and pathology; as well as how the use of the structure has responded to and reflects environmental, social, cultural and economic change through time;
- 'development and construction projects as well as ongoing facilities management' (the wider definition of BIM level 2 in the UK);
- the ephemeral and ambiguous, lacking set boundaries, consistently moving and changing, including infrastructure, trees, planting and water courses;
- the 'Internet of Sensors', responsive systems, and related actuators;
- location-specific 'intangible cultural heritage' and virtual heritage;
- secondary interpretation, community engagement and community-based bottom-up readings, and alternative interpretations, based on potentially ambiguous 'fuzzy' evidence;
- a structure and standards that support continual retrieval and reuse for decades if not centuries.

3.5 Some evolutionary stages for HBIM development

In the field of geospatial information, Maguire (1991) defined its development as evolving through three generations: in the first, data are entered to form an inventory; during the second, the continuing addition and updating of data to that inventory over time makes analysis of change possible; in the third, sophisticated spatial analytic tools enable various forms of 'what if?' analysis. Extrapolating on this theme Gardels (1997) defined a fourth evolutionary stage as a 'Web library model' where the data are 'just out there' and are found and deployed using special spatial search tools. Each of these generations supports more refined and broader analysis and decision making. Inventory supports filters and enquiries such as 'what is at . . .?', 'where is . . .?'. In 2D and to some extent in 3D this has become commonplace. OpenStreetMap has 3D prototypes; Google Street View and Google Earth take different approaches to Microsoft Bing in creating 3D cityscapes; and in-vehicle satellite navigation systems contain increasing blends of 2D and 3D. The outstanding challenge is navigation and discovery in multi-level complex built structures. There is also a prevailing lack of open standards and interoperability that keeps these in separate silos.

There are few heritage case studies that have reached the second generation and accessibly show 'what has changed?' The Walton Basin study (CPAT & Smith 2014), one of these few, is now unsupported and may shortly expire. It shows in web-based interactive VR (VRML) the changing appearance of cultural landscapes over millennia, with archaeological simulations of monuments and settlements in the Walton Basin in Wales, from the Late Glacial period to modern times. The study was generated from 3D data held in a GIS (MapInfo), using extensions in the MapInfo API to generate the VRML, so capable of ready amendment and update over time. The third GIS generation, deploying spatial tools for 'what if?' analysis, does not appear significantly in any heritage case studies, although elements appear in a range of funded research projects, and the London Charter (2009) is an implicit critique of much current practice.

The semantic web is predicated upon a more collaborative and interactive engagement by users. Alderson (2015) states that "location is the DNA that runs through all data". However, there is little open standard development to be found for cultural heritage buildings, let alone cultural heritage landscapes, intangible or virtual heritage, which evidences this, and few spatial search tools for 'just out there' data. The DURAARK (2016) project has recently prototyped applications of emerging semantic web standards such as the Resource Description Framework and the OWL web ontology language in order to provide a sound basis for the very long-term archiving, retrieval and reuse of architectural information, in particular point clouds and BIM. Similar approaches are analysed in the 'Linked Heritage' (2013) project report on 'Geocoded Digital Cultural Content'.

3.6 Taking inventory

HBIM research and development appears to be a very early point in an evolutionary tree, primarily focused on 'recording' and even more on 'accurate survey' (e.g. point clouds), which may be seen as taking inventory, the first of Maguire's (op. cit.) developmental GIS stages. Penttilä et al. (2007) retrospectively applied 'architectural information modelling' to the Alvar Aalto School of Architecture in Helsinki, prior to renovation. They concluded that an 'inventory model' was necessary to support the renovation process, as well as later facilities management, and that "the information structure of renovation projects differs remarkably from the structures of new buildings." In particular, unlike new construction where the ingredients are known, much of the detail of construction below the surface is unknown, and difficult to define using non-destructive techniques. Thus there is a need for more effective non-destructive techniques, methods of recording, and analysis of 'building pathology'. In their paper on the relation of recording to digital 3D modelling focused on the buildings in the Tower of London environs, Worthing and Counsell (1999) listed among reasons for recording: to enable the reconstruction of part or the whole of the building (or 'object') in the event of its damage or destruction (also examined in relation to the Welsh National Museum of Life laser scan project [Counsell et al. 2008]); to identify the changes that have occurred to an historic building over a period of time; where the information is used to inform decision making about the effect that proposed changes will have on the historic fabric; as an archival record of that which is inevitably lost or changed (even routine maintenance work can involve loss or change); or as a record of works actually undertaken to capture the chronological evolution of the building.

For HBIM, such (digital) record repositories require tools that support: capture/upload with data quality assurance and validation; storage (repository); retrieval including filters; tools for measurement, archaeometry and other analysis; display (in '3D' not just '2D'); with feedback loops for added semantics, e.g. significance, commentary and narrative. Significantly different data and processes may also be required depending on which outcomes are required. For example, Maxwell (op. cit.) points out that "conservation; restoration; rehabilitation; repair and maintenance [. . .] require different degrees of information and detail about the building, or area to be worked upon. This, in turn, will dictate the level of sophistication and depth of understanding that will be required from the survey activity".

One may also distinguish and perhaps even deploy different recording techniques by distinguishing between data that changes at different paces, from slow and steady to rapid. "While buildings are relatively slow to change and decay, so past records and now computer modeled analogues stay valid in the long term, yet their contexts, settings, gardens and grounds are open to rapid change. Effective recording of potentially rapid change is highly resource intensive, justifying exploration of automated data capture" (Counsell 2001). This is the particular challenge of recording cultural landscapes and heritage curtilages.

3.7 Near real-time data

"In 2016 Planet Labs will have enough satellites in orbit to image the entire globe, every single day" (Planet Labs 2016). While the 3–5 metres per pixel data that will be provided is not precise enough for BIM purposes, it may well be highly useful in detecting change in cultural heritage landscapes. To the developing range of precision remote-sensed data, including that offered by drones in the future, fusion needs to be achieved with the 'Internet of Sensors' or 'Sensor Systems' for near real-time data capture, which also offer great potential for automated analysis. "When you cut through the hype around the Internet of Things (IoT), it's the Internet of Sensors (IoS) and resulting data that matters most" (TTP 2015). "This means using modern technologies to continuously monitor the condition and operation of infrastructure and to intervene before problems arise and to develop better solutions for the future" (AGI Foresight Report 2020, 2015). At the Luxor workshop, real-time sensor data collections relating to Heritage Cottage (Cadw) and other case studies were described, and questions posed about the future relationship of HBIM to 'Smart Cities'. 'Once there are sensors, real-time data, and consequential analysis, will an added feedback loop and actuators or the like make a heritage building or structure into an automatically responsive (i.e. smart) building, and is this a necessary step towards developing a fully smart city?' For example, Kolokotsa et al. (2011) argue that "integrated control and optimization tools of sufficient generality using the sensor inputs and the thermal models can take intelligent decisions, in almost real-time, regarding the operation of the building and its subsystems." It may be argued that data automatically available from such 'Smart Heritage' and via 'participatory sensing' (see Chapter 17) may eventually lead us to Gardels's fourth stage of HBIM, where the data are 'just out there', often now described as 'Big Data', and where the challenges become those of automated filtering and analysis. This is similar to the "Internet of

Places" described by Plume (2015), but with the wider dimensions (including intangible heritage) needed for a fully developed HBIM. The developing trend towards this fusion is described as "cyber-physical" (Bosché 2016). Bosché further defined this for BIM as more than 'linked digital and physical elements', in that changes in the digital will be automatically reflected in the physical and vice versa.

3.8 Semantic information

The current state of the art appears to be substantially achieving accurate geometric surveys of heritage assets. It does not appear to offer effective approaches to complementing this with robust methods of establishing value and significance, or enhancing with semantic information. One of the authors has been informally discussing this with BIM software focused heritage architecture practitioners across Europe and beyond. Now many BIM software focused heritage architecture practitioners segregate the recording process, to its apparent detriment. In this current practice the laser scanner operator attends site, focused on achieving a fully comprehensive sequence of integrable 3D point cloud scans. The scans are then processed back at the office, off-site, usually merged into a single coherent combined point cloud survey. Errors in accuracy occur via calibration issues, the quantities of scans, the distance from the scanner, and even operator errors. For ease of use in the BIM software, the merged point cloud may well then be extensively thinned or culled. Experts describe, for example, using one-ninth of the original points, with consequent reductions in data quality, i.e. points at an original distance accuracy of up to ±2 mm are discarded prior to modelling. The BIM software is used to create a 3D parametric model that closely matches the thinned point cloud via a manual tracing process, but may well further simplify it.

In these recent discussions these practitioners explain that the modelling process helps them to analyse the materials, structure and pathology, and may assist in assigning value and significance to key elements. The benefit of an expert studying the physical reality and carefully noting each detail of value, significance, material form, structural condition, and potential pathology in situ is devolved. These BIM software heritage practitioners now do this off-site from a simplified partial geometric point cloud record, usually with RGB (red-green-blue) values for each point. This represents a substantial degradation of the on-site expert analytical recording process. A few are instead arguing for greater ease of use of a mobile in situ ICT process, an amalgam of automated geometric recording, mobile instrumentation and logging, enhanced by simultaneous expert semantic analysis and annotation, to digitally speed and enhance that qualitative manual recording process defined by ICOMOS (op. cit.).

3.8.1 The question can be asked, "How much to model?"

A related question that is relevant to the evolutionary process identified for the Sydney Opera House is: Why model everything rather than just what is required at the time? Models are often significant simplifications of the actual heritage building, with some consequential major challenges for data quality and assurance. In this respect Hichri et al. (2013) state that

> historical buildings are characterized by very complex and varied shapes, mostly not responding to classical geometrical laws. For example, walls are not always vertical and can be tilted in many cases. Some elements are even more complex such as capitals which have specific characteristics and different architectural styles. Modeling them becomes even harder because of their deterioration over time.

As with new construction focused BIM, the models might be formed from discrete elements, but due to the complexity and deterioration of many structures, the separation of the structure into such components risks even greater simplification. There is a recognised need in BIM for tagging with semantic information, and such components appear to many to be the only currently practicable approach to tagging with what is described as "non-graphical" information in many software solutions. Thus Murphy et al. (2009) focused on matching parametric elements from Georgian architecture in Dublin to laser point cloud survey data. This approach has been particularly challenged by Maxwell (op. cit.), who stated that

> its limitation in not having the precise individual architectural details recorded, as opposed to utilising a standardised catalogue of library based features, need to be borne in mind. As a result, the approach is perhaps best suited to streetscape and urban planning considerations, rather than building specific conservation requirements where an accuracy in understanding the building details is important.

Yet it becomes valuable to model when:

a   the actual building/artefact is not accessible in the present – whether covered up, demolished or destroyed, dangerous to access or subsequently changed;
b   for virtual access, providing an overview or comprehension is not achievable on the ground; that is for simplification of complexity (e.g. the UK National Trust applied unsuccessfully for millennium lottery funding to create interpretative visitor centres at the entrance to a number of their buildings, in order to provide the visitors with a comprehendible overview before immersion in the maze of rooms and levels thereafter);
c   for visual pattern matching, "both kinds of sources, original input and enhanced imagery, can be draped onto the 3D models to analyse thoroughly the whole object in three dimensions" (Lerma et al. 2011);
d   for analysis of historic construction processes (e.g. Guedelon 2015, Reeves & Paardekooper 2014);
e   for analysis of performance, for example, thermal/energy, structural, pathological;
f   for determining methods of reassembly, for example, following earthquake destruction, or anastylosis (Canciani et al. 2013b);
g   for predicting the future reconstruction, repurposing or, for example, predicting the future effects of climate change.

3.8.2 Models composed of (parametric) objects

There is still debate about the extent to which parametric product libraries of building components, the basis of most current commercial BIM packages, can be deployed for HBIM. There are still different schools of thought about the extent to which parametric modelling should be aligned to laser scan surveys. Murphy et al. (2009) began with a parametric architectural object best-fit approach to point cloud data. Yet Murphy more recently concludes (Dore and Murphy 2012) that there is a need to blend their best-fit 3D models into 3D GIS in order to enhance the process of tagging with semantic information, documentation and analysis. There is still much research to do to optimise the addition of semantic information to point cloud data, and no clear agreement on the best way forward. Automated approaches have been tried, but yet have often less than 90% success, or lead to gross simplification of the shape and form of the actual heritage structure, thus are seen by some as denying the history of the artefact over time, including the dimensional distortions that emerge as aging and settlement take place.

3.8.3 Minimum stratigraphic units – a form of 'cookie cutting'

BuildingSMART in the UK reports that "we generally found that while there is a good understanding of the benefits of 3D modelling, this is not converted into a deeper technical understanding of how to exploit all digital data as the lowest and reusable common denominator of information, beyond just the graphical output" (Kemp 2015b).

A more precise approach than adding non-graphical attributes to parametric object representations of deteriorated heritage materials and forms is a form of disaggregation into distinct elements that has been proposed based upon archaeological stratigraphic units. This appears to be the 3D equivalent of the 2D spatial selection 'data-clipping' tools used in GIS termed 'cookie cutters'. Canciani et al. (2013a) describe their prototype for heritage buildings, stating that "one of the crucial problems is the definition of 'base units', meaning base elements presenting homogenous characteristics, for the association of further data . . . the current study resolving the problem of the base unit by instead defining a minimum stratigraphic unit."

Much of the time and cost in modelling from point clouds is taken in painstakingly replicating the visual appearance of the building in segmented elements, before 'discarding' the point cloud. The point cloud data with associated RGB colour is already a very effective visual representation. The challenge remains how to segment it and associate semantic information with it. The EU-funded DURAARK (2016) project and subsequent investigation suggest that the time-consuming process of detailed modelling could in future be reduced by a much simpler, even crude, semantic BIM model (partially automatically generated) that is more loosely linked to the scanned point cloud data but serves as a filter to retrieve point cloud data.

3.8.4 Automated segmentation and semantic labelling of point cloud data

Other researchers point out how much can be achieved with laser point clouds combined with photographic and similar non-destructive testing records for analysis and understanding, without transformation into 'simplified models'. Emerging technologies such as CloudCompare (n.d.) permit new repair/refurbishment solutions to be checked for fit against point clouds without full digital modelling, although many seek to use the same 'clash detection' process to validate the accuracy of their fully constructed modelling against the original point cloud data. Bosché et al. (2015), in exploring automated analytical tools for masonry structures, point out that "point clouds (or generated meshes) are just 'raw' data, and the information truly valuable to experts needs to be extracted from that data, through its segmentation and semantic labelling." Bosché et al. report on development and testing of their algorithm that detects each individual masonry surface and distinguishes it from the adjacent mortar bonding surface. On a broader scale, the Felis Research Centre (Wang et al. 2008, Weinecker 2008) developed software that can automatically, without operator intervention, distinguish between evergreen and broadleaf trees, and extract simplified coplanar surfaces of building exteriors with reasonable accuracy from high-resolution aerial Lidar data. The DURAARK (2016) project has prototyped the automatic extraction and semantic identification of the planar surfaces of rooms (floors, walls, ceilings), with detection of doorways to relate one room to the next. Detailed models resulting from amalgamation of these approaches could support geo-tagged semantic information overlays, without deploying parametric constructional elements that make overprecise assumptions about the invisible construction deployed behind the surfaces.

3.8.5 Geo-tagged 'comments and narrative'

It should be noted that a visualisable model is not necessarily required in this process, and a case needs to be made if one is to be constructed primarily to support georeferencing. In their development of the Non-Domestic Building Stock GIS (solely a 2D GIS), Steadman et al. (2000) successfully captured, via street and internal surveys, adequate descriptions both 2D and 3D, with semantic information about construction occupancy and use, for subsequent analysis of energy efficiency, without recourse to a visualisable '3D model'. The Sydney Opera House as a facility was managed effectively for its first decades by a text-based inventory colour coded to elements of the building. Now augmented reality techniques might provide the same functionality.

Fai et al. (2011) used Navisworks commercial software to fuse data from a variety of sources, to geo-tag semantic labelling and to test the georeferencing of intangible heritage to their 3D digital modelled heritage reconstructions. Fai went on to state, "If the 'facts' change, the model can integrate this new material without having to be completely rebuilt." This does not however enable the user to determine the 'level of certainty' or ambiguity of data underlying semantic labelling, nor permit different and even conflicting interpretations supported by the same evidence. Both of these situations are likely in heritage, since it is often not possible to tell what lies below the surface without demolition.

While it does not yet sufficiently meet the wide range of needs for semantic heritage 'labelling' and narratives, there is now a more open source approach developing that may with extension and development effectively support semantic labelling, via the new construction focused BIM Collaboration Format (BCF). BuildingSMART (2015) has now released the second generation of BCF as a standard. Many of the concepts behind BCF were explored during the Virtual Environmental Planning (VEPs) project to fully address full community-based participation online, described as Comment Markup Language (CoML) (Schill et al. 2007). CoML was partially built upon an earlier heritage-focused georeferencing landscape prototype, explored in the Framework 5 EU project Valhalla. An interface was

> developed that permits archived video clips to be retrieved following a search and viewed alongside the model. The model viewpoint can then interactively be matched to the view or views in the archive clip to generate metadata about field of view, zoom and comments with which to tag the clip, for later search and retrieval by context or content. The same approach has also been implemented for static digital images, thus enabling scanned historic images or digital photographs to be incorporated in the database and retrieved using the same search parameters. Thus images that were not originally annotated with locational data can be rendered retrievable by searching the database for instances of plants or other objects.
>
> (Counsell et al. 2003)

Some of the cross-disciplinary concerns expressed in the 2006 UK research councils' workshop (EPSRC 2006) still have currency. Among potential digital solutions is a requirement for tools that enable open ad hoc semantic labelling of a variety of media in ways potentially not anticipated by ICT system designers. One may, for example, need to distinguish between semantic labelling that is planned by the author of a document, and post-publication labelling that is more difficult to predict. 'Hot-spot' tools would be needed to draw attention to each nexus where multiple 'labellers' have commented or expressed interest or views. Tools would also be needed to locate in sequence or conduct an animated tour of a series of related 'labels' with their associated views that might form a narrative, or even develop further, akin to a Web 2.0 'mashup'. Labels might therefore need to be audible. These aspects were also trialled in the development of CoML.

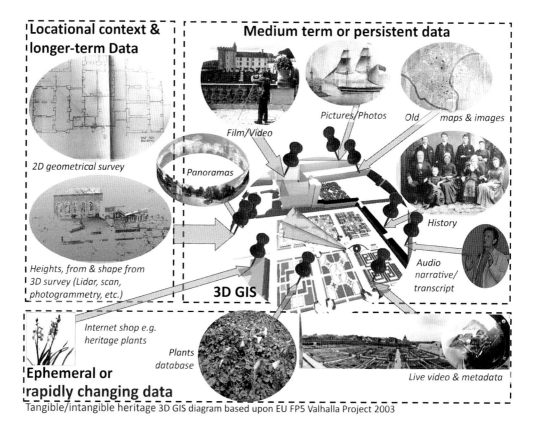

Figure 3.1 Tangible/intangible heritage 3D GIS diagram based upon EU FP5 Valhalla Project 2003.

3.9 Some recent indicators

The EU-funded Framework 7 DURAARK (2016) project, which has just ended, mapped 'Durable Architectural Knowledge', and particularly the long-term preservation of BIM models. The site contains a wide range of reports, with links to open source resources and tools that were developed within the project that the project partners hope will instigate a longer-term open source community development. Although not specifically focused on heritage, it defined a range of standards and approaches that would be valuable for heritage platforms or frameworks, particularly in addressing the substantive challenge of enabling an information resource for future generations without reentry of data or rework.

The Valhalla project (Counsell et al. 2003) remains one of a very few that have focused on the 3D GIS-based information needs of capturing and recording the diurnal and seasonal changes of planting in heritage landscapes. It used earlier versions of the GIS linked to VRML software later deployed for the Walton Basin (CPAT & Smith 2014) to blend streamed real-time data from security cameras with navigable web-based VR models. This approach was further enhanced in the VEPs project, where a case study was made of St Teilo's Church, rescued and reconstructed at the National Museum of Welsh Life. The church was virtually placed back into its original site (fusing Lidar data

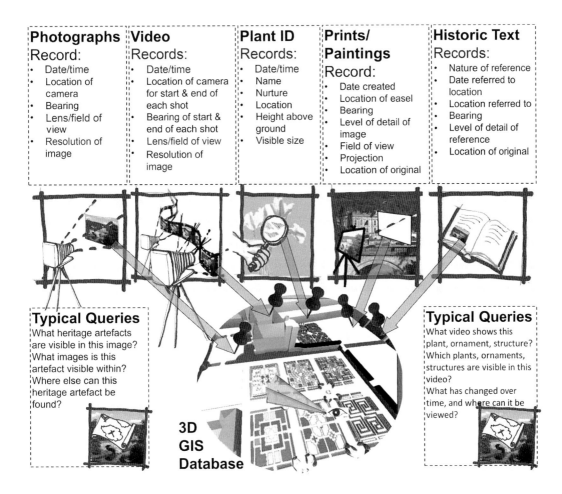

Figure 3.2 Recording and geo-tagging for retrospective data entry, based upon Valhalla Project 2003.

of the site with point cloud data from the reconstructed building), and a timeline tool deployed that allowed online users to explore back through time to its original medieval single cell structure, with adjacent motte and bailey castles and villages (Counsell & Littlewood 2008). While the focus of the VEPs project was community engagement, independent evaluation revealed major ease-of-use issues for all three prototype web technologies trialled. Richman and Holding (2008) found the "perception of how 'user-friendly' a 3D tool is depends very much on the experience a user has with 3D applications." Their report on VEPs evaluation revealed significant further research challenges, particularly regarding ease of use.

As a multidisciplinary design, engineering and project management consultancy, Atkins has extensive experience of applying BIM and geospatial techniques at all scales, from the restoration of important historical buildings up to major infrastructure projects and cultural landscapes. Significant projects, such as the development of a site management plan for the Bahla Fort and Oasis World Heritage Site and master plan for the National Botanic Reserve in Oman, can involve both extensive recording and modelling activities that are embedded in participatory processes. Collaboration platforms and visualisation tools are increasingly essential elements for HBIM practitioners, and Atkins has a strategic plan to migrate all the company's systems to the cloud by 2020 as well as trialling innovative technologies such as Microsoft HoloLens to enable virtual data to be experienced. Other innovations by Atkins in this area include the Digital Imaging for Condition Asset Management (DIFCAM 2013) platform that enables rapid acquisition of digital data for a structure that can be used for automating condition monitoring processes. This type of technology is inherently efficient, accurate and safe and may indicate the future for preventative maintenance of infrastructure with historical and cultural significance, i.e. it provides the capability for an asset to report its state to management systems that learn and adapt in response.

Batawa (Fai et al. 2011) is a recent case study of an early twentieth century factory and worker housing complex in Canada. The case study claims to have successfully prototyped the integration of accurate survey data with quantitative and qualitative virtual and intangible heritage, and the visual display of change over time. It demonstrated significant barriers: multiple expensive software packages were used, each requiring significant expertise. Indeed, without support from Autodesk Research, its limited success would probably not have been achieved. It successfully made a case for further funding intended to develop a far more accessible web development, in a project that appears to have ended in 2014. They describe the project, "Cultural Diversity and Material Imagination in Canadian Architecture (CDMICA)", as focused on "knowledge creation and mobilization for both tangible and intangible conservation, documentation, and digital archiving". However, the material now accessible on the web from this project (CDMICA 2014) is a limited gazetteer of some heritage cases, without heritage case-specific timeline (although there is a listing of initial construction dates for each case), georeferencing or intangible heritage. It provides highly limited interaction with 3D models of each case, if users are prepared to wait for the massive files generated by similar expensive 3D modelling software to download. When judged against the criteria of inclusive engagement of and ease of use by the stakeholder community, it does not display progress, and indicates a need for a research portal that collates development towards HBIM to avoid these regular reinventions of the wheel.

3.10 Conclusion

It can be argued that GIS and BIM appear still to be in silos, while HBIM, to be successful, will need fusion. Current BIM practice in particular still appears predicated upon the one-way push of information from expert to participant, inadequately supporting wide interactive participation and inclusion. On the other hand GIS practice has long enabled users to engage with data and perform their own analysis. With the increasing use of mapping services online, web users are becoming increasingly accustomed, but more in 2D than 3D. It may be that the step change forecast soon from the increasingly low-cost VR and AR technologies will at last result in widespread familiarity with 3D usage. Common standardised access to data remains a major barrier. Even at the inventory level, heritage data generally remains in silos, without common standards. For example, the National Trust in the UK publishes a gazetteer app, but the data is locked and proprietary, so difficult to fuse with information about heritage in other ownerships. Google specifically publishes 'Street Views' of National Trust properties, so legal registers of heritage can be linked via Web 2.0 'mashups' to create a much more substantially complete embryonic inventory of UK heritage. Yet the long-term persistence of such inventories is unlikely, and will not serve as an 'information resource for future generations', without clear standards for long-term retrieval and reuse. Few have downloaded the DURAARK project resources, but their approach to standardisation urgently needs to be built upon to deliver the goals of HBIM.

References

AGI Foresight Report 2020 (2015). Online at: www.agi.org.uk/about/resources/category/100-foresight?download=160:agi-foresight-2020.

AIA (2007). 'Integrated Project Delivery: A Guide Version 1'. Online at: http://info.aia.org/siteobjects/files/ipd_guide_2007.pdf.

Alderson, J. (2015). 'The GI [Geographic Information] Community: Open to a Bright Future'. AGI Foresight Report 2020. Online at: www.agi.org.uk/about/resources/category/100-foresight?download=160:agi-foresight-2020.

Australia ICOMOS (2013). 'The Australia ICOMOS Charter for the Conservation of Places of Cultural Significance – (The Burra Charter)'. Australia ICOMOS. Online at: http://australia.icomos.org/wp-content/uploads/The-Burra-Charter-2013-Adopted-31.10.2013.pdf.

Bosché, F. (2016). In discussion at Researcher Links workshop.

Bosché, F., Forster, A., & Valero, E. (2015). '3d Surveying Technologies and Applications: Point Clouds and beyond' a Technical Report for Historic Scotland'. Online at: https://web.sbe.hw.ac.uk/fbosche/publications/report/Bosche-2015-Report.pdf.

Boyes, G., Thomson, C., & Ellul, C. (2015). 'Integrating BIM and GIS: Exploring the Use of IFC Space Objects and Boundaries'. In Proceedings of GISRUK 2015. Online at: http://leeds.gisruk.org/abstracts/GISRUK2015_submission_106.pdf.

BuildingSMART (2015). 'BCF Open BIM Collaboration Format'. Online at: https://github.com/BuildingSMART/BCF-API.

Canciani, M., Ceniccola, V., Messi, M., Saccone, M., & Zampilli, M. (2013a). 'A 3D GIS method applied to cataloging and restoring: the case of Aurelian Walls at Rome'. *ISPRS-International Archives of the Photogrammetry, Remote Sensing and Spatial Information Sciences*, 1(2), 143–148. Online at: www.int-arch-photogramm-remote-sens-spatial-inf-sci.net/XL-5-W2/143/2013/isprsarchives-XL-5-W2-143-2013.pdf.

Canciani, M., Falcolini, C., Buonfiglio, M., Pergola, S., Saccone, M., Mammì, B., & Romito, G. (2013b). 'A method for virtual anastylosis: The case of the arch of Titus at the circus maximus in Rome.' *ISPRS Annals of Photogrammetry, Remote Sensing and Spatial Information Sciences II-5 W*, 1, 61–66.

CDMICA (2014). 'Cultural Diversity and Material Imagination in Canadian Architecture'. Online at: www.cims.carleton.ca:9011/cdmica/

Cloudcompare, n.d. Online at: http://www.danielgm.net/cc/.

Counsell, J. (2001). 'An evolutionary approach to digital recording and information about heritage sites'. In Proceedings of the 2001 Conference on Virtual Reality, Archaeology, and Cultural Heritage (pp. 33–42). Online at: http://dl.acm.org/ft_gateway.cfm?id=584999&ftid=91542&dwn=1&CFID=756668336&CFTOKEN=39262009.

Counsell, J. (2002). 'A role for GIS in capturing and unifying records of historic gardens and landscapes'. In UNESCO 30th Anniversary Virtual Conference on Heritage Management Mapping: GIS and Multimedia, Alexandria, Egypt. Online at: http://citeseerx.ist.psu.edu/viewdoc/download?doi=10.1.1.137.601&rep=rep1&type=pdf.

Counsell, J., & Littlewood, J. (2008). 'The Welsh medieval church and its context'. In Conference Presentation. Online at: www.academia.edu/3636602/St_Teilo15Nov08_compressed_FILEminimizer_

Counsell, J., Littlewood, J., Arayici, Y., Hamilton, A., Dixey, A., Nash, G., & Richman, A. (2008). 'Future proofing, recording and tagging historical buildings: a pilot study in Wales, UK'. In Proceedings of 2008 RICS Cobra Conference. Online at: http://usir.salford.ac.uk/11420/3/Future_Proofing%252C_Recording_and_Tagging_Historical_Buildings.pdf.

Counsell, J., Smith, S., & Aldrich, S. (2003). 'A role for 3D modelling in controlling webcams and interpreting real-time video'. In Geometric Modelling and Graphics Proceedings. 2003 International Conference (pp. 2–7).

CPAT & Smith, S. (2014). Online at: www.cpat.org.uk/vr/llandod/ (nb. requires Internet Explorer v11 or earlier, plugin functionality is disabled in most modern browsers. The Bitmanagement Contact browser plugin still supports this.)

DIFCAM (2013). 'The Project'. Online at: http://projects.npl.co.uk/difcam/the-project/.

Dore, C., & Murphy, M. (2012). 'Integration of Historic Building Information Modeling (HBIM) and 3D GIS for recording and managing cultural heritage sites'. In Virtual Systems and Multimedia (VSMM), 2012 18th International Conference (pp. 369–376). IEEE. Online at: http://arrow.dit.ie/cgi/viewcontent.cgi?article=1072&context=beschreccon.

DURAARK (2016). 'Durable Architectural Knowledge'. Online at: http://duraark.eu/.

EPSRC (2006). "Preserving Our Past – Workshop Report". Online at: www.epsrc.ac.uk/newsevents/pubs/preserving-our-past-workshop-report/.

Fai, S., Graham, K., Duckworth T., Wood N., & Attar R. (2011). 'Building Information Modelling and Heritage Documentation'. Online at: https://pdfs.semanticscholar.org/b76a/6be1ab4c5c148757eac8c73dc7a5c2995999.pdf.

Farrell, T. (2014). 'Our Future in Place'. Report for UK Dept of Culture, Media & Sport. Online at: www.farrellreview.co.uk/downloads/The%20Farrell%20Review.pdf.

Fassi, F., Achille, C., Mandelli, A., Rechichi, F., & Parri, S. 2015. 'A new idea of BIM system for visualization, web sharing and using huge complex 3D models for facility management.' *ISPRS-International Archives of the Photogrammetry, Remote Sensing and Spatial Information Sciences*, 1, 359–366. Online at: www.int-arch-photogramm-remote-sens-spatial-inf-sci.net/XL-5-W4/359/2015/isprsarchives-XL-5-W4–359–2015.pdf.

Gardels, K. (1997). 'Building environmental analysis systems with open GIS.' In Proceeding of the Third Joint European Conference & Exhibition on Geographic Information, JEC-GI'97 (pp. 28–297). Amsterdam: IOS Press.

Guedelon (2015). 'Newsletter, No 40, 7 October 2015'. Online at: www.guedelon.fr/fichiers/guedelon-chateau-en-vue-n-40_1444210232.pdf.

Hichri, N., Stefani, C., De Luca, L., Veron, P., & Hamon, G. (2013). 'From point cloud to BIM: a survey of existing approaches'. *ISPRS-International Archives of the Photogrammetry, Remote Sensing and Spatial Information Sciences*, 1(2), 343–348. Online at: http://cipa.icomos.org/fileadmin/template/doc/STRASBOURG/ARCHIVES/isprsarchives-XL-5-W2–343–2013.pdf.

ICOMOS (1990). 'Guide to Recording Historic Buildings'. London: Butterworth.

Kemp, A. (2014). 'Why BIM Is So Important to Our Industry'. Online at: www.atkinsglobal.com/en-GB/angles/opinion/why-bim-is-so-important-to-our-industry.

Kemp, A. (2015a). 'Open standards for open BIM'. In Planning and Building Control Today, 17th June 2015 issue. Online at: http://planningandbuildingcontroltoday.co.uk/bim-today-002/open-standards-open-bim-3/18512/.

Kemp, A. (2015b). 'Building Information Modelling – Are We There Yet?'. Online at: https://issuu.com/potion/docs/1chc_mag_p1–76_online.

Knutt, E. (2015). 'Aecom and BIM Academy Team to Create BIM/FM System for Sydney Opera House'. Online at: www.bimplus.co.uk/news/aecom-and-b1im-acade4my-t5eam-create-bimfm-/

Kolokotsa, D., Rovas, D., Kosmatopoulos, E., & Kalaitzakis, K. (2011). 'A roadmap towards intelligent net zero- and positive-energy buildings.' *Solar Energy*, 85(12), 3067–3084.

Lerma, J.L., Akasheh, T., Haddad, N., & Cabrelles, M. (2011). 'Multispectral sensors in combination with recording tools for cultural heritage documentation'. *Change Over Time*, 1(2), 236–250.

Linked Heritage (2013). 'Geocoded Digital Cultural Content'. Online at: www.linkedheritage.org/index.php?.

Linning, C. (2014). 'BIM – An Information Resource for Future Generations'. Online at: http://brisbim.com/wp-content/uploads/2014/12/BrisBIMx-Sydney-Opera-House-Presentation-FINAL.pdf.

London Charter (2009). 'The London Charter for the Computer-Based Visualisation of Cultural Heritage'. Online at: www.londoncharter.org/.

Maguire, D.J. (1991). 'An Overview and Definition of GIS', in Geographic Information Systems, eds. Maguire, D.J., Goodchild, M.F., and Rhind, D.W. Oxford, UK: Longman, Vol. 1, Chapter 1, p. 16.

Maxwell, I. (2014). 'Integrating Digital Technologies in Support of Historic Building Information Modelling: BIM4Conservation (HBIM) – A Cotac Report.' Online at: www.cotac.org.uk/docs/COTAC-HBIM-Report-Final-A-21-April-2014–2-small.pdf.

McTaggart 2015 quoted in news article on "BIM-consortium selected to help preserve and upgrade Sydney Opera House" online at: http://www.architectureanddesign.com.au/news/sydney-opera-house-selects-aecom-led-consortium-to accessed 7/4/16.

Molyneux, N. (1991). 'English Heritage and Recording: Policy and Practice', in Recording Historic Buildings, RCHME, London.

Murphy, M., McGovern, E., & Pavia, S. (2009). 'Historic building information modelling (HBIM)'. *Structural Survey*, 27(4), 311–327.

Niskanen, I., Purhonen, A., Kuusijärvi, J., & Halmetoja, E. (2014). 'Towards semantic facility data management'. In proceedings of the INTELLI 2014: The Third International Conference on Intelligent Systems and Applications.

Penttilä, Hannu, Marko Rajala, and Simo Freese (2007). 'Building information modelling of modern historic buildings: case study of HUT/architectural department by Alvar Aalto.' *eCAADe*, 25(Session 13), 607–613.

Planet Labs (2016). Online at: https://www.planet.com/data/#coverage.

Plume, J. (2015). "Integrated Digitally- Enabled Environment: the Internet of Places". AGI Foresight Report 2020 (op cit).

Reeves, F.J., & Paardekooper, R. (eds.) (2014). 'Experiments Past. Histories of Experimental Archaeology'. Leiden: Sidestone Press.

Richman, A., & Holding, A. (2008). 'Summary of Evaluation of VEPs Online Participation Tools'. Online at: http://www.veps3d.org/site/259.asp.

Schevers, H., Mitchell, J., Akhurst, P., Marchant, D., Bull, S., McDonald, K., Drogemuller, R., & Linning, C. (2007). 'Towards digital facility modelling for Sydney opera house using IFC and semantic web technology'. *ITcon*, 12, 347–362. Online at: www.researchgate.net/profile/David_Marchant4/publication/27483940_Towards_digital_facility_modelling_for_Sydney_Opera_House_using_IFC_and_semantic_web_technology/links/0c9605182e1ccc74b6000000.pdf

Schill, C., Koch, B., Bogdahn, J., & Coors, V. (2007). Public Participation Comment Markup Language and WFS 1.1. Urban and Regional Data Management, 85–92.

Schofield, L. (2015). 'BIM4Infrastructure: Rather than Answers, Do We Know the Right Questions Yet?"'. AGI Foresight Report 2020 (op cit).

Scottish Ten (2013). 'Sydney Opera House, Australia'. Online at: www.scottishten.org/index/partners/sydneyoperahouse.htm.

Silberman, N. (2015). 'Remembrance of things past: collective memory, sensory perception, and the emergence of new interpretive paradigms'. In Proceedings of the 2nd International Conference on Best Practices in World Heritage: People and Communities. Madrid: Editora Complutense, 2015 (p. 1172). ISBN: 978-84-606-9264-5. Online at: http://eprints.ucm.es/35080/1/Neil%20Silberman.pdf.

Steadman, P., Bruhns, H.R., Rickaby, P.A. (2000). "An introduction to the national Non-Domestic Building Stock database". *Environment and Planning B: Planning and Design*, 27(1), 3–10. Online at: www.ucl.ac.uk/carb/nondomestic/.

TTP (2015). 'Time to Think Internet of Sensors and Not Internet of Things'. Online at: www.ttp.com/files/download/458672a027a17bf.

UNESCO (2001). 'Intangible Cultural Heritage – Working Definitions'. UNESCO International Round Table. Online at: www.unesco.org/culture/ich/doc/src/05359-EN.pdf.

UNESCO (2003). 'Charter on the Preservation of Digital Heritage'. Online at: http://portal.unesco.org/en/ev.php-URL_ID=17721&URL_DO=DO_TOPIC&URL_SECTION=201.html

Wang, Y., Weinacker, H., & Koch, B. (2008). 'A lidar point cloud based procedure for vertical canopy structure analysis and 3D single tree modelling in forest.' *Sensors*, 8(6), 3938–3951.

Weinecker (2008). '3D Building Reconstruction from Lidar'. Final Reports of VEPs3D Interreg NWE Project. Online at: www.veps3d.org/site/files/20-jun-2008/16-32-39/E3_Building_recon_LiDAR.pdf.

Worthing, D., & Counsell, J. (1999). 'Issues arising from computer-based recording of heritage sites'. *Structural Survey*, 17(4), 200–210.

4 Heritage and time
Mapping what is not there

David Littlefield

This chapter examines ideas of authenticity and meaning, and considers how Heritage Building Information Modelling/Management may be used as a system which attempts to capture a site in its multiple conditions, enabling a comparison between its ideal and actual states, and permitting users to access the meanings implied by the differences between those states. HBIM implies, therefore, a recognition of what is not, as well as what is, there – while acknowledging that what is there may be intangible, such as association, identity, memory and cultural value.

4.1 Introduction

One of the principal questions within the emerging field of Heritage Building Information Modelling/Management (HBIM) is how such technologies and processes occupy the fuzzy ground of site as designed, site as constructed and site as found. There is inevitably a mismatch between site as imagined and site as fact – at both the macro scale (a site may have become ruinous) and the micro one (subtle erasures and shifts within surfaces). This paper examines ideas of authenticity and the presence of multiple authenticities, the importance of the intangible qualities of place, and the consideration HBIM should give to capturing the site as found, rather than the site as idealised.

Moreover, I argue in this chapter that the role of interpretation and preserving the potential meaning of a site must be a key element of HBIM, as compared with its BIM equivalent. With HBIM, which captures and enables the future management of a site which has already been constructed, it is imperative that social and cultural significance fall within the orbit of consideration. HBIM must embrace, if it is to be an effective heritage tool, interpretation, narrative, value and identity; it is these things, after all, which make heritage what it is. The term BIM implies technology, and is too often used as a synonym for 3D digital modelling or even a particular digital design tool. Multi-party collaboration within a digital environment provides a truer picture of BIM (Kara 2008: 36), and in this case HBIM is little different other than the consideration of the intangibles, the traces and the fuzzy areas from which ideas of heritage emerge. HBIM can, then, embrace digital technology in obvious and uncontentious ways:

- shared models, point clouds and site surveys can capture the condition of the site as found in a highly granular way;
- archival material (drawings, photographs, texts, recordings) and remembrance projects (the spoken word, song, ritual) can be captured and cross-referenced;
- social media, search tools and almost limitless data storage can facilitate useful networks of different stakeholders.

If, however, the above points are indeed obvious and uncontentious, it does have far-reaching consequences for a definition of HBIM. There is a question mark over editing and access rights; the reach of archival material may be almost limitless; and the surveyed condition of any site is always problematic (often estimated and even never complete) quite apart from the fact that the found site will differ from what was original or desired. This latter point is fundamental: the granularity of the heritage site is important in that surfaces can change abruptly due to weathering, staining, erosion and mark-making which cause the once continuous surface to become textured and nuanced. The survey of the heritage site can, then, include the recording of elements which are no longer there, due to un-designed or accidental interventions, inhabitation and the passage of time. Indeed, it is important to understand that much meaning is implied within such change and erasure, in that stories and other historical data (such as the occupation implied by the erosion of a stone step) can be detected through what is absent. The survey of a heritage site, therefore, is very much a process of noticing rather than merely looking, and acknowledging the difference between the site as originally conceived and the site as lived. It is the site as lived which often (though not always) catalyses an otherwise ordinary site into the heritage site; the chapel of the Mission San Antonio de Valero, Texas, would be an unremarkable Spanish-era church were it not for the 1836 battle of the Alamo which propelled the building and the site into the American imagination; 251 Menlove Avenue and 20 Forthlin Road, Liverpool, would be entirely ordinary post-war suburban dwellings were they not the childhood homes of John Lennon and Paul McCartney. It is often the case that places accrue significance, and even a sense of reverence and the sacred, not through any objective architectural qualities but as tangible reference points for far more important invisible qualities. What is critical, though, is the plurality of the heritage site and the consequences this has for capture within an HBIM tool. Heritage sites are nuanced, ripe for interpretation and even contested. Unlike BIM, which may reasonably seek to capture a proposed building/site in its entirety, technology simply cannot provide a definitive and finite model of the heritage site. The heritage site is too open and ambiguous for that, which is what makes HBIM quantifiably different from BIM. Heritage Building Information Modelling/Management entails, therefore, a mapping of (or a making room for) what is not there. This can be considered in terms of:

- what is no longer there (such as a worn or eroded surface) and
- what is intangibly there (the essence of, and associations with, a site).

Additionally, it is worth considering what is there but ought not to be. This paper concludes by suggesting a framework in which HBIM may be considered in manifesto terms.

4.1.1 The fuzziness of heritage and authenticity

Much has been written about heritage and authenticity, and the increasingly broad frameworks in which these terms can be framed. Starn describes well the inherent contradictions and ambiguities which dog present understanding, recounting the story of the ship of Theseus (which was repaired to such an extent that philosophers could not agree whether or not the ship remained original) as a parable for heritage practices. Such arguments persist when considering the preservation of things and sites.

> Preserve the design and the techniques for executing it and an original can arguably remain "the same". Preserve the historical material and the wear and tear of time will distinguish the original from its copy and indeed from its own original state. Stretched between alternatives such as

these, the case for historic preservation seems bound to go around in circles or turn downright incoherent. Critics like it that way. So do the movers and shakers with reasons – power, profit, presumption – for wanting to trump the claims of the past on the present.

[Starn 2002: 2]

Starn and others (such as Jukka Jokkilehto and David Lowenthal) have described the efforts undertaken to provide a degree of certainty within this persistent grey area. ICOMOS' 1994 *Nara Document on Authenticity*, the *Declaration of San Antonio* which followed two years later, and the 2008 *Quebec City Declaration on the Preservation of the Spirit of Place*, for example, all give considerable importance to the role of the intangible qualities of place. These qualities include values attributed to particular sites (which may vary from culture to culture) and a recognition that "the spirit of place is made up of tangible (sites, buildings, landscapes, routes, objects) as well as intangible elements (memories, oral narratives, written documents, festivals, commemorations, rituals, traditional knowledge, values, odors) . . ." [ICOMOS 2008]. Critics have treated such efforts at definition with a range of responses, from exasperation and derision [Lowenthal 1985: 231] to one of liberation from the "tyranny" of a fixed and fictitious past [Kidd 2011: 24]. Glendinning asks whether such openness in definitions is one of inevitable progress or temporary crisis, while recognising that conservation and associated terms occupy shifting theoretical territory:

If the Conservation Movement has always, despite its own propaganda, been a child of Progress and Western modernity, a product of the ideal of using the past for useful modern progress, then where does it stand today, in an era of postmodern relativism that has deconstructed all the old certainties of grand narratives and normative values?

[Glendinning 2013: 6]

We therefore find ourselves at an awkward point in history. The terms of reference are blurred just at the point when technology is enabling ever greater specificity. The challenge for HBIM is responding to that openness; if HBIM privileges the known and the measurable, it may find itself uncomfortably distanced from the profound questions being asked of heritage itself. Without a seismic shift in contemporary cultural norms, possibly with some sort of return to those "old certainties", HBIM must embrace (or at least acknowledge) ambiguity, uncertainty, interpretation and conflict.

A case study far beyond the realms of architecture and place is instructive – that of Scarlett O'Hara's dress. In David O. Selznick's 1939 production of *Gone with the Wind*, Vivien Leigh wears a vivid green "curtain dress"; so iconic has this dress become that a replica was made in 1986 in order that the precious original be preserved, leaving the more robust copy to stand in for touring exhibitions. In 2010 funds were raised to conserve the original (and four others worn on set). This work, undertaken at the Harry Ransom Center, University of Texas at Austin, has led to a series of testing questions being asked about the nature of the life of an object and the status of changes wrought over time. The makers of the 1986 replica dress found the original to contain baffling details which they did not believe to be the intention of the highly skilled studio seamstresses, unless the product of hasty on-set repairs; in such instances, the replica makers created details differently, creating a copy which was deliberately not faithful to every last stitch. During the work of 2010, conservators were troubled when having to tackle that "original" poor and weak detailing, especially when it could not be known when the details/repairs had been made: "Were these repairs made later by less skilled hands after months of touring the nation? Or were they intentional, reflecting the desperate state of Scarlett's financial affairs and dwindling resources in the screen narrative?" [Morena 2014: 123]. After much soul-searching, conservators preserved the found detail, while adding additional support. The result is a pair of dresses: a replica which demonstrates the original (imagined) intention of the dress, and an original which preserves poor workmanship which may or may not have been part of the designed narrative of the garment. Additionally, the original dress contains a wire support that does not appear to be present on screen, and may well date from a 1974 MoMA exhibition, "inserted for display purposes, perhaps to add drama and to convey a sense of movement" [Morena 2014: 124]. Moreover, even the colour of the original dress is ambiguous; it is relatively certain that designer Walter Plunkett deliberately faded the dress to provide it with a degree of realism, however there is broad agreement that the fading is more extreme today than in 1939, while the Technicolor processes of the filmmaking added an on-screen vibrancy absent from the actual artefact. Indeed, the original dress was displayed at the Victoria & Albert Museum in 2012 under a green-coloured filter in order to present the garment in such a way that it would match the screen image held in visitors' imaginations. According to Morena, one might entertain the idea of there being three versions of the original dress: the version seen on-screen; the actual fabric worn by Leigh; and the reproduction which matches more faithfully the original the expectations of the viewer. Certainly, she argues ideas of originality can shift "if it is accepted that an object's only authentic state is its present known state, not an assumed previous state" [Morena 2014: 126].

Marks and traces of use and occupation (often unplanned and accidental) can, then, test the notion of authenticity – and this applies equally to both artefacts (such as dresses) and architectural sites. Bath Abbey, which this author has described elsewhere [Littlefield 2016], is an interesting case study in that the present floor comprises original stones, none of which are in their original places. This leads to a structural ambiguity in that the stones (in fact, "ledger stones" which record the details of those buried beneath) have borne no relation to particular human remains since they were shifted by George Gilbert Scott in the 1860s as part of a programme to treat subsidence. The ledger stone recording the epitaph "Here lie the remains of Frances Fairman Ayerst, who died 15th Novr 1840" almost certainly does *not* mark the location of the mortal remains of Frances Ayerst (see Figure 4.1). Further slabs are so eroded through footfall (see Figure 4.2) that it is becoming difficult, even impossible, to read the inscriptions, leaving visitors to wonder at texts such as:

. . . *ith this Stone*
. . . *deposited*
. . . *ortal Remains*
. . . *William*

I argue that, as Morena suggests, in order to understand Bath Abbey in its most authentic state, we must understand the building as it exists now. That comprehension of the "now-ness" of the abbey includes a recognition of change as well as an acceptance of the fact that we cannot know (because records do not exist) the position of the ledger stones prior to the nineteenth-century stabilising work. This repositioning of the stones (as well as the erosion, cracking and peeling which characterise them) comprises a surface which is a long way from any chronologically original surface in terms of texture, clarity, datum and name/named relationships, yet its heritage is undiminished. "Like the child's threadbare comfort blanket, it becomes transformed, while remaining very much its authentic self" [Littlefield 2016]. The floor, therefore, is plural: it can be considered authentic in its present, while not denying the authenticity of the floor as it once was (see Figure 4.3). Any HBIM system which wishes to capture the abbey must acknowledge this plurality, leading, perhaps, to a model of layers and difference without privileging one layer over another.

This is what makes BIM and HBIM fundamentally different: both must seek to encapsulate the whole of the building or site, but the embrace of HBIM is far more complex and extensive than its BIM counterpart. BIM is a tool optimised to capture the whole design intention, while for its

Figure 4.1 Bath Abbey Frances Ayerst, Bath Abbey, UK. The ledger stones which comprise the abbey floor no longer indicate the presence of named individuals. It is unknown where the remains of Frances Ayerst are now located.

Figure 4.2 Bath Abbey erosion, Bath Abbey, UK. Circulation routes have concentrated erosion to the ledger stones, in many cases erasing text completely. Ledger stones are neither in their original positions, nor complete.

Heritage and time 37

Figure 4.3 Bath Abbey ledger edge, Bath Abbey, UK. Many ledger stones were cut in order to make them fit during nineteenth-century repositioning. Their incompleteness is not the result of slow erasure over time, but of sudden, brutal decisions.

heritage cousin the concept of wholeness is far more diverse and unpredictable. The consideration of sites as "living witnesses" to traditions, values, events and cultural significance lies at the heart of the idea of heritage [ICOMOS 1964: 1] and, like any living witnesses, historic places exist as dynamic rather than static forms. This dynamic nature of heritage can be troubling, especially when history disrupts the continuity of place, such as the destruction wrought upon Warsaw during World War II, or the descent of Le Corbusier's 1929 Villa Savoye into a state of ruination in the 1960s.[1] Unplanned and unauthorised change can easily be dismissed as inauthentic and a diversion from core messages, yet those core messages (the official view of what makes a site significant) can themselves be contested. I would argue that UNESCO's consideration of Gerrit Rietveld's 1924 Schröderhuis as "an outstanding expression of human creative genius in its purity of ideas" as opposed to an occupied space is not above critique. "The significance of the Rietveld Schröder House as a historical material document does not lie in its occupational history. It was and still is considered to be a constructed manifesto," says the UNESCO report on the building's cultural value [UNESCO 1999: 16]. Thus the changes made to the house during Mde Schröder's (and Rietveld's) occupancy are judged to have been inconsequential to the principal narrative of the space, which is its power as an idea; thus changes over half a century of occupation have been reversed and nullified. In terms of a quite different structure, Inigo Jones, commissioned by James I to undertake a study of Stonehenge, concluded that the structure was of Roman origin, possibly to help justify his own introduction of classical architecture into Britain (Tait 1978). Others, aware that interpretation is not always neutral, simply suggest that every age reads Stonehenge to suit its own purposes [Hill 2008].

Heritage can be a powerful political and cultural tool, and the uses to which sites are put can tell only partial tales. HBIM, configured as an open source repository of information and interpretations, can offer a more comprehensive and living account of a site's *authenticities*. This is not to say that anything goes – that the site should be democratised to such an extent that it embodies every and no meaning. Rather, the notion of the plurality of site and interpretation, even if contested and counterintuitive, can combine to create a more whole model than the unidimensional schemas described above.

Again, observations from beyond the field of architecture are relevant; emerging thinking in biology offers a useful model. It has long been recognised that "friendly" bacteria assist in digestion, a fact exploited by the manufacturers of some food products,[2] but new research demonstrates that bacteria plays a far more integral role in human performance than aiding the digestive system. Microbes can assist in matters as diverse as triggering the immune system and manipulating mood, and thus could be considered less as alien species tolerated by humans and more as "a forgotten organ, as essential as the liver" [Jones 2011]. Indeed, Seth Bordenstein and Richard Jefferson argue for the body to be considered not as a pure and discrete form but as a collection of species which combine to create a "super-organism", "micro-biome" or collective "performance unit" [Arnold 2013]. "The universality and significance of microbial symbionts in multicellular life is now unmistakable . . . microbial symbiosis is a process by which two or more distinct organisms interact as one entity" [Bordenstein and Brucker 2012: 443]. This research conjures images of the human being as a cloud of symbiotic life forms; significantly, this cloud or combination of symbionts is unique to each of us. "We all leave traces of our microbes behind. We are haloed by an invisible nebula of bacteria, fungi and viruses . . . It's our vaporous calling card" [Smith 2016].

What such research demonstrates is that notions of dirt and contamination can be recast, and what might be overlooked or dismissed as inconvenient detail might well be read as an integral part of a whole, unified system. I have proposed that the wear, erasure, absence, loss and impurity of the ledger stones at Bath Abbey is as much a part of those stones, as original and authentic, as the stones laid down centuries ago. In other words, the marks of the passage of time, including the removal and casual repositioning of the stones by Gilbert Scott, ought not be considered unfortunate ruptures in the continuity of their authenticity but as very real moments in time to which the stones bear witness. That the stones no longer tell whole truths (the resting place of Frances Ayerst is not marked by the stone which bears the name) is part of the story of the stones, just as the erosion caused by visitors' footfalls is also part of that story. That is, the change wrought by time is not dirt or impurity, but an essential condition of the authentic artefact, just as our own microbial symbionts are an integral part of our own human functioning.

Absence (of the body, of the text) becomes powerfully present at Bath Abbey, and worthy of record. Similarly, the absent faces in reliefs of Queen Hatshepsut within the Egyptian temple of Dayr al-BaḤrī have become remarkable historical records. In this latter case, the obliteration (undertaken c1430 BCE) is far older than that exhibited in Bath; indeed, the absence of Hatshepsut's face has been a historical fact for far longer than it was ever present (see Figures 4.4a and 4.4b). The reason for such vandalism, likely undertaken after her death on behalf of her stepson, Pharaoh Thutmose III, is uncertain – possibly part of an effort to feign an unbroken lineage of male kings [Tyldesley 2015]. Today, the absence of the queen's face is arresting: the state-sanctioned vandalism appears crude and savage, with no attempt at remodelling or deploying the skills of masons to create a featureless void – a clean, officious redaction. Here, the removal of Hatshepsut's face from her numerous representations appears more as rushed attacks than careful incisions. The sudden shift in surface texture within the temple, from the craftsmanship of the original surface to the brutality of the defacing, is as remarkable as the defacing itself, and provides a fine example of the physical granularity of heritage. The shifts in surface conditions are important, as they imply stories – and without stories, heritage has nothing to say.

Figure 4.4a and Figure 4.4b Hatshepsut. Dayr al-BaḤrī, Egypt. The face was removed from likenesses of Queen Hatshepsut c1430 BCE, likely on the order of her stepson, Pharaoh Thutmose III, possibly to feign an image of an unbroken line of male kings. The difference between the original state and the found state is stark.

Figure 4.4a and Figure 4.4b (Continued)

Less brutal, but no less invasive, are the marks left by Alexander and later Roman administrations on the temple at Karnak, close to Dayr al-Baḥrī but on the other (east) bank of the Nile. One of the antechambers within the temple complex contains Egyptian-style depictions and cartouches of the Macedonian, while evidence for Roman occupation is less subtle, comprising more recognisable European columns and friezes (see Figures 4.5a and 4.5b). Again, and this is especially the case with places of antiquity, the heritage site can be seen as a series of accumulated layers. The challenge for HBIM is in capturing the sedimentary character of this time-based composition, instead of the igneous quality of the sudden completeness of the BIM model. Any HBIM resource will need to acknowledge the multiple, or parallel, authenticities of a site, resisting the temptation to privilege one particular reading or ideal origin over another. The BIM model shifts, of course – it, too, is a dynamic tool which is designed to respond to the inputs of its contributors, and even when complete it will be subject to amendment if its operators take care to modify it as its analogue mirror is modified. A key difference with HBIM, though, as I have attempted to demonstrate above, is that there will never be a definitive state to begin with – only a tentative beginning upon which increasing layers of understanding can be applied. The assembling of an HBIM resource will always be an ongoing project.

4.2 Conclusion – towards a framework for HBIM

BIM is about certainty – modelling the new, anticipating problems, reducing (ideally eliminating) risk and uncertainty. Within a heritage context, the building or site already exists and therefore the exact condition of what lies beneath the surface cannot always be known – not without invasive investigation. HBIM, therefore, implies a degree of uncertainty. The sources of materials will not

Figure 4.5a and Figure 4.5b Temple at Karnak, Egypt. The temple bears the marks of multiple eras and power structures – Egyptian, Macedonian and Roman. These periods manifest themselves in layers, some of which are peeling away to reveal the original intent. Such examples also reveal the difference between originality and authenticity.

Figure 4.5a and Figure 4.5b (Continued)

always be known and neither will the building's original condition. The idea of the original is itself problematic: the historic building has a past; it has been occupied; it has very probably been the subject of change; and, importantly, the site will have accrued cultural value (stories, associations) which may become of greater significance than the built fabric itself. Places cherished for their event value (something once happened there) or their personality value (someone once lived there) are such examples – they have become heritage sites for reasons other than any intrinsic, measurable qualities of the site itself.

As described above, sites classified as heritage are problematised through the consideration of the term "authenticity" – an elastic concept through which an object's or site's truthfulness can be tested. No site remains static, and this is especially true of occupied sites. Buildings are touched, amended, maintained, repaired and subjected to the winds of fashion. They become eroded, smoothed, stained and renewed. Windsor Castle today is not the same Windsor Castle of 1992, or 1692, or 1092.[3] Yet it remains Windsor Castle, and authentically so. Authenticity, then, is plural. The problem for HBIM, which implies an intention to capture a site via a shared digital model or resource, is this: What, exactly, is being captured? This question implies further questions: Who does the capturing? Who owns what is captured?

I propose that the digital tools and processes which underpin building information modelling be considered in the context of heritage in the ways set out in Table 4.1. This table is set out as a set of oppositions, in much the same way as land artist Robert Smithson compiled his "Dialectic of Site and Nonsite" table in 1972 [Doherty 2009]. Through this work, Smithson articulated a framework for considering one's experience of being on-site compared with the experience of viewing artefacts, such as drawings and photographs, which represent the site. Although Heritage BIM will inevitably describe any given site in exacting detail, and the resolution of digital scanning is certainly

Table 4.1 BIM/HBIM

BIM	HBIM
captures the building as it is intended to be	captures the building as it is
a resource and test of a structure yet to be constructed	a resource describing a building which has already been constructed
a resource which ensures a new structure conforms to codes and other parameters	a "non-judgmental" resource describing an inherited structure which will inevitably deviate from the ideal
a design, construction, coordination and FM tool	a tool which may record a structure prior to its loss or damage
the known – the "idea" of the intended structure will never be more complete	will embody much of the unknown, including competing and evolving ideas
off-site – assembled to predict how a site will perform once changed	on-site – a response to the site as found
ambition of exactness, predictability, perfection	makes room for imperfection, the accident, the undersigned, the unauthorised
looking, testing	noticing, narrative
singular	plural
integrated	layered
definitive	interpreted
enables site as designed to match site as built	enables comparison between site as designed, site as found and site as imagined
directed	negotiated
ownership is defined	ownership is loose

impressive, I argue that HBIM must embody an attitude of openness, interpretation and ambiguity that standard BIM does not. The model must be accessible to a range of audiences and contributors, and any mapping of a site must consider the nature of mapping itself – as well as the physical contours, both idealised and actual, there are memories and stories which can be mapped. This makes HBIM an open source and open-ended project, embodying inputs ranging from the quantifiable techniques of conservation to the vagaries of politics and identity. Heritage is, after all, a tool through which societies tell stories about themselves. In the case of well-known people and places, "the associative potential would be almost without limit in volume, richness or complexity" [Evans, in MacDonald 2006: 567]. Heritage Building Information Modelling, in order to reach maturity in its own right rather than position itself as a subset of BIM, must include a careful consideration of *information*. Within the context of HBIM, information includes not just fact but site readings and a recognition of the fuzzy qualities of historic places – fuzzy in that their physical and metaphysical edges shift. Their surfaces are not always where we expect them to be; and these places, as culturally constructed artefacts, continually reposition themselves in society's collective imagination.

Notes

1 Bernard Tschumi, incidentally, celebrated this period: "The Villa Savoye was never so moving as when plaster fell off its concrete blocks." "Architecture and Transgression", published in *Oppositions*, Princeton University Press, New York, 1998, page 360.
2 French manufacturer Danone describes its Actimel yogurt drink thus: "As a yogurt drink Actimel contains two traditional yogurt cultures: Lactobacillus bulgaricus and Streptococcus thermophilus. But it also contains a third, the culture L. casei Danone®. This is exclusive to Danone and has been extensively studied for more

than 15 years." http://www.actimel.co.uk/all-about-actimel/the-story-of-actimel/ [Accessed 15 March 2016]. Japanese firm Yakult describes its products similarly: "What's in the bottle? Lots and lots of very good stuff. Well, 6.5 billion bacteria to be precise. Called *Lactobacillus casei* Shirota, they're named after the scientist Dr Shirota who cultivated this unique strain that is scientifically proven to reach the gut alive. The other few ingredients make up the fermented milk drink you see today. You see, with Yakult, it's all about the bacteria." http://www.yakult.co.uk/?gclid=Cj0KEQjw5Z63BRCLqqLtpc6dk7gBEiQA0OuhsPLdLVV33sPrUkLubKIZqqVRmrND033Yv4qBKoh97bwaApEA8P8HAQ [Accessed 15 March 2016].

3 On 20 November 1992 more than 100 rooms at Windsor Castle were damaged or destroyed by fire. "The areas that were most badly damaged, such as St George's Hall, were redesigned in a modern gothic style, while the other parts were restored to the condition in which George IV had left them." https://www.royalcollection.org.uk/sites/default/files/Windsor_Castle_Fact_Sheet.pdf

References

Arnold C. 2013. "The hologenome: A new view of evolution", *New Scientist* issue 2899. 12 January 2013. Reed Business Information.
Bordenstein S. and Brucker R. 2012. "Speciation by symbiosis", *Trends in Ecology & Evolution* 27(8): 443–451.
Doherty C. 2009. *Situation: Documents of Contemporary Art*. London and Cambridge, MA: Whitechapel Gallery and MIT Press.
Glendinning M. 2013. *The Conservation Movement: A History of History of Architectural Preservation*. Oxford: Taylor & Francis.
Hill R. 2008. *Stonehenge*. London: Profile Books.
ICOMOS. 1964. The International Charter for the Conservation and Restoration of Monuments and Site (The Venice Charter).
ICOMOS. 2008. *The Quebec Declaration*. Paris: ICOMOS.
Jones S. "Is dirt really such a bad thing?" *Daily Telegraph*. 22 March 2011
Kara H. 2008. "Discourse networks and the digital: structural collaboration at the Phaeno science centre". In: Littlefield D., *Space Craft: Development in Architectural Computing*. London: RIBA Enterprises.
Kidd J. 2011. "Performing the knowing archive", *International Journal of Heritage Studies* 17:1.
Littlefield D. 2016. "The living and the dead: an investigation into the status of erasure within the floor of Bath Abbey", *Interiors: Design, Architecture, Culture* 7:1 (Taylor & Francis).
Lowenthal D. 1985. *The Past Is a Foreign Country.* New York: Cambridge University Press.
MacDonald (2006), "Digital Heritage: Applying Digital Imaging to Cultural Heritage", Routledge, Jordan Hill Oxford, ISBN: 13 9780750661836.
Morena J. 2014. "Definitions of Authenticity: a study of the relationship between the reproduction and the original *Gone with the Wind* costumes at the Harry Ransom Center". In: Gordon R., Hermens E. and Lennard F. (Eds.), *Authenticity and Replication: The 'Real Thing' in Art and Conservation* (pp. 119–130). London: Archetype Publications.
Smith J. 2016. "Microbe CSI: How to read the air for clues at crime scenes". *New Scientist*, issue 3063. 5 March 2016. Reed Business Information.
Starn R. 2002. "Authenticity and historic preservation: Towards and authentic history", *History of the Human Sciences* 15(1): 1–16.
Tait A. 1978. "Inigo Jones's Stone-Heng", *The Burlington Magazine*, 120(900): 154–159.
Tyldesley J. 2015. http://www.britannica.com/biography/Hatshepsut [accessed 28 September 2016].
UNESCO. 1999. WHC Nomination Documentation, June 1999, p. 16. http://whc.unesco.org/en/list/965/documents/ [accessed 16 March 2016].

5 From history to heritage

Viollet-le-Duc's concept on historic preservation

Khaled Dewidar

For architects who pioneered and developed the modern movement, the architectural theory of Viollet-le-Duc was useful, precisely because it helped them to break free of the historical past and the theoretical tradition of earlier architecture. In our own time, when preservation has assumed as great an urgency as the continued development of modernity, Viollet-le-Duc's ideas about preservation of old structures have acquired a new import and offered valuable insights to patrons and builder alike. Perhaps it may be as well to endeavor at the outset to gain an exact notion of what we understand by restoration. I shall simply try to present in my argument, based on the concept of a model as realized by Viollet-le-Duc's Gothic example, an apt instrument for different restoration projects of an edifice. This principle, with its adaptable phenomenological description, would provide a special significance in the case study chosen to be presented in this chapter: the Egyptian Society of Political Science, Statistics and Legislation.

> There are two ways of expressing truth in architecture: we must be true according to the program of requirements, and true according to the method and means of construction.
>
> Eugène-Emmanuel Viollet-le-Duc (2011)

It was only since the first quarter of the present century that the idea of restoring buildings of another age had come into focus. Perhaps it is appropriate to endeavor at the beginning to gain an exact notion of what we understand about restoration. The Romans replaced, but did not restore; a proof for this is that there is no Latin word corresponding with our term "restoration". *Instaurare*, *reficere*, *renovare* do not mean to restore, but to reinstate or to make new. Our age has adopted an attitude toward the past in which it stands quite alone among historical ages. I had undertaken to analyze the past to compare and classify its phenomena and to construct its veritable history.

Viollet-le-Duc is generally acknowledged to be the premier theorist of modern architecture. Because he approached the theory of architecture in terms of principles rather than rules, the application of those principles required the formulation of a method, an explanation of which would be presented. On March 11, 1844, both Viollet-le-Duc and Jean-Baptiste-Antoine Lassus were awarded the commission to restore the Gothic cathedral of Notre-Dame de Paris. This commission marked the moment when Viollet-le-Duc became Mérimée's protégé and was put on the track to a central position in the Parisian cultural establishment.

Henceforth, his thought and work in the field of architectural restoration were to be grounded in the Gothic tradition, as opposed to the classical orientation of prevailing taste and practice. The Notre-Dame project led Viollet-le-Duc to view the field of architecture and to prophesy its future in terms of what he had learned from Gothic architecture. Viollet-le-Duc developed his architectural theory based on the rationalized interpretation of architecture in general, and of Gothic architecture in particular, interpreting it as a logical structural system. The Notre-Dame project involved a church much altered since its creation in the late twelfth century and much neglected since the revolution. Both Viollet-le-Duc and Lassus had to decide which aspects of the fabric design to retain and which to alter. Viollet-le-Duc changed the form of the flying buttresses along the nave, the earliest universally acknowledged buttresses in Gothic architecture. We can still see, at the end of the west side of the south transept arm, a trace of the original design. Observing in the same area a series of virtually identical scars in the masonry of the clerestory, he deduced that the traceried windows of the thirteenth century had replaced a shorter plain window. Accordingly, he decided to "restore" the original wall configuration in the bays adjacent to the crossing. He was, in my opinion, removing genuine thirteenth-century work in order to make manifest the elevation he thought had been there at the outset.

His decision involved lowering the roofline; as a result, the roof would no longer cover the roundels, which originally opened into the dark area over the gallery vaults. He filled the roundels with stained glass, thereby treating them in a manner in which they had never been seen in the Middle Ages. More was to come: all statues of a biblical king had stood on the west façade; he ordered replacements for those vandalized in the anti-royal revolution. Ironically, other features he created have become among the most cherished aspects of the cathedral.

The grotesques on the tower balustrades had weathered to unrecognizable nubbins. He replaced them with original figures that are perhaps the most memorable on the entire cathedral. Similarly, the crossing had been accented by a spire that was replaced by a new one. Finally, he overwhelmed the final design by adding new figures for the apostles and archangels at the angles and corners of the roof.

These modifications on the cathedral were presented in order to emphasize that Viollet-le-Duc's theory on restoration was neither to bring a building back to the end of its evolution – in this case, its state at the end of the Middle Ages – nor to restore it to a pristine version of its twelfth-century design. Rather, it was to rebuild it in a condition of completeness that could never have existed at any given time. For Viollet-le-Duc, restoring ruined structures to full order was not simply a case of a bright completion. Rather, he had a remarkably detailed knowledge of history and a wonderfully intuitive imagination with which he peopled the monument and envisioned it in use. Viollet-le-Duc is chiefly remembered as a restorer based on his own systematic interpretation of the intentions behind medieval architecture. For him the beauty of Gothic architecture was the result of a systematically rational approach to building rather than of aesthetic or iconographical considerations. He proposed that a restoration must take changes into account and usually retain them. While neutral components in the structural elements, such as ashlar blocks, might require renewal or outright replacement, decorative features should generally not be recurved because it is impossible to reproduce their authentic character. In other words, an old building should not be made like a new one, but should retain wear and damage. One of Viollet-le-Duc's cardinal rules was to respect the historical integrity of the building as it has come down to us. In this we can see the reverence for the contextual guidance that has become one of the principal tenets of present-day restoration. His argument, in my opinion, is based on redefining the structure in a better condition, a stronger and more perfect way. As a result, the restored edifice would have a renewed sense of existence, capable of withstanding the ravages of the time.

Viollet-le-Duc's method of inductive analysis and interpretive reconstruction based on comprehensive knowledge would be the cornerstone for the case presented. His name is associated with certain views on the art of restoration which, despite their close links with the picturesque inventions of the Romantic imagination, have not been entirely discredited in the obsessional, objective twentieth century. If Gothic architecture had any didactic value for Viollet-le-Duc, it lies not only in the echoes that it had raised in the imagination of the time, nor in its role as a model to be imitated, but also in its function as something to be explained and analyzed. No other architectural style lent itself more readily to this intellectual exercise than the rigorously logical structures of the twelfth and thirteenth centuries. His seminal article "On Restoration" in the *Dictionnaire raisonné de l'architecture* is a straightforward appeal towards integrity, founding a new system on the hierarchy of historic preservation of an edifice. His thematic model for a Gothic cathedral is an example for completeness, expressing his intention towards an inductive analysis for historic preservation. The logical spirit that pervades this model endows it with a sense of finality and reality.

In the Egyptian Society of Political Science, Statistics and Legislation, one cannot operate with too much prudence and discretion. To state it plainly, a restoration can be more disastrous for a monument than the ravages of the centuries. In this case, one truly does not know what is to be feared more: negligence that truly allows what is threatened with ruin to fall to the ground or ignorant zeal that adds, suppresses, completes and ends by transforming an old monument into a new one, devoid of any historical interest. We should also understand perfectly in this project that the architect made every effort to restore to the building through prudent repairs the richness and brightness of which it has been robbed. It was not a matter of making art but only of submitting to the art of an epoch that no longer exists. Figure 5.1 shows the Egyptian Society of Political Science, Statistics and Legislation.

In restoring this edifice, it was necessary to decipher texts and consult all existing documents on the construction of this building, descriptive as well as graphic. To give our beautiful building all its splendor and richness, this was the task imposed upon us. Eight major points, as revealed by Viollet-le-Duc's theory on historic preservation, governed the preservation and remodelling of the case presented.

1   The courageous use of new materials and their method of application.
2   The application of new structural elements for the overall stability.
3   Structural integrity (new and old).
4   Study of the architectural style based on the relationship among material, structure and space.
5   The final utility of the edifice.
6   The emphasis on novel features and new requirements.
7   Sincerity and exactitude of the final work, based on respecting ancient architectural vestiges.
8   The final completion of the edifice, leading to a model to be imitated. Every portion that was removed was replaced with better materials, in a stronger and more perfect way.

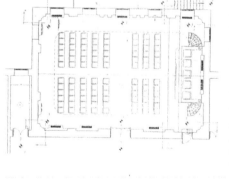

Figure 5.1 The Egyptian Society of Political Science, Statistics and Legislation.

As a result, the restored edifice should have a renewed sense of existence, longer than has already elapsed. It cannot be denied that a process of restoration was a severe trial to the structure. The scaffolding and the partial removal of masonry shook the whole work from its base to its top. This diminution of solidity was counteracted by increasing the strength of the renewed parts – by introducing improvements on the structural system, the introduction of different sizes of structural steel I beams, well-contrived tie rods to support the floor of the structure and by providing better appliances for structural resistance. Based on its structure, its anatomy and its temperament, every detail of the building was mastered. Similarly, it was our task to be acquainted with forms and architectural styles belonging to the edifice, and to the school to which it owed its origin.

Since all edifices whose restoration is undertaken have a particular use, the role of the restorer cannot be assumed to change the actual utility of the building. The edifice should not be less convenient when it leaves the architect's hands than it was before restoration; moreover, the best means of preserving a building is to find a use for it, to satisfy its requirements so completely that there shall be no occasion to make any changes. Based on this concept, the case presented was restored for a final utility (as a major conference hall and meeting rooms). We must admit that we were on slippery grounds as soon as we deviated from literal reproduction. We proceeded, as a master of the Middle Ages would have done, not to hide novel features – since the ancient masters, far from dissembling a necessity, sought on the contrary to invest it with a becoming form, even making decorative features of these new requirements. The case of the central heating system, air-conditioning and air handing units were emphasized and treated with special care, not to hide these new additions but, on the contrary, to reveal their usage as a new work of art on the boundaries of the main conference room.

A very important principle to be observed is to pay regard to every vestige indicating an architectural arrangement. The architect should not be thoroughly satisfied until he or she has discovered all the combinations that accord with the vestiges of the ancient work. Based on this, and on the vestiges of the previous existing chairs and their special method of fixation on the ground, we were able to decide on their arrangements, their number and the inclination of the floor in relation to the sight lines of the audience. This furnished a guarantee for the sincerity and exactitude for the work to be done.

Based on the idea of "completing the structure" to give back to our beautiful building all its splendor and to restore to it all the richness of which it has been robbed, this was task imposed upon us. Externally, the building is surely beautiful enough that it would be pointless to add anything to it. Internally, this was not the case. The two major stained-glass windows that overlooked the main conference hall and the main staircase required a complete restoration and redesigning. The restoration of these windows required a scrupulous examination of their construction, for which it was necessary to devise some architectural modifications. Perhaps without changing their interior and exterior profiles, we gave them a much greater strength by thickening their tracery. These two windows were completely ruined. They needed to be renewed almost entirely. We believed that the execution of stained-glass windows would be one of the most splendid means of interior decoration. Nothing equals the richness of these transparent pictures, which are an indispensable component of monuments of this epoch. The engravings and the precious drawing on the ceiling also required a complete restoration. A minute swabbing was the first operation to be utilized, in order to know the state of these engravings and to recover the traces of these paintings. It was evident in this case that this cleaning gave us some positive proofs of the general system formerly adopted. Up till now we have admitted painting only as decoration of the ceilings. In a restoration project like the one presented, it was impossible not to try to harmonize all the accessory objects with the final design of the edifice, above all when they are really important. Nothing was removed or replaced unless it was to the benefit of the final design of the whole project.

Finally, the principles that we have set forth, based on Viollet-le-Duc's concept of restoration and historic preservation, should not be forgotten when it is a question of a monument as important as the one presented. In my opinion, we have to remain constantly faithful to the principles that we have set. It is necessary not to hesitate and make excuses but to proceed with confidence, take no risks, and to be sure to succeed. We think that each part, added in whatever epoch, ought in principle to be preserved, strengthened and restored in the architectural style appropriate to it. This has to be done with a reverent discretion.

Reference

Eugène-Emmanuel Viollet-Le-Duc 2011. *Lectures on Architecture*. New York: Dover Publications.

6 Integrating value map with Building Information Modeling approach for documenting historic buildings in Egypt

Soheir Hawas and Mohamed Marzouk

The concept of Building Information Modeling (BIM) is implemented in this chapter to develop a value map by applying recent technologies for recording and documenting historic buildings in Egypt. The concept of the value map has emerged to support and maintain aesthetic, architectural, cultural and environmental values, and highlights the natural elements. This chapter proposes the use of BIM technology to collect, compare and share and manage all the heterogeneous data related to the geometry and state of conservation of historic buildings in Egypt. The proposed methodology considers scanning historic buildings using laser scanners to generate numbers of geometric features (point clouds). The processed surveyed data is then used as a mapping framework for plotting interactive geometric information and parametric objects. These parametric objects represent architectural elements, which are then combined together to generate a virtual model of the historic building. The resultant BIM model is a rich digital model that can be used to support decisions related to the analysis and conservation of historic buildings in relation to surrounding environments.

6.1 Introduction

Historic Building Information Modeling (HBIM) is defined as a library of parametric objects based on historic architectural data, in addition to a mapping system for plotting the library objects onto laser scan survey data (Murphy et al., 2013). In the architecture, engineering and construction (AEC) sector and among heritage communities there is increased importance on modeling, documenting and representing buildings with intelligent information-enhanced models as opposed to previous digital representations that contained only geometry. The main motivation for smart information-enhanced models is the wide variety of applications that the information-rich models can be used for. This includes applications for documentation and management of buildings along with great capabilities for energy, structural and economic analysis of buildings.

The concept of Building Information Modeling (BIM) is a major development in 3D CAD modeling and provides the ability to represent buildings with smart information-enhanced components. BIM incorporates object-oriented, parametric and feature-based modeling concepts combined with the addition of a dynamic 3D database for storing information relating to buildings. Due to its many benefits, BIM has received a lot of attention in both industry and academia. As BIM was designed for modeling and representing new and modern buildings, the focus of this attention to date has mainly been on the use of BIM in the planning, design and construction stages of a building (Volk et al., 2014). Recently, however, there has been a shift in BIM-related research from early life cycle stages to maintenance, refurbishment and management of buildings throughout their complete life cycle (Volk et al., 2014). The benefits of BIM make it a very suitable solution for modeling and managing information relating to a building after the construction stages and throughout the building's life cycle. A BIM for an existing or historical building can be used as a documentation tool for conservation work, retrofitting or renovations, or as a tool for performing building analysis.

The automatic generation of structured BIMs from point clouds is a primary focus for a lot of research in the area of as-built BIM. Although progress has been made in this area, the current practice for creating as-built BIMs requires manual modeling techniques which can be time-consuming, labor-intensive, tedious and require skilled workers. Because BIM is mainly used for design of new buildings, commercial software is also currently limited as far as tools for modeling complex architectural elements and irregular geometries that often occur in existing buildings due to deformation and damages over time. The time-consuming and tedious nature of the as-built BIM process motivates the need for new, more automated solutions designed specifically for modeling existing and historical buildings from survey data.

The creation of an as-built BIM or HBIM requires accurate measurements to be taken on a building, which can be used to create a 3D BIM representation. This data can be collected using laser scanning, photogrammetry or other traditional survey technologies. The current problem with these technologies is that they produce very large unstructured datasets such as 3D point clouds that need to be converted to structured 3D models for useful further applications. The creation of an as-built BIM or HBIM from survey data can be divided into four main stages. This includes data acquisition through an image survey campaign, generating a 3D mesh model, creating a point cloud model and finally modeling with BIM. The fourth stage of modeling with BIM is the longest stage of an as-built BIM project. This is a reverse engineering process where BIM components are created and mapped to the survey data to create the BIM model.

Many research efforts have been made to model historical buildings. Murphy et al. (2009) outlined in detail the procedure of remote data capture using laser scanning and the subsequent processing required in order to identify a new methodology for creating full engineering drawings (orthographic and 3D models) from laser scan and image survey data for historic structures. Murphy et al. (2013) started with remote collection of survey data using a terrestrial laser scanner combined with digital photo modeling. The next stage involves the design and construction of a parametric library of objects. In building parametric objects, the problem of file format and exchange of data has been overcome within the BIM ArchiCAD software platform by using geometric descriptive language (GDL). The plotting of parametric objects onto the laser scan surveys as building components to create or form the entire building is the final stage in the reverse engineering process. The final HBIM product is the creation of full 3D models including detail behind the object's surface concerning its methods of construction and material make-up. Guarnieri et al. (2010) presented a case study on the development of a web-based application for user access and interactive exploration of three-dimensional models by providing integrated geometrical and non-geometrical information into an intuitive interface. The main feature of this interactive system is to provide the user with a completely new visit experience based on a free interactive exploration interface of the object. Remondino (2011) reviewed the actual optical 3D measurement sensors and 3D modeling techniques, with their limitations and potentialities, requirements and specifications. Examples of 3D surveying and modeling of heritage sites and objects are also shown.

Oreni et al. (2013) illustrated the utility of switching from a 3D content model to a Historic Building Information Modeling (HBIM) in order to support conservation and management of built heritage. This three-dimensional solution is based on simplified parametric models, suitable for industrial elements and modern architecture that can be usefully applied to heritage documentation and management of the data on conservation practices. In this sense, the potentials in starting the definition of an HBIM targeted library are investigated, towards the logic of object data definition, beginning from surface surveying and representation. Vault and wooden beam floor analysis show how an HBIM for architectural heritage could be implemented in order to assemble different kinds of data on historical buildings. Fai et al. (2011) focused on adapting automated data processing and pattern recognition that leverages 3D point cloud data towards the generation of a simple building envelope. They described the Batawa project model, which is a redevelopment proposal for approximately 600 hectares of land that includes a former factory in Toronto (a cluster of three nineteenth-century heritage buildings) with its rich history of modern architecture and town planning. The purpose is to document the heritage assets of Batawa and to develop a BIM model using available software packages that are appropriate for specific applications (AutoCAD, Civil 3D, SketchUp, Revit). That will serve as a digital archive to help in conserving the extant heritage buildings and planning and to test future development proposals within the context of these historic buildings and plans.

Baik et al. (2013) introduced the Jeddah Historical Building Information Modeling approach (JHBIM). This approach is used to represent Hijazi architectural elements based on laser scanner and image survey data. The approach is divided into three steps, which are: 1) capturing data using image/range-based methodologies, 2) processing of data and 3) modeling of historical components as parametric objects using different building information modeling applications. Kersten et al. (2014) presented a methodology to record buildings using 3D acquisition methods. They applied their approach to a Segeberg townhouse in Germany. A historical documentation of six construction phases was presented in the model. They also generated a virtual tour from panorama photographs, which allows visualization from both inside and outside the building. Megahed (2015) introduced an overview of concepts as well as surveying and representation techniques used in historical building information modeling (HBIM). She also presented a framework that helps in defining different aspects of historic preservation and management. The introduced framework described how policy, process, technology and people should interact throughout the project life cycle.

6.2 Value map

The concept of urban harmony has emerged in order to preserve the cultural, aesthetic, architectural and environmental values of different urban areas, especially after what happened in the last thirty years of inappropriate appearance of many urban and heritage areas. Therefore, the definition and documentation of areas with natural and architectural values is the main step to face current problems. A value map is an urban tool that incorporates a methodological framework that enables identification and profiling areas with values. Moreover, it helps in defining their properties such as volume, area, axes, paths, locations, and so forth. Maintaining and preservation of heritage areas depends on a sequential and integrated series of procedures. The first procedure is to study and analyze the characteristics and features of heritage areas. The second procedure is to identify the aspects of the richness and originality and the values that characterize heritage areas. The value map introduces a reference framework for identification and formulation of intervention policies to coordinate the cultural and aesthetic appearance of historical areas as well as natural areas such as sanctuaries, parks, palm trees, and so forth. Figure 6.1 summarizes the importance and purposes of a value map, whereas Figure 6.2 depicts the framework of a value map.

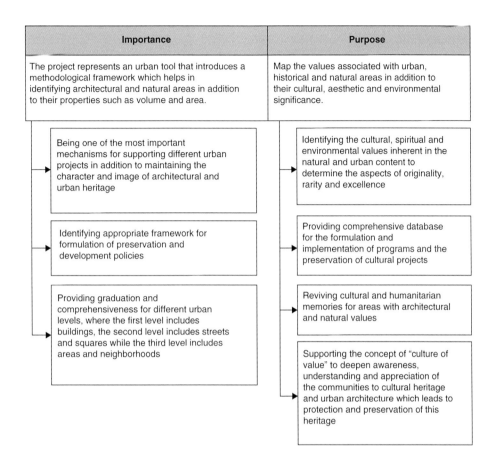

Figure 6.1 Importance and purpose of a value map.

6.3 Proposed methodology

This section presents the methodology for image processing using point clouds to generate an HBIM model. Once the HBIM model is created, the next stage involves the integration of the 3D model into 3D GIS for further management and analysis (see Figure 6.3).

6.3.1 Image survey campaign

A historical window and entrance located in downtown Cairo have been captured by default camera with close equal distances to refine the model with the best accuracy for details. The more photos captured for the model, the more accuracy we would get. Through open source software Autodesk Memento, the captured photos have been imported, and before modeling it, we checked the validation for all the photos and excluded the one that doesn't match with others. Finally, the validated captured reality photos have been converted into high-definition 3D meshes. Figure 6.4 shows the close stands for capturing images. The image survey campaign on left shows the positions and numbers of stands to capture the targeted model and on right shows entering captured photos and checking validation before creating the model.

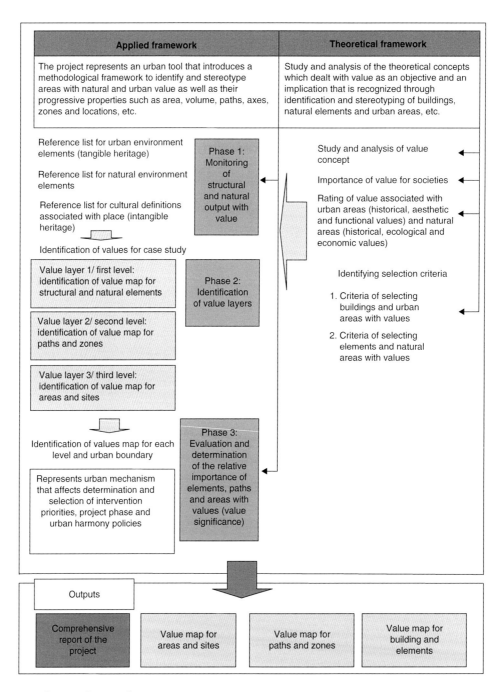

Figure 6.2 Value map framework.

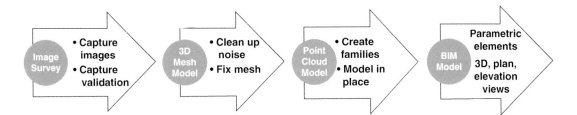

Figure 6.3 Stages of HBIM generation.

Figure 6.4 Image survey campaign.

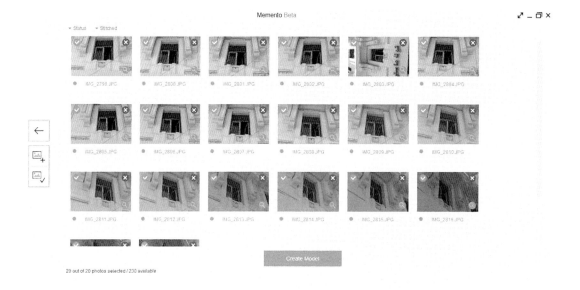

Figure 6.4 (Continued)

A historical window and entrance located in downtown Cairo have been captured by default camera with close equal distances to refine the model with the best accuracy for details. The more photos captured for the model, the more accuracy we would get. Through open source software Autodesk Memento, the captured photos have been imported, and before modeling it, we checked the validation for all the photos and excluded the one that doesn't match with others. Finally, the validated captured reality photos have been converted into high-definition 3D meshes. Figure 6.4 shows the close stands for capturing images. The image survey campaign on left shows the positions and numbers of stands to capture the targeted model and on right shows entering captured photos and checking validation before creating the model.

6.3.2 3D mesh modeling

The first stage in generating a solid mesh from the scanned pictures is sub-sectioning and cleaning. Generally the point cloud will contain a fair degree of noise. Excess noise will prevent an accurate solid mesh being created, for the most part, unable to discriminate between the clusters of noise and solid surfaces. This noise can come from a variety of sources, particularly in urban environments where people, vehicles and even precipitation will cause partial shadows or point clusters within the cloud data. Figure 6.5 illustrates the impact of the noise present within the scene. Excess noise comes from different sources, such as sun reflection on the building, or an excessive number of points, or from a bad digitalization.

After generating a 3D model, the model will be cleaned up from excess noise by smart selection and cleanup tools provided in the platform. Also with smart fixing tools, the model will be fixed with cutting-edge mesh analysis/diagnostics for finding and fixing mesh issues such as particles, holes,

Figure 6.5 Noise impact in the scene.

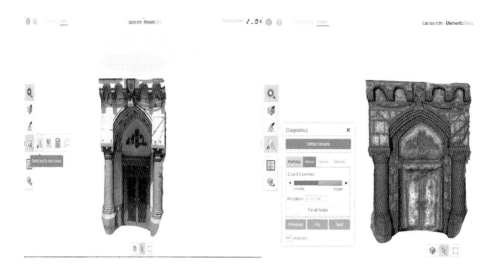

Figure 6.6 Noise cleaning and fixing mesh issues.

spikes and tunnels. Using smart re-topologies tools such as re-triangulate, subdivide, and detect and fix mesh issues will finally help to get the best 3D mesh model. Figure 6.6 shows the models during cleaning and fixing from excess noise.

6.3.3 Point cloud and HBIM model

The 3D point cloud, after cleaning, fixing and decimation phases, demonstrated as a good model for the case study. In order to verify reliability and usability of methodology for point cloud processing in a BIM environment with high accuracy, we have exported the model with a .rcp extension and imported it in the BIM platform; we use Autodesk Revit 2015. Figure 6.7 shows the two models after cleaning and transferring both of them as point cloud models.

Figure 6.7 Generated point cloud models.

Generating models from point clouds aid in minimizing the number of steps and avoiding loss of accuracy, data quality and details. Revit also shows the point cloud in each view (maps, front and 3D views). Some objects were made immediately, using the basic commands of the Revit menu. The modeling was performed using parametric elements already in the Revit internal library; some objects were created as families outside, and then they were imported into the model. Not many libraries are available for free in online databases. The more complex elements were made in place using the Revit model form: Extrusion, Blend, Revolve, Sweep, Sweep Blend and Void Forms. Figure 6.8 depicts the BIM modeling process using Autodesk Revit 2015.

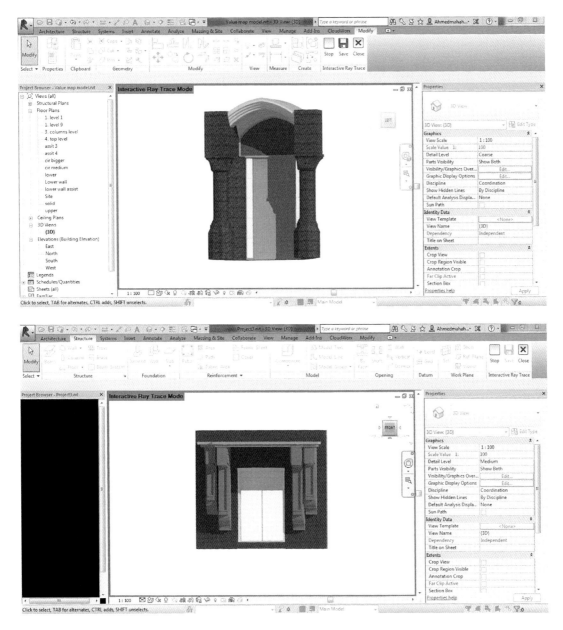

Figure 6.8 Generated HBIM models.

6.4 Conclusions

This research presented a methodology to develop a high-quality 3D model that is semantic-aware, able to connect geometrical-historical survey with descriptive thematic databases. In this way, a centralized HBIM will serve as a comprehensive data set of information about all disciplines, particularly for restoration and conservation. The use of laser scanning and photogrammetry can aid in recording a very high level of details in the field for heritage documentation. The HBIM process can be automated using accurate parametric objects that can be altered and mapped to heritage survey data. As a result of using HBIM, several documents such as plans, sections and elevations can be produced. The integration to GIS can provide further capabilities for linking the 3D heritage model to information systems. The integrated models help in performing efficient management and analysis that is required for maintaining important cultural heritage sites. As such, preventive maintenance modeling can be done regularly for more preservation of our heritage sites.

References

Baik, A., Boehm, J., Robson, S., 2013. Jeddah historical building information modeling "JHBIM" Old Jeddah – Saudi Arabia. *International Archives of the Photogrammetry, Remote Sensing and Spatial Information Sciences*, 40, pp.73–78.

Fai, S., Graham, K., Duckworth, T., Wood, N., Attar, R., 2011. Building Information Modeling and Heritage Documentation. *CIPA 2011 Conference Proceedings: XXIIIrd International CIPA Symposium, (CIPA)*, Prague, Czech Republic.

Guarnieri, A., Pirotti, F., Vettore, A., 2010. Cultural heritage interactive 3D models on the web: An approach using open source and free software. *Journal of Cultural Heritage*, 11(3), pp.350–353.

Kersten, T.P., Hinrichsen, N., Lindstaedt, M., Weber, C., Schreyer, K., Tschirschwitz, F. 2014. Architectural Historical 4D Documentation of the Old-Segeberg Town House by Photogrammetry, Terrestrial Laser Scanning and Historical Analysis. In *Digital Heritage. Progress in Cultural Heritage: Documentation, Preservation and Protection*. ed. Ioannides, M., Magnenat-Thalmann, N., Zarnic, E.F, Yen A-Y, and Quak, E, 5th International Conference, EuroMed 2014, Limassol, Cyprus, pp. 35–47.

Megahed, N.A., 2015. Towards a theoretical framework for HBIM approach in historic preservation and management. *International Journal of Architectural Research*, 9, pp.130–147.

Murphy, M., McGovern, E., Pavia, S., 2009. Historic building information modelling (HBIM). *Structural Survey*, 27(4), pp.311–327.

Murphy, M., McGovern, E., Pavia, S., 2013. Historic building information modelling – adding intelligence to laser and image based surveys of European classical architecture. *ISPRS Journal of Photogrammetry and Remote Sensing*, 76, pp.89–102.

Oreni, D., Brumana, R., Georgopoulos, A., Cuca, B. 2013. HBIM for Conservation and Management of Built Heritage: Towards a Library of Vaults and Wooden Bean Floors. *SPRS Annals of the Photogrammetry, Remote Sensing and Spatial Information Sciences, Volume II-5/W1, 2013 XXIV International CIPA Symposium*, Strasbourg, France, pp. 215–221.

Remondino, F., 2011. Heritage recording and 3D modeling with photogrammetry and 3D scanning. *Remote Sensing*, 3(6), pp.1104–1138.

Volk, R., Stengel, J., Schultmann, F., 2014 Building information modeling for existing buildings – literature review and future needs. *Automation in Construction*, 38, pp.109–127.

7 Capturing heritage data with 3D laser scanners

Antony Pidduck

Through the use of technology, the opportunity is here to accurately capture in minute detail the built heritage around us, to record for the future, develop our knowledge and understanding and use the information to support sustainability. This chapter examines the process and seeks to determine if there is a place for these methodologies.

Tens of thousands of years ago, our ancestors began recording the world around them in the form of cave paintings, and these have provided insight into our past, our heritage. We continued the tradition through the ages, often in a stylised and politicised manner, in paintings, etchings and sketches. Recently this has been achieved using the photograph, originally on paper but now digitally, and with the mobile phone now containing a digital camera, chronicling our lives has never been easier, either through image or video. There are a number of initiatives using this methodology to create models of monuments to archive a record for the future of our past. Why, then, do we need 3D scan data to record our built heritage – how is it obtained and what are the benefits?

The pace of development in the world of digital continues unabated, with incrementally and wholly new technologies that were initially driven through military application but now by the insatiable requirement for immediacy. Everything is in a process of miniaturisation, coupled with routines and subroutines, algorithms and unlimited fast access to data via 4G+ or Wi-Fi. Currently in development is the Internet of Things, the interconnection of all objects (buildings, cars, electronics, white goods etc.) to collect and exchange data without human intervention. Can this improve the conservation of our heritage?

Ultimately, this is where heritage will move, to become the fusion of analogue with a digital interface. The basic fabric will still need monitoring and interventions but these will be informed and managed through the use of data continuously gathered and analysed autonomously. In the twilight years, as the development of a built asset slows to a shuffle under the force of conservation, preservation becomes the order of the day, valued for heritage, human endeavour of the past. The record is important, determining in part the lengths taken to preserve these assets; hearsay today is less accepted as access to digital repositories is now unlimited. Digital data capture supports these records, extending the knowledge and evolving nature of the asset. Furthermore, we share the data locally and throughout the world without derogation using the Internet, providing access to our knowledge, our heritage. Is this quality data, and how do we capture it and ensure that future generations retain access to it? Figure 7.1 shows a Leica Geosystems laser scanner capturing data in the undercroft of Rufford Abbey in Nottinghamshire.

The historical architectural solution was to take a myriad of accessible measurements, recording by hand with possibility for error. Those areas out of reach were photographed, and where there was a regularisation of component, the dimensions were calculated. For example, a panel of bricks would be measured at low level, an object selected at high level and the distances extrapolated from the

Figure 7.1 Leica Geosystems C10 laser scanner in the undercroft of Rufford Abbey, Nottinghamshire. The ribbed ceilings and columns determined the positions of the scanner; however, the exhibition objects hindered a complete record being obtained.

brick dimensions. There is the assumption that within a tolerance there was accuracy, but that did not allow for non-regular components, revised methodologies and so forth. Furthermore, any deviation of plumb is not recorded, and dimensions for a radius surface in elevation or section requires a set of tools or templates to accurately obtain for record. Archaeologists sketch and include scales in addition to dimensions that can help in situations where inaccuracies develop in the records. The benefit to the consultant, and therefore by extension the heritage asset, is as the record is built by eye and hand there is an intimacy which supports greater understanding and aids assessment. This analysis wanes when the computer screen mediates the relationship.

There are a number of situations where data acquisition can be difficult, such as at the side of transport routes, or because of the fragility of the structure, perhaps due to damage or natural detrition. Recording in this situation may need additional measures which may add cost or place others in difficult, unsafe positions to secure the structure. An alternative solution needs to be considered, one where time or the risk to the data gatherer can be constrained.

It is vitally important to understand at an early stage what the data is for, as this will determine a number of the parameters in the capture. Data as a record requires completeness and quality. If the data is to build a model, however, then it is feasible to capture only the general dimensions to build objects from. For example, when building a model with windows, the structural opening for a window needs locating in the wall object, along with a height and width dimension. Beyond this, a collection of referenced images can provide a record and be linked to the object. Therefore, in this situation extensive acquisition of data may be a step too far.

Laser scanners are part of the hardware being continually developed to accurately record our built features in a reasonable time frame. This data forms the basis of 3D interactive models, accurately dimensioned and overlaid with images to inform both the expert and the layperson alike. Regular data capture leads to model evolution and continual knowledge development, but also forms the basis of revolutionary expansive understanding of the heritage asset for future generations.

Originally lasers, or 'light amplification by stimulated emission of radiation', were developed by the military as range finders before being used in construction for line and level maintenance. The first scanners using a spinning mirror were placed in aircraft for ground mapping before terrestrial applications were considered. When fixed on top of a robust tripod, today's laser scanner has the ability to capture millions of points per second from which to build 3D point clouds, and in the same 'black box' digital images can be captured too. These individual data sets are merged in a post-process operation to provide dimensionally accurate 3D models.

Manufacturers provide specialist versions of instruments dependent on required accuracy (sub-millimetre is possible) and distances (>250 mts) using a number of range-finding technologies. Suffice to say the resultant information is a point in space fixed by x/y/z coordinates from the reference location at the centre of the instrument. Greater accuracy is achieved by pulsing the laser at the same point and, after omitting the spurious results, recording the optimum value. These quality adjustments can counter the difficulties of radius and non-reflective surfaces to promote accurate results as discussed later in the chapter.

In the terrestrial scanner the mirror located on a horizontal axis spins at speed, reflecting the laser at the object to be recorded; all but the data in the bottom quadrant (90° approximately) of the circle is obtained. The mirror also directs the returning laser reflection to the receiver. The body containing the mirror also spins on the vertical axis, rotating through 180° to capture a full hemisphere of points where they are located within the range of the instrument. The data is generally stored on an internal hard drive and displayed on an integral touchscreen for immediate verification. If required, a second 180° body spin captures the images in the field of view and saves them as .jpg file extensions. It is worth noting at this point that these are individual files and can form a database of high-quality images for future use.

The methodology follows a similar scenario for all scanning instruments. A surveying tripod is set on stable ground and approximately levelled by eye across the machined top face. The instrument is securely attached to the top face before being accurately levelled using three screws in the baseplate, guided by the circular vial level located in the casing. Once this process is complete, the internal compensators allow for any slight deviation. Using the touchscreen, the final process is to adjust the settings to determine the quantity and quality of the data to be captured. It is vitally important to have stable ground for the tripod, for example, not on a poorly bedded paving slab – if during data capture somebody were to step on the slab, the movement would cause the instrument to go off level and stop.

The instrument is a line-of-sight device: if objects cannot be seen they will not be recorded, and this is one of the limitations of scan data. Furthermore, it is a 'skin' of points co-located without links to any other point – it has no depth – the thickness of the material generally cannot be determined from the data. In order to build a complete 3D model it is necessary to complete more than one scan to fill those unseen areas.

If there are objects between the building and the scanner, for example, currently the laser cannot see through or around the object, and the data will remain incomplete. This is rectified by careful choice of another scanner position recording a second set of data which, when brought together with the first, will provide the missing parts. It is important to remember there may still be missed data which may leave visible holes in the model. It is wise to consider the quality of the asset or final use of the model before completing additional scans to record every aspect and fill the missing parts. Undertaking this is time-consuming and may result in an unwieldy model which will cause issues as the project progresses.

Moving towards the building can assist the capture of data, especially in protruding areas such as the ceilings of porches or window reveals, although this can be at the expense of roof areas which cannot then be seen by the laser. The data at roof level is generally difficult to obtain, especially where the line of sight is acute to the building. Heritage buildings with parapets and wide overhangs leave invisible areas wherever the instrument is located. Higher ground, adjoining buildings or the object itself can be used to provide locations to gather additional data sets. However, the issue is finding sufficient accessible and safe positions to set up the tripod where it will not subjected to excessive movement due to the exposed position.

When considering locations to capture the data, it is worth spending time at the site considering the advantages and disadvantages of particular positions. Best practice is to plan and record on a site drawing the scan locations as the project proceeds; knowledge of these locations can assist in the process of bringing the scan data together in forming the model. Choice of location is based on a number of factors including traffic, proximity to risks and so forth, but a risk assessment can benefit the process as safety of personnel, the public and the equipment should be paramount. When looking for the optimum locations for the instrument, overlap between the scan data should be built in, preferably on the object. A third point of reference could also be used, such as another building away from the object of interest, especially if there is limited scope for connecting the scans. An example of this would be a terrace of buildings with no connection between the front and rear; a tall tower could link the two elevations.

For any heritage asset, the site, location and setting are all valuable aspects to the object being recorded. Any listed object can have significant yet remote elements which are considered important to the setting or to the use. As the scan data is a dome covering the whole view 360° horizontally, there is an additional layer of information being recorded which should be considered too. The result may be to add scans to fully capture the surrounding situation, although these may be considered secondary to the main object and the quality may be revised to suit this hierarchy of data.

In an ideal world, the area between the scanner and the building becomes an exclusion zone for people, vehicles, signage and so forth, as this improves robustness of the data. Furthermore, whilst a number of the instruments are rated at Ingress Protection 54 (IP54: 5, limited protection against dust ingress, and 4, protected against splash water from any direction), the data can be affected by rain and dust in extreme cases. When it is raining, the area above the scanner especially becomes awash with data from raindrops, which may need removing when building the model. The latest firmware for scanner instruments coupled with modelling software is able to deal with this situation in the algorithms. It is possible to pause and resume the collection without impact on the data set for large high-sided vehicles and so forth, although this extends the time period of the project on-site. There are solutions which may need to be enacted such as early start/late finish, weekends, holidays, or night working if images are not required (very important if the asset is on a major transport route) or winter when deciduous trees have no leaves or fruit. This may also have an impact on the recorded accuracy in the model stage, as discussed later in the chapter.

Whilst giving thought to the scanning data collection it is also necessary to consider the image acquisition. The instrument has some basic control settings to cope with overcast or sunny situations, but high contrast will cause inconsistencies in the model information. If there is a dependency on these images for building surveying and recording, it may be necessary to complete the data capture at times when there is an overcast sky or on façades not subjected to direct sunlight. There can be benefits in the presence of direct light when using standard cameras as it adds shade to enhance the 3D understanding, but this is not necessarily the case when using a scanning instrument. Complete scan data has infinite viewing possibilities to aid examination and understanding, especially valuable in complex models, and it is therefore beneficial to have uniform lighting conditions to improve the model rendering and the information in the image. Figure 7.2 shows scan data captured and displayed on the instrument screen; functions such as pan, zoom and so forth allow the data to be checked for integrity.

As previously touched on, the quality of scan data, and therefore accuracy, can be considered as two functions: the distance between each point on a single plane and the accuracy of determining the position of the point in space. Both of these functions can be adjusted on the instrument. Aspects such as complexity of the object surface to be scanned or uniqueness may impact on the quantity of information to be gathered. The quality can be defined as the spacing between the points; by decreasing the distance, the number of points increases and the quality of the data rises. However, the consequence is extending the data capture time and larger file sizes. This is defined on an instrument as a function of distance from the object – a large area to be scanned, easily visible without intervening trees and so forth, may benefit from fewer scan positions at a greater distance away, capturing more of the object and therefore the need for the higher resolution.

Another facility on the instrument is the ability to define the area to be scanned, thereby reducing the time for data capture and the file size. There are possible consequences to this action, generally only identified in constructing the model, and it could be argued the time saving is of little significance. With certain instruments the time taken to complete a full dome scan and image capture with reasonable accuracy and medium quality can be less than 10 minutes, and with a skilled operative the whole process, including setting up, can be as low as 15 minutes. A benefit of this area definition may be to scan small, detailed areas such as statues, friezes, tympanum and so forth at a higher resolution. To ensure the data is complete for highly detailed surfaces there is a need to move the instrument more regularly to get to the unseen areas at the higher resolution. When completing this workflow it is important to note the warning earlier and ensure there is enough overlap to bring the scans accurately together. Therefore, it may benefit the model to do a low-resolution scan to assist in the location of the object and a high-resolution scan to capture the object to provide the data for the model. An alternative solution is to capture the detailed areas using a camera and build a 3D model in photogrammetry software. This will provide a high-definition surface model which may be more satisfactory to record the condition and easier to build a 3D printed model from. The accurate measurement data may be unnecessary; if it had been scanned with the façade, the dimensions are already recorded.

Data for internal space can be captured using a laser scanner; the benefits and limitations are similar to those external to a building. Foremost, the stability of the floor can provide difficulties, and in this situation remote control can be used to start/pause and identify when the process is complete. Furniture, fixtures and fittings can inhibit data captured, especially at floor level, which may provide difficulty in assessing a building or recording details. With a building of exceptional interest, it may be necessary to remove some or all of these to record the significant elements. The one major difficulty is in linking the data between rooms, and therefore it can be necessary to factor in an instrument position in the doorway between each space.

Due to the shorter distances involved, there is the possibility to reduce the number of points captured without negatively affecting the accuracy of the data. That said, if radius corners or soft surfaces are prevalent, a high-quality setting would improve the accuracy of the data recorded. Image capture for internal space is another aspect for consideration; the scanner may have some adjustment for artificial light, however, this is generally coarse and will not compensate for all lamp types, especially with the complex range of LED sources. Where there are features of particular importance it may

Figure 7.2 Scan data captured and displayed on the instrument screen will show the intensity of the returning laser light using the colour; functions such as pan, zoom and so forth allow the data to be checked for integrity.

suit to define the area and, without repositioning the instrument, rescan at the higher setting with shorter distance between the points and/or the higher quality, although reimaging the area would not be required. Certain instruments are more suited to data capture in small spaces by being lighter and more compact, but it is important to ensure there is sufficient overlap between the scans to accurately position them when building the model.

It may be of benefit to establish a relationship to the external areas in a small number of locations throughout the scan of the internal space. A complex model will be the sum of a large number of scans, and the ability to relate more than one to others in the series will provide an opportunity to check for accuracy and ensure quality of data. The methodology in this situation will be determined by a number of factors around accessibility and proximity. In a simple situation, a room with a view from the instrument of an adjacent structure recorded from another scanner location will provide all that is required to check the accuracy. Where this is difficult it may be necessary to introduce a target into the window area visible from both scanner positions (see Figure 7.3), where the accuracy of the model can be tested at the target. The target location should be recorded on the plan and remain in position between the scans and should be scanned at a higher resolution from each of the instrument locations.

Geolocation is the accurate positioning of objects on the planet surface. Global Positioning Systems (GPS) or Global Navigation Satellite Systems (GNSS) use a number of satellites to define coordinates, and these can be placed in the point cloud model. The methodology requires one or more locations for the instrument. After capturing the data, the tripod remains in position and an antenna is attached, linked to the GNSS unit. The resultant millimeter-accurate position of the scanner instrument is recorded and downloaded into the data. For a number of years geographic information systems have been used to store, process, study, manage and present data, and links can be drawn to Building Information Modelling (BIM). As more information is captured and uploaded to these systems there is greater opportunity for analysis to improve the management and conservation of our built heritage.

Once the data has been captured it can be downloaded to a computer ready for processing. In consideration of data security, it is good practice to download a copy to a second device for storage without process. Finally, for the surveyors' purpose it would be advisable to copy all image files to a third location labelled with the position cross-referenced to a drawing and a direction for the first image. It is at this point consideration should be given as to how to collate the model. There may be advantages in segregating the external from the internal data sets and breaking them down floor by floor for large buildings. This will reduce the size of the files being handled by the computer, which may provide benefits as discussed later in the chapter. It is important to refer to the drawing to ensure all data sets are included and to note where there are overlaps between each data set.

Another aspect to consider is the computer hardware being utilised to process the data – whilst speed and size of RAM available is important, the graphics card can be a deciding factor, especially when moving the model on the screen leads to time to rebuild the view. To give some idea as to file sizes, eight scans of medium quality can result in a file size of 8.5 GB. Figure 7.4 shows internal point cloud data.

Once the data is downloaded and stored, the workflow can go a number of ways depending on the information required. A command in most software will import the data from each of the locations with the images before auto-stitching using algorithms to place the data in a single 3D model space. On occasion, if the auto-stitch fails to find the overlap, the scans can form disparate groups or will be left as single point clouds for manual manipulation, alignment and grouping. Once complete, parts of the cloud no longer required can be deleted. However, if a number of data sets are to be brought together in a single model this may be best left until the process has been completed, as the discarded data may be needed for alignment. There may have been instruments from more than one

Figure 7.3 A target is located in a window to meld internal and external scan data.

manufacturer used to gather the data, due to size or weight considerations, but most software can cope with this when building a model.

Within the model it is possible to clean unwanted data or noise, which may include vehicles or the 'ghosts' of people moving through the point clouds. For example, if the internal areas are to be merged with external, there will be an amount of data collected through the window glass, usually ceilings or internal walls, which should be carefully cut out. Alternatively, there is the possibility of

Figure 7.4 Internal point cloud data reveals the structure in the surface.

opening each point cloud and removing these unwanted points before stitching the clouds. The benefit to this is the time taken in manipulating the cloud to carefully identify and delete the data (larger clouds can take more time in regenerating the view). A secondary aspect is in a single point cloud there is less possibility of removing important data. However, the opening and closing of each cloud can take time depending on the quality of the data recorded. Once completed, the point clouds are auto-stitched as before and finally 'merged' to a single model. Again, for security of the data it is worth backing up the model, copying it into another model space and appropriately naming the file. If the data has been split into more than one set, these need to be drawn together into a single model. A word of warning, though: it is important to retain file links to build the model, otherwise if you move the files or save them to a new location it may be necessary to rebuild the model.

There is a possibility of considerably reducing the data, and therefore the file size, by removing all data for the surrounding areas. This may leave the object without context, but without the interference of the excess data the object can be analysed to the fullest extent. It may be noted at this time that there are areas of the model without data, holes where there were 'screens' before the object interrupting the laser, even taking account of more than one set of scan data. Figure 7.5 shows a 3D model of a Grade II listed campus building. The image data colouring the points reveals the 'holes' – black areas where data has not been captured.

As with any other 3D computer model, it is interactive, providing the ability to move around in the model and pan/zoom in to examine areas in greater detail. As the model is dimensionally accurate, any distance can be sought between two points in the cloud; these can be provided as a product in three axes or a direct length. Furthermore, areas and volumes can be identified and defined and the data stored in the model.

The data can be viewed in orthographic or perspective mode and 2D images extracted from the data. The colouring of the cloud via 'rainbow' colours generally helps to view the data and is based on the intensity of the reflected laser light; there is often a function to view in grey scale too. If the images were captured, the software can interrogate these and assign the recorded colour to the

Figure 7.5 This 3D model of a Grade II listed campus building with the image data colouring the points reveals the 'holes' – black areas where data has not been captured.

point, thereby recreating the model as seen on the day. As previously discussed, however, this will be somewhat dictated by the lighting conditions, leaving the views impacted on by contrast or colour rendered by the light source. This is a problem previously encountered in standard colour photographic records. It must also be noted that the images do not always represent the scanned situation. For example, if the scan data shows an empty highway and the image contains a vehicle, it will attach the colour from the image to the road surface data, leaving a 'flattened' car; similar situations will occur in the records of heritage assets.

One solution to the colour or contrast issue is to use a second high-end camera, substituting the integral camera of the instrument. This is placed on the tripod to record the images and import them into the point cloud. Whilst this will take time to align with the point data, the resulting object information may benefit from the expenditure; again, this may also depend on the rationale for the data acquisition and the status of the object.

Fly-through films can be made directly from the point cloud for export to provide informative narrative for tourism or tutorials on the heritage asset. It must also be noted that there may be holes in the data and these can distract the viewer, causing confusion, especially where other parts of the cloud can be viewed through the hole.

The data can be sliced very thinly in any plane, making plans and sections easy to construct and record. These may also be cleansed of noise and be prepared to extract for record with labels or dimensions as appropriate. Another opportunity exists in these slices: if there was suspected movement in the object, it would be possible to rescan after a period of time and compare the slices of data to determine the direction and extent of any deformation. This could aid understanding and support a solution which, when complete, could be monitored in a similar way. This situation presented itself when a local landmark was defaced by a lorry, removing part of the external masonry skin to an oriel tower. The building had been scanned for interest prior to the accident and will now be compared to the current situation to determine the extent of the damage, before providing support in the specification of the remedial work.

It is possible to determine the accuracy of the data from the model once it has been registered, although this may not reflect the true situation. When scanning in external locations, the movement of fauna or flora, detritus, overhead cables, light fittings and so forth will cause variances in the data, and these will be recorded as an error which does not reflect the accuracy in the solid object. Excessive errors may show there has been a problem in the registration; further investigation is then required to determine if the clouds are accurately aligned or refinement is required. The software may also perceive overlap in the scans where there was nothing – in this situation, the links between the data sets may require breaking to improve the accuracy.

Currently there is no software available to go from point cloud to a solid object-based model in a single step. It is therefore necessary to use a workflow to transform the data by tracing the objects for inclusion in a BIM model. There are algorithms to extract certain standardised objects, for example, pipes and sectional steel structure, from a point cloud, and to show these as solid objects. If this is a requirement of the data set, it is important to scan from multiple locations to fill where the data is insufficient to show the pipe run. It is often necessary to carefully check the objects and the data from which they were drawn to ensure the software has correctly identified the object. These objects will need extruding to complete the run and flanges or fittings added to create a complete model. Where there are valves and pumps, it may be necessary to use a high-quality setting to capture the data for the object due to its complex nature.

Finally, for modifications in plant areas, clash detection is available by merging a proposed 3D model of service runs with a scan of the existing situation. The software can identify clashes and produce the schedule for interrogation by the operator. However, not all identified situations are clashes as it detects points in the same location as the solid object; therefore, if the points are a reflection from a surface, it may provide spurious results. This leads to the conclusion that the operator needs to carefully clean up the scan data before merging it with the model to reduce such results and quicken the process.

Developmentally, the scanning of simple walls and ceilings is not far behind, but openings are currently proving elusive. With a standardised object in a building such as a window it is now possible to use a pick box, define it and the software will locate all similar objects in the point cloud. The object recognition in scan data is a developing field, and the future will see rapid development, although it will rely on standardised parts. Heritage assets, however, especially those of the older order, are anything but regularised. This will provide much difficulty, and we may have to determine alternative workflows to compile complete 3D representations for inclusion in the BIM.

As previously discussed, the point cloud is just that – a point cloud; other than that the points occupy the same model space, there is no link between them. It is possible to place a skin over the data, and there are some good examples of these types of software available. This can help when trying to interpret the results, although where the data has holes in it there will be spurious features. This can be especially difficult where flora is close to the heritage asset. The data, however, can bring forward some interesting additional features which may not have been identified in other surveys. It is possible to see fluctuations in areas of wall where there previously had been window openings, doors, staircases or details such as taller skirting boards, dado or picture rails, providing insights into the evolution of a property. If the data is sliced for section or plan, the thicknesses of wall and floor/ceiling voids can be determined, leading to prediction of materials or details used in the construction.

The cloud can be littered with flags which are visible during interrogation, and additional information such as an asset register or the installation/service manual for the equipment can be hyperlinked to these. With bespoke products and unique design features in heritage assets, the registers can be extensive documents. However, these can often have vital information missing, as the data is not available without completely dismantling and rebuilding the asset. The additional information can only be collected during maintenance, repair or refurbishment. This notwithstanding, the main thrust of the BIM benefit is collaboration, leading to efficiency, and the production of the 3D model as a tool to improve the interrogation and dissemination of information. Leading on from this, research is required to determine the need for extensive object modelling or if the simplified acquisition of complex data in the form of point clouds is adequate or if it can be a mixture of the two.

The ability to use flags in the point cloud could also be used to benefit the storage and retrieval of the heritage asset records. Hyperlinks can be set up to take the user on journeys of discovery. Once scanned, documents can be catalogued and referenced to the point cloud, and the two data sets can be synergistic, bringing additional knowledge and understanding. Either data set can be searched or the results analysed. There is, however, an immediacy brought by the visual model; the surrounding flags can be a starting point for these discoveries. Comparisons between the point cloud and the visual records can provide clues to the development in the asset without having to visit the building each time. Furthermore, construction records may provide some of the answers to develop the object data to build into the BIM model.

It is possible to reveal details which were previously difficult to appreciate. The Gothic-style Arkwright Building, part of the Nottingham Trent University City Campus, was damaged during the Second World War when a bomb exploded at the west end of the building. To the front of the building is a grassed area scattered with trees, and across the street buildings stop the observer from receding sufficiently to observe the wide façade (see Figure 7.6). A scan data model was built and the objects obscuring the façade cut away, leaving an unencumbered orthographic elevation. Comparing the east and west ends immediately reveals the inconsistencies, showing the extent of the original damage and the simpler details used in the reconstruction.

The point cloud is exported in a standardised file format ready for importing into BIM software, and some software has a specific plug-in to translate the data as it is imported. Generally there is a range of export formats in the software package to choose from; if the data is being supplied to a third party for modelling, a check will need to be made as to the best format. The data can be viewed and basically manipulated, although there will be limitations should errors in alignment or missing data be identified. The next stage is to locate the edges and start to 'trace' over the point cloud to produce the solid model. There are restrictions in the process, and these may see simplified objects being built (for example, a wall will exclude some of the intricate detail or patterns). The cloud can also be sliced to show plans and sections to trace to build into the model.

From the supplied data it is possible to build a relatively accurate object-based model. However, for a number of reasons it may be necessary to return to site to manually obtain data where none

Capturing data with 3D laser scanners 73

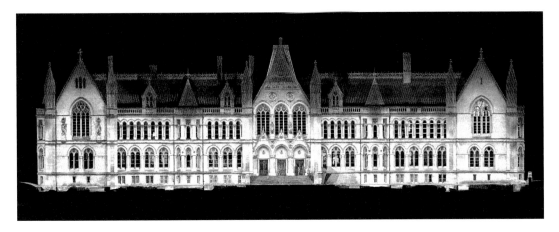

Figure 7.6 This point cloud data orthographic image of Nottingham Trent University Grade II* listed Arkwright Building reveals the asymmetric façade due to rebuilding after WW2 bomb damage.

Figure 7.7 Image taken from film made whilst flying a UAS over land adjacent to the building.

could be recorded with the scanner. There are some situations where either there was an immovable article on site or where the data from a small room could not be linked accurately with the main group as there was little overlap in the data between the clouds. However, as there is accuracy in the surrounding model, merging the manually obtained data into the BIM model will maintain a complete data set with minimal opportunity for oversights.

There is a wide range of methodologies available to record the heritage asset to support a sustainable future, although all have their limitations. The laser scanner is a tool to gather and present data in a 3D format, has great accuracy and removes some of the human error. However, there are a number of problems which currently beset its use in all circumstances.

The inability to record features out of the line of sight often leaves some of the more vulnerable yet very important areas without data, such as the roof. However, this could soon be resolved using unmanned aerial systems (UAS) to fill in the missing areas once the issue of flying in an urban area is resolved. Figure 7.7 shows data gathered using UAS technology to film in 4K format and extract images such as this one. A full roof and high-level masonry survey can be completed without having to use access systems, thereby reducing the associated risk of injury and at a cost benefit.

The model building stage provides expansive amounts of information in an accessible format to laymen and consultants alike, but yet again there is difficulty in transferring the data to an object-based model for BIM. There are solutions coming through to address this, especially for standardised components, which transform the laborious nature of this stage.

Likewise, the issue of data size may now be resolved in hosting the model in the cloud and accessing it via the web. As with all digital data, there is the question of security – how can we ensure the data does not become corrupt, and how can we ensure we can still read it in the future? These are questions that not only have to be addressed for point clouds but equally for BIM models.

As with all things digital, the cost of this technology is reducing and, at the same time, the experience is expanding to improve efficiency and therefore the cost benefit to a client. There will come a time when all conservation plans will include a point cloud model as the basis for a record. Like the building, this model should evolve, providing accurate records for future custodians to consult when making decisions in managing the asset. As with the cave paintings, the historians will be able to gain insight to our past and the development of our surroundings with levels of accuracy never before achieved.

8 Evaluation of historic masonry
Towards greater objectivity and efficiency

Enrique Valero, Frédéric Bosché, Alan Forster, Lyn Wilson and Alick Leslie

Internationally, historic masonry buildings constitute a significant proportion of the traditional built environment. In the UK alone, there are 450,000 listed buildings and 10.6 million pre-1945 structures, a high proportion of which have been constructed in traditional masonry. The value of the historic built environment is significant and as such it requires regular, well-considered and cost-effective survey and maintenance.

Survey for the sensitive repair of these buildings has generally been undertaken adopting visual methods. Whilst this is an important mechanism for determining construction characteristics, associated materials, condition and subsequently the scope and cost of repairs, it has been shown to have limitations. Survey subjectivity has presented itself as an issue of concern, and variability in evaluation is the starting point for cumulative errors that lead to unnecessary, costly and philosophically contentious fabric interventions. Survey accuracy also requires adequate access, which is not always easy to facilitate and is a significant cost that can ultimately reduce frequency and scope of inspection. Scaffolding in particular represents a large portion of survey costs; and it may also cause irreversible physical damage to the masonry substrates, if tied into them. In addition, working at height can be dangerous for operatives during erection and operation.

Innovation in survey technology, now often referred to as *reality capture* technology, is prevalent. Reality capture technologies, including ground-based and aerial laser scanning and photogrammetry, promise more comprehensive and accurate data collection. This should ultimately support greater levels of objectivity in determination of fabric condition, in addition to making data acquisition safer, faster and more cost-effective. This chapter builds a case for the deployment of these technologies to support building surveying and maintenance decision making, with a focus on historic masonry. The arguments particularly resonate when considering proactive maintenance strategies that rely upon accurate, coherent information, as well as Building Information Modelling (BIM) for the organisation and management of these data. The chapter also highlights the emerging awareness of restrictions encountered in the use of these modern survey technologies, that is the extensive time and cost of processing the acquired data to extract information of ultimate practical value to surveyors. To show that these constraints can be alleviated even in complex situations, preliminary results are reported on the development of a novel system for analysing three-dimensional data from rubble stonework (acquired by laser scanning and photogrammetric means). The system delivers valuable information, arguably supporting more objective determination of regions associated with undercutting masonry via excessive joist erosion, and quantification of pinning (or galleting) and repointing intervention required. This work shows how this kind of solution can be used to improve the efficiency and accuracy of measurement and costing works for rubble stonework that has traditionally been evaluated on an approximate square-metre basis.

8.1 Introduction

Internationally, the importance of reactive and increasingly proactive building maintenance is well recognised as an essential mechanism for preserving the value associated with historic building fabric. In addition to designated heritage buildings, there are innumerable traditionally built structures that do not have legislative protection but require appropriate high-quality maintenance to ensure satisfactory long-term performance and aesthetic continuity. Whilst maintenance is clearly pivotal from a cultural perspective, it is also economically important. Reflecting this, it has been estimated that 50% of national wealth across Europe is contained within the existing built environment and that maintenance is a major contributor to the gross domestic product (GDP) of the majority of European countries [7]. In the UK, maintenance accounts for half of the total expenditure on construction, and there are nearly 450,000 listed buildings and 10.6 million pre-1945 buildings, constituting the largest part of the built environment [56]. The financial value of repair works was estimated at £36 billion in 2002 [3, 27]. Of this value, masonry is not an insignificant sum, and it has been estimated that in Glasgow alone, the cost of masonry repairs required over a twenty-year period is approximately £600 million [89]. Other major international cities with a tradition of masonry construction may likely require similar investment. Logically, if these works were undertaken they would contribute significantly to the GDP of national economies and simultaneously support indigenous traditional craft skills.

The importance of maintenance for historic buildings is not new and was emphasised by William Morris in 1877, who stated, "stave off decay with daily care" [81]. Since then, maintenance has gradually become a 'first order' consideration that has been embedded in the principal building conservation legislative frameworks and charters that are utilised to direct best practice [11, 16]. Indeed, most recently the revised edition of BS 7913 "Guide to the Conservation of Historic Buildings" [16] dedicates a complete section to maintenance, highlighting the importance of proactive maintenance and indicating that "maintenance is the continual care of a historic building and is the most common and important activity in their conservation and preservation" [16].

Yet, despite its importance and the sector size, maintenance appears to be poorly regarded by the public and construction industry alike due to various interconnected and disparate factors [38]. Principally amongst these is the relative cost of access and bespoke survey services that individually do not benefit from economies of scale. Compounding this, costs for repair projects are well understood to be difficult to accurately predict due in part to the specialist nature of the work and their supporting materials [38]. Inaccuracy in determination of suitable repairs, their associated pricing and rational evaluation of project duration is commonly encountered. This situation cannot be countenanced and, as Cassar emphasises, "with dwindling resources and escalating cost, pragmatic choices are often necessary and inevitable" [20]. Inaccurate pricing compounded by excessive inspection and access costs practically mean a reduction in the scale of project operations. The relative expense of the essential preliminary and enabling works consume the budget, leaving diminished scope for practical fabric interventions.

Adding to the complexity of prioritisation within technical, philosophical and economic parameters, a fourth and emerging factor in the evaluation is environmental sustainability. This is theoretically and practically encapsulated in the growing field of 'green maintenance' [35]. A correlation has been made between effective evaluation of defects and deterioration, repair options and the cumulative effect of routine maintenance operations and their associated environmental impact [35]. As survey technology adoption increases it will undoubtedly help to meaningfully support advances in efficiency and in accuracy of reporting and influence decisions made. Pragmatically, better reporting has the potential to yield better understanding of the degree to which fabric has decayed, triggering intervention in the form of minimal works (ideally doing nothing) or major works. The arresting of decay and retention of original fabric is a primary consideration for building conservation that simultaneously enables greater retention of embodied energy and carbon imbued in the structure. Efficient and accurate survey can therefore be correlated with support for minimal intervention decision making that underpins broader cultural, sustainable and economic aspirations.

The aforementioned issues are discussed in Section 8.2, which provides a review of the current practice and protocols in masonry survey and maintenance and highlights typical defects and their diagnosis. Section 8.3 then investigates how novel advanced technologies, including rapid 3D reality capture and Building Information Modelling (BIM), could provide surveyors with objective and informative data on building fabric, and how these data could support them in ensuring greater objectivity and consistency in a cost-effective manner. The reviews and discussions of Sections 8.2 and 8.3 are summarised in Section 8.4, which concludes with the identification of an emerging yet critical restriction to the use of modern survey technologies: the extensive time and cost of processing the acquired data to extract information of ultimate interest and practical use to surveyors. To illustrate how this restriction can be alleviated, Section 8.5 reports preliminary results on the development of a system for efficient and effective processing of 3D point clouds of rubble stonework (obtained using reality capture technologies such as laser scanning or photogrammetry) to extract valuable information such as the location and quantity of undercutting mortar, and subsequently the amount of repointing and pinning required. Section 8.6 concludes this chapter, encouraging survey specialists to expand their collaboration with computer science and ICT experts in recognition of the benefits from adopting the novel survey technologies reviewed.

8.2 Review of current practice for masonry condition survey, repair and maintenance: Defects and investigative protocols

Condition surveys are used for the analysis of structure and fabric but have a broader scope, such as directing those entrusted with the repair of historic buildings to prioritise interventions. The many purposes of surveys include: to record fabric; to form the basis of defect diagnosis and pathology; and to support preliminary and detailed costing, for the establishment of specifications and broader repair strategies. Condition survey data typically include description of the element or component, location, conditions and associated defects. They can importantly include basic information regarding identification and rudimentary measurement of affected deteriorated materials and components. More specifically, when specialist masonry surveys are employed, the costing and pricing processes are typically based on the accurate measurement of dimensional stones to be replaced (these measurements form the basis of cutting schedules sent to quarries) or, in the case of rubble stonework, the approximate amount to be replaced. Measurement must also be conducted for an array of complementary works that include plastic repair, repointing, pinning and consolidating delaminated surfaces, adopting doweling materials, deep void packing and pinning (or gallet) [36, 39]. The quantification and pricing of those works should ideally be based upon a detailed building elevation with 'marking up' of individual stones. Delivering such detailed, accurate documentation has been challenging due to the traditional utilisation of visual surveys as a primary mechanism for the evaluation of condition. Visual surveys of masonry, particularly in random rubble, are very complex, time-consuming, and subject to subjectivity, hindering the timely production of meaningful and effective repair strategies.

This section initially reviews common defects encountered in masonry (Section 8.2.1), leading to the evaluation of current methods used for the documentation/measurement of those defects (Section 8.2.2). A discussion is provided pertaining to the frequency of surveys (Section 8.2.3). The impact of survey accuracy is then discussed – in particular, issues surrounding the subjective interpretations of surveyors (Section 8.2.4) – prior to concluding the section with a discussion on the limitations of current approaches to costing and the impact of access provision (Section 8.2.5).

8.2.1 Common masonry defects

Masonry defects are extremely varied and deterioration processes are correspondingly complex. Notwithstanding, a surveyor evaluating deteriorating masonry fabric will often be confronted with recurring themes. The most prevalent defects are summarised in Table 8.1.

The defects highlighted in Table 8.1 are problematic and often suggest underlying issues, the most alarming of which is ultimately structural movement and significant erosion leading to progressive collapse. Whilst it is recognised that not all of the defects are readily visually identifiable, adopting scanning and photogrammetry technologies has the potential to advance fabric diagnostics. Examples include using laser scanning technologies, infrared thermography and/or penetrating radar for assessing materials and detecting construction anomalies. The integrated utilisation of technologies is taking us beyond traditional methodologies, and this concept is central to this work. It is recognised that 3D reality capture technologies can enable significant efficiencies in data acquisition and greater objectivity in representations of buildings that are utilised as primary working documents (i.e. plans, elevations and sections).

Table 8.1 Masonry defects: Manifestation, identification and diagnosis [48, 90, 92]

Defect	Manifestation, Identification & Diagnosis
Structural movement	Rotation of wall faces/out of plumb. Buckling of walls. Evaluation of planar surfaces. Inconsistencies in level/plumb/deflection.
Cracks	Cracks associated with structural movement – differential and uniform settlement, moisture and/or thermal movement. Volumetric instability and movement associated with chemical instability and responding materials change/conversion.
	Load path alteration and force redirection leading to localised stress concentration (overloading) in masonry and mortars. Both compressive and tensile fracturing in materials associated with overloading – stress concentration especially in fenestration, buttresses and piers etc.
	Fundamental importance of size, location of fractures and whether they are compressive or tensile in nature. Analysis of fracture/crack patterns as a primary mechanism for supporting pathology/prognosis and diagnosis.
Erosion and fabric loss of masonry units	Friable materials, surface dissolution, alveolisation (honeycombing), voids, recessed materials, faces blown in brickwork etc. Large or small magnitude spalling events. Binder migration and wall core voiding.
Eroding pointing and bedding mortars	Undercutting of masonry unit. Increased prevalence of moisture ingress and penetrating dampness at depth. Deterioration from original wall line.
Delaminating natural stone	Face-bedded stone ('on cant' – deterioration along bedding planes), contour scaling, fabric detachment.
Contaminated masonry	Efflorescence and crypto-efflorescence salt blooms, salt crystallisation deterioration and surface dissolution. Crusts and blisters associated with pollutants. Discoloration – both from internal and external environment (NOx and SOx pollutants) and hydrocarbon sulphate deposits (SOx) from unlined chimneys etc.
Biological activity	Mould and moss, growth often consistent with high and sustained moisture contents and low potential evaporation. Lichen growth and surface interaction.

Clearly, the production of visual survey information (elevations, plans and sections etc.) is the starting point for more advanced technical processes. The identification, evaluation and determination of suitable interventions are interconnected, and inaccuracies in survey documents compound problems in later application. Advanced survey processes can be seen as a complex 'chain of events' or decisions that are subject to cumulative error (including the production of the original survey drawings that underpin specialist evaluation). The ability to implement effective and rigorous building pathology or forensic engineering strategies relies upon accurate identification via examination (diagnosis) and an ability to predict likely outcomes (prognosis). Survey therefore meaningfully underpins diagnosis and prognosis, but reporting objectivity cannot necessarily be guaranteed and has been shown to be of concern to the sector [39, 83].

8.2.2 Survey protocols for investigation

Numerous approaches to building surveys and associated protocols have been developed in attempts to enhance accuracy, consistency and objectivity in investigation and reporting. The 'HEIR' approach [76] is both tacitly and expressly utilised to guide the surveyor in rational decision making. When contextualised within a stone survey capacity, the assessment of 'Historical' documentation and former survey reports forms the basis of a comprehensive desktop study. 'Evaluation' relates to the survey that is commonly undertaken visually due to its low cost, absence of 'hi tech' equipment and surveyor familiarity [21, 47]. Visual surveys have often traditionally been supplemented with fully measured hand-drawn surveys (later site drawn and transposed to CAD) with annotated information relating to necessary repairs [5, 43]. 'Investigation' requires more complex analytical approaches, additional techniques and the use of multi-hypothesis generated questioning to determine construction, condition and pathology [24, 42]. 'Report' requires the logical structuring of often complex information relating to the building form and fabric being evaluated.

McDonald [58] highlights protocols in which expectations for architectural investigation may be more specifically outlined for masonry fabric assessment. Within his approach, key stages are established, including reconnaissance; surface mapping; non-destructive testing; destructive testing; and laboratory analysis. This offers a greater level of investigation objectivity, supported initially by surveys and increasing levels of sophistication and complexity in scientific analysis techniques. Logically, increasing hypothesis generation requires additional supporting scientific investigation to prove or disprove the questions posed; all have cost implications for the client.

Visual surveys and hand-measured production of drawings for masonry are common but have been prone to error and are extremely expensive to undertake. As a reaction to this situation, survey technologies have been employed to obviate the former issues outlined [21, 47]. Techniques such as rectified photography and photogrammetry have been extensively utilised and have been shown to yield effective results in recording and documentation in comparison with hand-drawn surveys [64]. These working documents enable scalable images to be produced that can be subsequently 'marked up' to identify the required interventions. The utilisation of these technologies gained some traction, but they were principally associated with specialist, 'high' importance structures with relatively uniform masonry styles (i.e. ashlar and squared coursed rubble). It is clear that a paradigm shift is occurring with novel 3D reality capture technologies and broader HBIM considerations, offering the possibility of better documentation and reporting, evaluation and management of repair projects.

8.2.3 Decision triggering for fabric repair: When to intervene?

The time between fabric interventions is influenced by many variables, including materials durability, degree of exposure, building detailing, quality of repair, and specification [37, 94].

Triggering of maintenance interventions can occur as a result of various proactive and reactive survey mechanisms. Regular preventative and planned maintenance is commonly adopted [21] but is associated with inefficiencies. Establishment of budget requirements for works has been typically based upon maintenance profiling (longevity modelling). These models are mainly supported by data relating to the life expectancy of materials and components, architectural design form and detailing and exposure levels [74]. Inspection-based maintenance is more nuanced, with regular surveys being used to form the basis of required interventions irrespective of expected life expectancies. This can help obviate unnecessary maintenance interventions that may be dogmatically applied in profiled operations. This also importantly creates an environment that achieves better conservation outcomes due to its potential to attain 'minimal' or least intervention results.

The determination of the minimal acceptable condition may vary considerably from inspector to inspector. Overly favourable or 'light-touch' evaluation and interpretation of fabric may lead to inactivity that whilst retaining historic patina may ultimately lead to an increased likelihood of more significant, cumulative defects such as potential falling masonry. Alternatively, overzealous evaluation, interpretation and reporting of masonry substrates may result in unnecessary repairs that can also have a negative philosophical impact, as more of the original material will be removed than is required. A loss of historic patina would occur in this situation, and a potential reduction of architectural integrity may result. Additionally, consideration must be given to loss of embodied energy in the form of 'spent' carbon that is 'locked up' in the historic masonry fabric. Premature intervention clearly has negative implications for broader carbon reduction strategies required by heritage organisations.

8.2.4 Utility, quality and accuracy of reporting

The accuracy of a survey is clearly essential to communicate intervention requirements and repair strategies. Yet, these surveys have often been inadequately produced by surveyors (lead professionals) who do not fully understand the requirements of masonry contractors and the nature of the information that they utilise to meaningfully cost works [5] or develop repair strategies. In light of this situation, several authors have identified the need to achieve a more collaborative approach to the project's preliminary development [5] [39] where surveys are undertaken jointly with professionals and specialist contractors. The result of such an approach offers the promise of increased accuracy in reporting and efficacy of the supporting information that has shared meaning to all those involved in the fabric repair scheme.

8.2.4.1 Subjectivity

Theoretically, the more objective the starting point, the better the evaluation of required repairs and their underlying strategies. Yet, current assessment techniques rely heavily on the judgement of the surveyor evaluating these defects, a process that is often based upon rudimentary heuristics such as the 'mid 1/3' rule for wall rotation, basic slenderness ratios and erosion depth in relation to mortar bed height [10] [32]. Whilst bed joint erosion calculations (and by extension the need to intervene) are relatively easily applied in brickwork, squared coursed rubble and ashlar, these calculations are much more complex in the case of random rubble masonry due to variability in stone and mortar dimensions. Poorly informed or inexperienced surveyors can be naturally and understandably prone to extreme caution in suggesting remedial works that are overzealous in nature. Conversely, failure to fully grasp the potentially significant and dangerous situation that may confront them is problematic.

The determination of defect and suitable repair is thus hindered by *subjectivity* in fabric evaluation requirements [39]. Straub [83] highlights this situation, stating,

> the practice of condition assessment by building inspectors yields variable results due to subjective perceptions of inspectors. Surveyor variability is defined as the situation where two or more surveyors, surveying the same building, arrive at very different survey decisions. This variability is caused by a variety of factors such as previous experience, attitude to risk, and heuristics – the use of 'rules of thumb' – a leaning towards a particular opinion regardless of the available evidence.

Attempts to enhance reporting uniformity have led to the utilisation of system-based approaches or protocols to surveys [72]. Whilst helpful, these do not reconcile the building inspector's inexperience and variable interpretation and may therefore yield considerably different results that have tangible implications for the repair strategies, cost and aesthetics.

8.2.5 Costing/pricing

Determination of quantities and subsequent costs for the repair of masonry substrates is complicated by the variation of masonry types. Measurement of works is particularly difficult for random rubble masonry that lacks regularity. This is compounded by an inability to accurately identify specific areas for repair and the dimensions.

Specific cost data is not well defined for conservation works, and only recently has the National Building Specification (NBS) attempted to rectify this via the rudimentary inclusion of more specialised clauses. Indeed, in the Standard Method of Measurement (SMM), limited consideration is given to specification techniques, and therefore these are undertaken in a non-uniform, bespoke manner. Traditionally, SMM7 [71] included basic 'measurement rules' for natural stone rubble walling, and highlights that "stonework is measured according to mean dimensions." Other salient details are included, and supplementary information indicates that 'type of pointing' should be stated, but no specifics are stated beyond that. *Griffiths Price Book* [40] includes data on the pricing of alterations and repairs for masonry, with a generic section for 'repointing stonework' that includes "rake out and repoint; rubble walling"; the units utilised for this process are square metre (m^2) and there are no considerations given for pinning requirements, highlighting a lack of understanding in repair processes in traditional masonry and in lime works. *Spon's Architects' and Builders' Price Book* [26] includes a section entitled "Natural Stone Rubble Walling" with various masonry materials types also being quantified in m^2. No specific unit is given for pointing/repointing. Importantly, no mention is made of associated common repair processes such as pinning, consolidation or deep void packing.

More recently, New Rules of Measurement (NRM) [73] established the measurement protocol for masonry work under "Renovation Works" and more specifically "Masonry Repairs". NRM [73] specifically identifies masonry repair techniques under "Measurement Rules for Components", including specialist treatments such as cutting out decayed, defective stones and inserting new (whether isolated repairs, stitching and the like); plastic stone repairs; re-dressing stonework to new profiles; grouting; and rejointing/repointing existing masonry. These were formerly omitted from measurement protocols. This illustrates a greater awareness of the complexity of costing repair works for specialist masonry fabric repair in the NRM. Notwithstanding, the NRM loosely defines units of measure for evaluation of many works, for example, suggesting repointing can be measured in m or m^2. It is simply understood that the inability to rapidly and accurately determine repointing on a linear m basis necessitates the use of traditional approximation in m^2. Additionally, all of the aforementioned still insufficiently describe historic masonry, devaluing consequent production of project documentation. The greater the lack of meaningful applicability, the greater the potential for measurement and cost error.

The above-mentioned measurement requirements along with the measurement techniques currently available to surveyors collectively lead to inaccuracies that are particularly significant for random rubble works compared to those encountered in regular brickwork, squared coursed rubble or ashlar masonry construction. Time-consuming, labour-intensive repair processes such as pinning (or galleting), and deep void packing are thus estimated on a m^2 basis instead of using more accurate metrics, such as the depth, width and length of joint mortar pointing (m).

8.2.5.1 Access

Inspection systems and, more specifically, provision of access are an onerous cost for reactive and proactive maintenance interventions. It has been estimated that 10%–20% of maintenance costs are associated with the provision of access for inspection. Standard scaffold solutions may need to be tied through holes screwed into the masonry, which causes irreversible damage to the masonry. Alternatives do exist, such as buttress scaffolds. These and other complex scaffold solutions are costly and create higher levels of erection risk. It is not unrealistic that such complex scaffold solutions count for up to 40% of the total project cost. This situation is reflected in all types of construction, but is particularly important for historic buildings. This is due to the relative age and complexity of the fabric and associated degradation of the materials that may be many centuries old. Furthermore, alarmingly, almost half of construction fatalities and a third of major injuries result from falls from height [45], many of which are associated with scaffolding. Clearly, cheaper and safer access methods for accurate and complete inspection of building fabric would contribute to a significant reduction in maintenance and repair costs.

8.3 Progressive survey techniques – reality capture and data management

It is reasonable to assert that traditional surveying techniques will always be required, forming the 'cornerstone' of evaluation. However, new technologies can facilitate data recording and the integration of information, presenting the results in a unified, well-structured, compact manner, providing a better understanding of the fabric and reducing unnecessary interventions. In 1995, Ogleby [64] reviewed the advances made in techniques and technologies for creating records of significant sites and monuments, with particular focus on photogrammetry. While a valuable contribution at the time, the techniques presented, e.g. 'réseau plate film flattening devices', would now be considered 'outdated'. This illustrates the significant progress that has been made over the last twenty years, both in reality capture technologies to acquire data and information and communication technologies (ICTs) to manipulate, organise and visualise the acquired data. This section will successively review modern reality capture technologies, with focus on 3D imaging, and then ICTs that are being considered for data processing (i.e. information extraction) and data/information management.

8.3.1 Reality capture technologies

The literature clearly shows that two main modern technologies are now considered by cultural heritage (CH) experts for obtaining 3D records of historic monuments: laser scanning and photogrammetry. While the value of each is studied in more detail in Sections 8.3.1.1 and 8.3.1.2 and contrasted in Section 8.3.1.3, they both have line-of-sight limitation, which means that they can only attain surface 3D data and cannot 'see through' objects to obtain data from their inner parts or from other objects located behind them. For this, other technologies must be used, like ground-penetrating radar (GPR). Also, other 'invisible' parameters, such as temperature, cannot be obtained by means of 3D data acquisition technologies. Therefore, devices like thermographic cameras are increasingly used to improve the information acquired from buildings. The value of GPR and thermographic cameras for historic monument condition assessment is rapidly reviewed in Section 8.3.1.4.

8.3.1.1 Laser scanning

Laser scanning is very recent (it was unknown to Ogleby in 1995) but constitutes a revolution to land and monument surveying. Modern laser scanners sweep their surrounding space with a laser beam to obtain dense and accurate point clouds. While laser scanners can be used airborne, mounted on planes or helicopters to capture land elevation (in this case experts refer to the technology as *LiDAR*), in the context of monument recording it is more commonly used ground-based, in which case it is referred to as *terrestrial laser scanning (TLS)*. To illustrate the great value of TLS to monument survey, the recent Leica ScanStation P40 can acquire hundreds of millions of points in a 360° × 270° field of view with a single-point accuracy of ~1 mm, in a few minutes.

Within the context of historic monument surveys, noteworthy examples of the use of TLS include the work of Wilson et al. (2013), who illustrate the distinct benefits of TLS for the survey of large and complex historic monuments via case studies of UNESCO World Heritage Sites, including New Lanark in Scotland, Rani ki Vav stepwell in India and the Sydney Opera House in Australia. Cardaci et al. [18] show that TLS provides significant value compared to traditional manual surveys, and also demonstrate how CAD models generated from the data can be successfully used for structural analysis using the Finite Element Method (FEM). Nettley et al. [63] use TLS and LiDAR to obtain a photorealistic geospatial model of the historic quayside at Cotehele Quay, integrated in an accurate digital elevation map (DEM) in order to assess the potential impact of rising sea levels resulting from climate change. Temizer et al. [86] show the value of TLS to survey underground structures like the Byzantine cistern situated under the court of the Sarnicli Han building. They also investigate the impact of various levels of point filtering on the accuracy of the mesh produced from the data. De Matías et al. [57] investigate the use of TLS for surveying historic structures like walls, pillars and vaults. In particular, using the Coria Cathedral as a case study, they show how TLS can be used to measure structural displacement (e.g. wall collapse) and link those to cracks manually extracted from the data.

8.3.1.2 Photogrammetry

As the review of Ogleby [64] shows, photogrammetry (PG) is actually a well-established method for obtaining 3D records of historic monuments. Nonetheless, significant progress has been made in the last two decades both in terms of hardware and software to rapidly and accurately obtain such records. Most importantly, high-resolution and portable digital cameras are now widely available at a relatively low cost. Furthermore, the development of robust automated techniques for feature detection and matching in digital images (e.g. SIFT [54] or SURF [9] features), as well as dense matching approaches [84], have dramatically improved the image processing stage, rendering it essentially entirely automated.

Within the context of historic monument surveys, noteworthy examples of the use of PG include the work of Cappellini et al. [17], who employ PG to produce 3D models of monuments that are used to generate 2.5D orthophotos of walls. Using the example of Roman walls, the orthophotos are used to conduct the semantic annotation of the *opus* of different sections of the wall. Santagati et al. [78] demonstrate the value of software designed for photogrammetric 3D reconstruction of historic monuments. Despite the accuracy of the model (~1.2 cm on average), they note that a great advantage of this kind of software is its use of cloud computing that expedites the otherwise time-consuming processing for reconstructing the 3D model from the images. However, they also highlight that photogrammetric reconstruction is less robust than TLS, as precision is rather variable and some reconstructions are sometimes erroneous – an issue already acknowledged in the wider literature (see Section 8.3.1.3). They conclude that, although an excellent tool, photogrammetry should be used with care. Finally, we note the recent work of Sánchez-Aparicio et al. [80], who conduct structural damage assessment of an interior dome using a 3D CAD model obtained through photogrammetric reconstruction; FEM is used for the structural analysis and understanding of the cause of defects (deformations and cracks) identified in the 3D model.

In a somewhat different context, it is interesting to note the work of Bruno et al. [14], who have developed a system that combines stereo-imaging and structured-light sensing to conduct underwater surveys. The integration of these two techniques is motivated by the poor conditions (both water visibility and conditions of the artefacts) that can be encountered in underwater exploration. However, it seems that stereo-cameras have not been considered in other contexts such as the survey of the fabric of historic monuments.

8.3.1.2.1 AERIAL PHOTOGRAMMETRY

Photogrammetric reconstructions need oblique and perpendicular images from strategical points to obtain a good resolution and accurate model. A pole-based system can be considered to take pictures higher than human height. However, poles can only be comfortably used up to 5 m or so. For greater heights, aerial solutions must be considered. In fact, planes or helicopters have long been used with under-mounted camera systems for photogrammetric applications (similar to LiDAR systems). However, recent and rapid developments of unmanned aerial vehicles (UAVs) are providing a new platform for photogrammetric systems that further resolve access issues.

The value of UAVs to surveying has already been demonstrated in various contexts such as for ecological surveys [88, 95] or structural surveys [60, 75]. All these works show how avoiding the use of planes or scaffolding systems is beneficial to reduce acquisition time and budget.

In the context of historic monuments, UAV-based photogrammetry has been studied to provide alternative solutions to TLS. For example, Puschel et al. [68] propose the use of terrestrial and UAV pictures to create an accurate 3D model of Castle Landenberg, using tie points as a reference to accurately model the building. Remondino et al. [70] review the different stages of data acquisition and processing, such as planning, camera calibration, 3D reconstruction and applications. Lately, Koutsoudis et al. [50] proposed a photogrammetric system combining UAV and terrestrial pictures and compared the resulting reconstruction with TLS, obtaining promising results. Xu et al. [91] actually combine 3D data from TLS and a UAV-mounted camera for the reconstruction of a historical monument in Fujian, China. TLS point clouds are used to model the façades, and photogrammetric information is used to complete the roof area.

8.3.1.3 Comparison and complementarity of techniques

The works above show that both TLS and single-camera PG are widely considered for historic monument surveys, and (heritage) professionals often ask the question: Which one is better? Various researchers have tried to answer this question, and the answer is nuanced.

Brunetaud et al. [13] integrate and compare TLS and PG for creating 3D models for historic monument preservation (their study is conducted at the Castle of Chambord in France) and conclude that laser scanning provides advantages in terms of modelling speed since it directly provides information in 3D; furthermore, the accuracy and density of the laser scanned data appeared higher, which would better suit deterioration monitoring.

Andrews et al. [1] conducted TLS and dense PG surveys of the great barn at Harmondsworth (UK) to compare the two techniques. The comparison highlights issues related to access for the laser scanning: due to various obstacles, parts of the roof could not be adequately scanned. In contrast, the (PG) camera could be mounted on a 5 m pole to ensure adequate access. Fassi et al. [34] also note that photogrammetric data acquisition is significantly faster than TLS. Their comparison of the 3D point clouds obtained with the two techniques achieved quite good results, with variances between the two outputs (meshed) being approximately 2 cm for areas where both methods acquired data in similarly good conditions. This indicates that, in certain contexts, both methods could be used interchangeably.

Fassi et al. [34] also performed an in situ comparison of TLS and dense PG (using Agisoft Photo-Scan) for historic monument survey using three case studies (Roman ruins, a church façade and a chapter interior). The comparison led to average differences between the two models below 2 cm for an aerial reconstruction and below 5 mm for the façade and chapel interior scenes. They also note that the portability of digital cameras is a great advantage to acquire 3D data in difficult-to-reach areas.

Lerma and Muir [53] carried out a comparison of laser scanning with photogrammetry, reconstructing a 3D model by means of Visual Structure for Motion (VSfM) and Patch or Cluster based Multi View Stereo Software (PMVS/CMVS), and concluded that for accurate documentation of carved detailing on stone, photogrammetric techniques provided the greatest level of flexibility and reliability, although the two techniques used in combination may provide the best results.

Despite the very encouraging results reported by the different comparisons above, researchers have noted that the quality of photogrammetric reconstructions depends strongly on the camera lens quality (and its internal calibration) and environmental conditions during data acquisition. The lack of predictability of the photogrammetric reconstruction thus balances against the otherwise significant advantages in terms of data acquisition and cost that PG has over TLS [44].

The apparent complementary strengths of TLS and PG have actually led many to integrate them. For example, Grussenmeyer et al. [44], Vianna Baptista [8] and Bryan et al. [15] all employ TLS to get 'overview' scanned data (which is still very accurate and dense) and complement it with dense photogrammetric models typically acquired at closer range to obtain even higher resolutions. As noted in Fassi et al. [34] and Grussenmeyer et al. [44], PG is also useful to provide local 3D models that fill the holes in laser scanned point clouds resulting from access limitations, or when the quality of the colour information is particularly important, for example when the data is later used to analyse various degradations [19].

Note that researchers have not stopped sensor integration to the combination of TLS and PG; they also consider other reality capture technologies such as LiDAR [44] and thermal imaging [19].

8.3.1.4 Other technologies: GPR and thermographic imaging

TLS and PG are not able to record physical properties other than geometry and visual appearance. Other modern devices, such as ground-penetrating radars (GPR) or thermographic (i.e. infrared) cameras, are thus nowadays used to capture data about other 'invisible' yet valuable properties.

Infrared cameras can be used to analyse fabric surface temperature distribution. The thermographic information facilitates the observation of areas where problems may not yet be visible [6]. The evaluation of thermal bridges in building façades [4] as well as insulation works and air leaks [85], moisture ingress and condensation [93] are increasingly studied by means of thermographic cameras.

GPR devices have been used for years to identify buried architectural elements, in most cases to measure and map buildings previously erected in that place [52]. Nonetheless, structural components can also be evaluated by means of GPR to identify internal deterioration issues. Lai et al. [51] present the use of GPR to evaluate material properties, such as steel bar corrosion in concrete, concrete hydration and water content distribution in porous concrete and brick walls. With a focus on historical buildings, Ranalli et al. [69] propose to use GPR to monitor the state of conservation of façades (e.g. identifying internal cracks, voids or degraded mortar) and for the evaluation of the thickness of walls.

8.3.2 From data to information, visualisation and decision making

8.3.2.1 Historic BIM

In the architectural, engineering and construction and facility management (AEC/FM) sector, Building Information Modelling (BIM) is rapidly developing as a means to more efficiently and effectively build, operate and maintain building assets. This management strategy also applies to historic monuments, with the aim of achieving a more collaborative approach, with models that integrate all relevant data and information to enhance interpretation and visualisation. In these contexts, experts use the term Historic BIM (HBIM) [1] [55] – the term 'Heritage BIM' is also sometimes employed.

At the core of an (H)BIM model is a parametric, semantically rich 3D model. When a BIM model is to be generated for an existing model, it is now commonly undertaken from a point cloud obtained by TLS [2] [59]. Hichri et al. [46] provide a general review of the stages to go from point cloud to BIM, also discussing preprocessing and data representation matters. Macher et al. [55] similarly present a semi-automatic approach for creating 3D models of historical buildings from point clouds. They divide the building in sub-spaces, model surfaces and fit primitive shapes to architectural elements. Since a BIM model is supposed to be parametric, Murphy et al. [62] propose to reconstruct BIM models of historic buildings using a geometric description language (GDL) that is used to parametrically encode the building and component structures (non-primitive shapes are modelled using NURBS [non-uniform rational basis spline], meshing and Boolean operations). In a similar manner, Oreni et al. [65] present a workflow using NURBS and vector profiles to model the geometry of historic monument components. In addition, Garagnani and Manferdini [41] present a Revit plug-in (GreenSpider) that matches a spline to imported 3D points.

Regarding the modelling of building components in BIM models, an important difficulty is the relative unicity of objects and shapes, which is much more pronounced in historic buildings than in contemporary examples. Fai and Sydor [33] thus argue for distinguishing the 'typical' and the 'specific', and create 'typical' BIM objects that can be deformed to obtained 'specific' ones for a given historic monument. While this approach addresses the concerns of heritage experts regarding the unicity of objects in historic buildings (and therefore in HBIM models), it raises the need to create libraries of 'typical' BIM objects that can be deformed [12]. A library of interactive parametric HBIM model façade components and a parametric building façade with floors and windows is presented by Dore and Murphy [28].

8.3.2.2 Data structuring and visualisation

These processes, from data acquisition to 3D model generation, produce a significant volume of information. In fact, this information often needs to be combined with additional information (e.g. weather and climate, prior records and repair documentations), which further increases the size and heterogeneity of the information that needs to be understood and managed.

Various authors have worked on different tools aiming to offer solutions to this integration problem. Salonia et al. [77] present a software package that enables the integration of heterogeneous data (geometric information, text, images, etc.) represented in GIS format. The system enables the exploration of the information in the form of layers, such as decay layers, and conducting query operations on the graphical representations and database records. A web version of the system is also presented. Pauwels et al. [67] integrate BIM and semantic web technologies, and Smars [79] presents open source software tools developed by the author to facilitate in situ documentation of architectural and archaeological heritage.

Specifically for stone surveying, Drap et al. [31] propose a web application for digital photogrammetry in which the user can measure structural elements (ashlar) and a database containing different kinds of blocks is also available to represent the walls. Brunetaud et al. [13] and Stefani et al. [82] present a software tool for semantic segmentation of 3D data, scientific monitoring and decision making for the conservation of stone walls.

8.3.2.3 Semantic labelling/segmentation

Section 8.3.1 demonstrated the interest of heritage experts in using TLS and PG data to better record the state of historic monuments. However, it is important to realize that TLS or PG point clouds (or generated meshes) are just raw *data*, and the *information* truly valuable to experts needs to be extracted from that data, through its segmentation and labelling. This is particularly important to the study of material and structural defects. This subsection is devoted to reviewing works related to the study of TLS and PG data to detect, segment and label material or structural defects.

Current practice for the evaluation of masonry defects, whether using TLS/PG data or more traditional surveying methods (as discussed earlier), is based on visual observations, and subsequent manual data segmentation and labelling. For example, numerous authors show how orthophotos obtained from 3D reality capture data can be used for manual labelling and mapping of degradation (e.g. using the ICOMOS glossary [8, 13, 19, 48]). Photogrammetry appears particularly appropriate for this kind of analysis because of the typically higher quality of the images, which is valuable to the labelling task.

Manual segmentation and labelling is however time-consuming and not necessarily very accurate and repeatable/reproducible (i.e. it varies with surveyors and experience). Researchers have thus aspired to develop (semi-)automated segmentation and labelling algorithms. For example, colour can be analysed to characterise materials and degradations in fabric. Moltedo et al. [61] investigate compression, entropy and gradient analysis for characterising materials in images. Cossu et al. [25] similarly investigate histogram threshold and edge detection techniques to identify, segment and label stone degradation regions, characterised by holes or cavities, from colour images. Thornbusch and Viles [87] analyse colour images of roadside stone walls over a six-year period using basic image segmentation tools from Photoshop (e.g. histogram analysis) to semi-automatically detect soiling and decay, as well as their evolution over the period. They also note that "variation in external lighting conditions between re-surveys is a factor limiting the accuracy of change detection."

Others base their works on more complex data analysis techniques. In Kapsalas et al. [49], the authors compare various segmentation algorithms, such as region growing or difference of Gaussians, to detect the topology and patterns of stone corrosion (black crust) in close-range 2D colour images. Cerimele and Cossu [22, 23] apply region-growing algorithms, more precisely the fast marching numerical method, to detect decay regions in sandstones (sediment or cavity decays) in 2D colour images, given seed pixels selected by the user. And finally, a semi-automatic delineation and masonry classification is carried out by Oses et al. [66], where they use artificial intelligence techniques (k-NN classifiers) to identify stone blocks in 2D images.

88   Enrique Valero et al.

Most works focus on using 2D colour information (e.g. coloured orthophotos) to identify and evaluate defects. However, 3D data can also be valuable to such analysis. In fact, there is value in conducting such analysis by collective use of 3D and colour information. With this strategy, Cappellini et al. [17] propose a semi-automatic approach to semantically label 2.5D data (colour and depth profile) of walls obtained using photogrammetry. The authors focus on labelling the wall opus using image processing techniques and parametric models of the opus. The opus model parameters include the "geometrical shape and dimensions of the stones or bricks, [and] the presence of the intervals amongst them and their installation".

Most works to date have been demonstrated for regular masonry façades constructed in uniform styles such as ashlar and squared coursed rubble [30] [29]. These techniques would however be unsatisfactory when attempting to evaluate random rubble walls due to the absence of regularity and meaningful delineation of individual stones.

8.4 Need

The review of the literature clearly shows that novel technologies and information management techniques offer great potential for surveying, assessing and managing historic monuments in general, and (historic) masonries in particular. Novel sensing technologies, including TLS and PG, ICTs, cloud services and virtual and augmented reality, and deployment of BIM practice will altogether transform current practice surrounding the management of, and public engagement on, historic monuments. In the authors' opinion, the aforementioned technologies will significantly facilitate proactive maintenance schemes for historic monuments.

However, the literature review also indicated that, while the prior constraint to effective historic monument management was the acquisition/recording of reliable data from the monuments, this constraint has now been alleviated by the development of the technologies above (TLS, PG, databases and other data management systems). As a result of this (r)evolution, a new restriction can now be identified from the literature review: the lack of effective and efficient means to extract valuable information from the very large quantity of data now made available at low costs. Indeed, focusing on stone masonry surveys, the literature shows that analysis of the data in order to extract the information that is actually valuable to surveyors and other heritage experts is still essentially undertaken manually, with 'naked eye' evaluations (e.g. segmentation of walls into different areas and measurement of those areas, for example, for presenting and estimating the amount of repair). Such processes are time-consuming and thus expensive. Furthermore, they are prone to subjectivity and are thus not repeatable.

There is thus a growing need for efficient and effective methods for (semi)automatically processing data from modern survey technologies (using these as well as other data sources) to facilitate the job of surveyors and heritage experts and enable them to focus on more value-adding activities such as conducting pathology analysis from identified defects, and the development of specifications of appropriate and sensitive repair strategies.

Beyond identifying this important need, this manuscript also aims to demonstrate that this restriction can be addressed by developing novel computer algorithms for effectively processing reality capture data, in particular large 3D point clouds. This manuscript particularly reports preliminary results pertaining to ongoing works for the extraction of 'first-order' information from survey data from TLS and PG, with a focus on stone masonry data.

8.5 Illustrative results

This section reports results from ongoing work on the development of a computer system for the analysis of textured 3D point clouds of rubble stone walls (acquired using TLS or PG). The system aims to automatically segment rubble stone walls (with random stone shapes and patterns) into their individual stones and mortar joints. This segmentation/labelling supports the extraction of further information of value to surveyors. Focusing on mortar joints, the system aims to report the mortar's linear depth profile, which is important to accurately estimate the quantity of repointing to be undertaken. It also aims to automatically calculate the amount of pinning (galleting) that should be conducted.

Detailing the algorithms developed to achieve the aforementioned processes is beyond the scope of this chapter. But the reporting and analysis of the results obtained shall contribute to support the idea that well-crafted algorithms could constitute powerful tools to support surveyors in their work, enabling them to achieve higher levels of survey completeness, objectivity and efficiency. We do not suggest that such algorithms could in any way replace surveyors altogether. Instead, we argue that they can reduce the amount of painstaking, often low-value-adding tasks currently conducted by surveyors (e.g. measurements), thereby freeing time for them to focus on value-adding tasks, such as defect assessment, repair strategy establishment, and conducting more frequent surveys (e.g. within proactive schemes).

The experimental results reported below are obtained with data acquired in the east garden of the medieval Craigmillar Castle, in Edinburgh, UK. A dense 3D point cloud of the complete garden has been captured using a TLS Leica ScanStation P40, via three scans co-registered and altogether georeferenced using an existing georeferenced control network (see Figure 8.1).

The following three subsections report the results obtained for a subset of that point cloud corresponding to a section of the main castle rampart. The section (see Figure 8.1 [bottom]) is approximately 5 m wide and 2.5 m high, and has been selected for its overall level of complexity. Indeed, as can be seen in Figure 8.1 (bottom), this wall section presents varying textures resulting from rainwater flowing through the machicolations at the top of the rampart. Furthermore, one can see a fenestration that has later been blocked up, which adds complexity in terms of the size and shape of the stones encountered.

8.5.1 Stone/mortar segmentation

We developed an algorithm that simultaneously considers 3D information both globally and at local level to segment the wall data into regions corresponding to individual stones and mortar joints. Without going into any further details, this algorithm is based on the analysis of the data in the frequency domain, and mortar joints are detected as 'narrow valleys' surrounding compact regions (i.e. stones).

Figure 8.2 shows the result obtained for the selected wall section. The algorithm effectively deals with stones and mortar joints of various shape and profile. Each 'compact' coloured region is a detected stone, and even though the same colours may appear to the reader to be assigned to several stones, these are actually different tones.

A quantitative assessment of the segmentation accuracy has been undertaken by comparing the results obtained by the algorithm against an accurate segmentation conducted manually in the orthophoto. Figure 8.3 illustrates the results for the segmentation of stones and mortar joints. Figure 8.3a illustrates the labelling performance results. Overall, the results are good, although the size of the stones seems slightly overestimated by the current algorithm. Figure 8.3b summarizes the results in Figure 8.3a by colouring each stone according to the percentage of its area (in the orthophoto) that is properly labelled as stone. As can be seen, the area of most stones is typically well segmented, with ratios over 75% for the majority of stones. The few errors appear for some (but not all) of the smallest ones.

Figure 8.1 Textured laser scanned point cloud of the east garden of Craigmillar Castle in Edinburgh, UK. *Top:* the entire point cloud. *Bottom:* a zoom onto a 5 m × 2.5 m section of the castle's rampart wall facing the garden.

8.5.2 Study of mortar joints

The automated segmentation and labelling algorithm separates stones from mortar joints. While this can be useful on its own (e.g. to automatically track the movement and/or erosion of individual stones over time), additional analysis of these areas can provide further valuable information.

For example, the mortar joint data can be further processed to elicit recessed zones, and therefore the amount of maintenance works needed. Indeed, as noted in Section 8.2.5, numerous wall repair processes are currently estimated on a square-metre (m^2) basis, which leads to significant approximations in repair costs. More precise repair work quantification and therefore costing would be obtained if more accurate measurement metrics were used, such as length, depth and width of joint mortar pointing. While such accurate measurements are too onerous to conduct manually (which is why m^2 approximations have been utilised), we demonstrate here that point cloud data provided by TLS and PG can be automatically processed to obtain such measurements by further processing the data within the segmented mortar joint areas. In this chapter, we demonstrate the results obtained with an algorithm we have developed to automatically measure the length, width and depth of

Figure 8.2a Wall stone segmentation results: (a) an orthophoto and 3D view (point cloud) of the wall section; (b) the segmentation results in the same orthophoto and 3D view.

Figure 8.2b (Continued)

Figure 8.3a Labelling performance results. (a) Black and magenta (dark grey) regions are pixels that are correctly recognized as stone and mortar respectively. Yellow (light grey) regions are 'false positives', i.e. mortar areas that are incorrectly labelled as stone. White regions are 'false negatives', i.e. stone areas that are incorrectly labelled as mortar. (b) Each stone is coloured according to the percentage of its area in the orthophoto that is properly labelled as stone.

Figure 8.3b (Continued)

mortar joints. The algorithm aims to first detect the mortar area centrelines (skeleton) and then studies the depth and width of the mortar areas along these lines. Results for the selected wall section are shown in Figure 8.4. From the centrelines (Figure 8.4(a)), the length of this material in the wall can be obtained exactly. For the selected wall section, the length is automatically calculated to be 110 m, which compares with a length of 125 m obtained from the manually segmented data. This result is positive, particularly when set within the perspective of the complexity of the case considered here and contextualised within the current lack of objective method for conducting such measurements. The depth of the mortar recess (i.e. distance from stone edge to the maximum depth of pointing) along the centrelines can also be automatically calculated, which enables the detection of mortar areas that require repointing. In Figure 8.4(b), the greater red the line, the more recessed the mortar joint. The width of the mortar joints along their centrelines can be also be derived to enable the evaluation of the volume of mortar that would be required for repairs. In Figure 8.4(c), the more red the line, the wider the mortar joint.

Furthermore, it is also possible to automatically estimate the quantity of pinning (galleting) that should be necessary to repair a given wall. Indeed, traditional mortars cannot be used to fill large joints between stones without pinning stones due to excessive mortar shrinkage and subsequent failure. Like other repairs, the quantity of pinning is typically derived approximately from the wall surface (on a square-metre basis). Instead, we have developed an algorithm (preliminary) that further processes the results above to automatically detect likely locations of pinning stones. Figure 8.5 shows the automatically detected locations of pinning stones for the selected wall section.

Additionally, the detection of large (non-recessed) mortar areas between stones without appropriate pinnings could indicate the use of cement in the mortar (as indicated above, lime works without pinnings would almost certainly fail if utilised in this manner). Therefore, the proposed algorithm could also be used to detect whether cement is likely to have been used for earlier repairs and will direct the surveyor to potentially cut out and repoint in lime even if the cement mortar is sound.

Figure 8.4a Automatic measurement of the length (a), depth (b) and width (c) of mortar joints in the selected wall section.

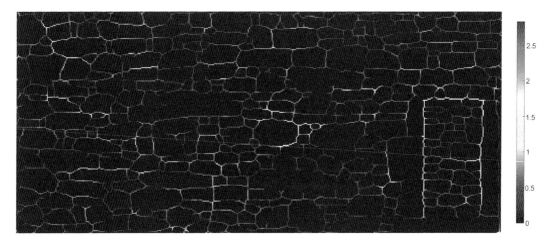

Figure 8.4b (Continued)

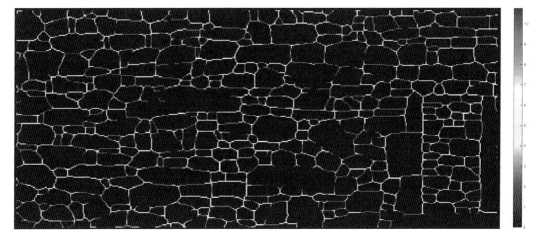

Figure 8.4c (Continued)

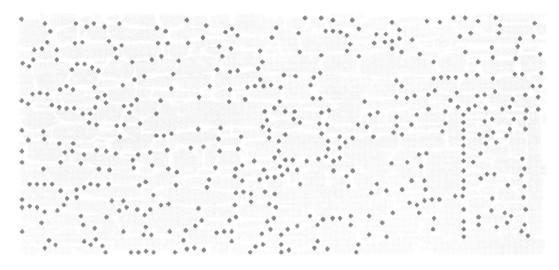

Figure 8.5 Automated detection of proposed pinning stone locations. The stones are shown in green (light grey), the mortar regions in white, and the pinning stone location in red (dark grey).

8.6 Conclusions

This chapter has reviewed current practice in surveying and the evaluation of historic masonry, with a focus on random rubble masonry and lime works. This showed that measurement of these forms of construction is made difficult by various factors including:

- The lack of regularity in this type of structure historically led to the use of ad hoc, approximate work measurement methods.
- The lack of access compounds measurement difficulties and increases survey costs. Access problems are common for these structures due in part to their complexity of evolution in building form and materials.
- The current methods of assessment rely significantly on the experience of surveyors. Surveyor variability and subjectivity in evaluation can lead to further cumulative error in assessments.
- Condition assessment requires the consideration of multiple data and information sources obtained on-site but also from third parties and from desk studies. Structuring and managing this information over time using traditional (paper-based) practices is challenging.

To address these issues, surveyors are now increasingly considering novel technologies in two areas: (1) novel reality capture technologies, in particular TLS and PG; and (2) ICTs for the structuring and management of this data, including (H)BIM. This chapter offered a review of the most recent work in these areas, with a focus on the survey of historic buildings. Despite the clear potential of these technologies to support greater efficiency and data accuracy, one can also notice the relatively slow uptake within individual and multi-agency organisations entrusted with the repair and maintenance of historic buildings (except for a few leading organisations). This chapter identifies that, while the aforementioned technologies are removing barriers to efficient, detailed and reliable data acquisition and structuring, restrictions to the overall process have shifted to the data processing stage, where the information of actual value to surveyors needs to be extracted from these large amounts of data.

Significant progress is required in this area if the above technologies are to be widely embraced as a new standard form of practice. To demonstrate the feasibility of addressing this new challenge, the last part of this chapter summarised recent research work conducted by the authors on the processing of dense 3D point clouds of rubble masonry works. It was shown that valuable information could be automatically extracted from such data, including the detection of individual stones, better estimation of the amount of mortar in a linear-metre form, and the quantification of the amount of pinning required.

Naturally, these are only preliminary results. But they show how well-crafted algorithms can provide significant value to surveying practice, delivering greater objectivity in measurement and freeing surveyors from painstaking measurement works, allowing them to focus on more value-adding interpretation and decision-making activities. The authors would thus encourage survey specialists to expand their collaboration with computer science and ICT experts to develop efficient and accurate solutions for survey, adopting novel technologies such as those discussed here.

Acknowledgements

This chapter and more particularly the work presented in Section 8.5 were made possible thanks to a research project grant from Historic Environment Scotland. The views and opinions expressed in this article are those of the authors and do not necessarily reflect the official policy or position of Historic Environment Scotland. The authors would also like to acknowledge CyberHawk Ltd. for their generous in-kind contribution to the above project and the Historic Environment Scotland Digital Documentation, Science and Estates Survey team for their support.

References

1. D.P. Andrews, J. Bedford, and P.G. Bryan. A Comparison of Laser Scanning and Structure for Motion as Applied to the Great Barn at Harmondsworth, UK. In *International Archives of the Photogrammetry, Remote Sensing and Spatial Information Sciences, Volume XL-5/W2, 2013 XXIV International CIPA Symposium*, pages 31–36, Strasbourg, France, 2013.
2. Yusuf Arayici. Towards Building Information Modelling for Existing Structures. *Structural Survey*, 26(3):210–222, 2008.
3. Arup. Maintaining Value-Module 5: Demand and Supply, Building the Business Case for Planned Maintenance, 2003.
4. Francesco Asdrubali, Giorgio Baldinelli, and Francesco Bianchi. A Quantitative Methodology to Evaluate Thermal Bridges in Buildings. *Applied Energy*, 97:365–373, 2012. Energy Solutions for a Sustainable World – Proceedings of the Third International Conference on Applied Energy, May 16–18, 2011 - Perugia, Italy.
5. John Ashurst. *Conservation of Ruins. Butterworth-Heinemann Series in Conservation and Museology*. Oxford: Routledge, 2006.
6. C.A. Balaras and A.A. Argiriou. Infrared Thermography for Building Diagnostics. *Energy and Buildings*, 34(2):171–183, 2002. {TOBUS} – a European method and software for office building refurbishment.
7. Constantinos A. Balaras, Kalliopi Droutsa, Elena Dascalaki, and Simon Kontoyiannidis. Deterioration of European Apartment Buildings. *Energy and Buildings*, 37(5):515–527, 2005.
8. Maria Lucia Vianna Baptista. Documenting a Complex Modern Heritage Building Using Multi Image Close Range Photogrammetry and 3D Laser Scanned Point Clouds. In *International Archives of the Photogrammetry, Remote Sensing and Spatial Information Sciences, Volume XL-5/W2, 2013 XXIV International CIPA Symposium*, pages 675–678, Strasbourg, France, 2013.
9. Herbert Bay, Andreas Ess, Tinne Tuytelaars, and Luc Van Gool. Speeded-Up Robust Features (SURF). *Computer Vision and Image Understanding*, 110(3):346–359, 2008.
10. P. Beckman. *Structural Aspects of Building Conservation*. London: McGraw-Hill, 1995.

11 D. Bell. Technical Advice Note 8: The Historic Scotland Guide to International Charters. Technical report, Historic Scotland, Edinburgh, 1997.
12 Raffaella Brumana, Daniela Oreni, Anna Raimondi, Andreas Georgopoulos, and Angeliki Bregianni. From Survey to HBIM for Documentation, Dissemination and Management of Built Heritage. In *Proceedings of Digital Heritage International Congress 2013*, volume 1, pages 497–504, Marseille, France, 2013.
13 Xavier Brunetaud, Livio De Luca, Sarah Janvier-Badosa, Kévin Beck, and Muzahim Al-Mukhtar. Application of Digital Techniques in Monument Preservation. *European Journal of Environmental and Civil Engineering*, 16(5):543–556, 2012.
14 F. Bruno, G. Bianco, M. Muzzupappa, S. Barone, and A.V. Razionale. Experimentation of Structured Light and Stereo Vision for Underwater 3D Reconstruction. *ISPRS Journal of Photogrammetry and Remote Sensing*, 66(4):508–518, 2011.
15 P.G. Bryan, M. Abbott, and A.J. Dodson. Revealing the Secrets of Stonehenge through the Applications of Laser Scanning, Photogrammetry and Visualization Techniques. In *International Archives of the Photogrammetry, Remote Sensing and Spatial Information Sciences, Volume XL-5/W2, 2013 XXIV International CIPA Symposium*, pages 125–129, Strasbourg, France, 2013.
16 BSI Group. Bs 7913:2013 – guide to the conservation of historic buildings, 2013.
17 Valeria Cappellini, Chiara Stefani, Nicolas Nony, and Livio De Luca. Surveying Masonry Structures by Semantically Enriched 2.5D Textures: A New Approach. In Marinos Ioannides, Dieter Fritsch, Johanna Leissner, Rob Davies, Fabio Remondino, and Rossella Caffo, editors, *Progress in Cultural Heritage Preservation*, volume 7616 of *Lecture Notes in Computer Science*, pages 729–737. Springer, Berlin Heidelberg, 2012.
18 A. Cardaci, G. Mirabella Roberti, and A. Versaci. From the Continuous to the Discrete Model: a Laser Scanning Application to Conservation Projects. In *Proceedings of the Congress of the International Society for Photogrammetry and Remote Sensing (ISPRS)*, pages 437–444, 2011.
19 T. Cardinale, R. Valva, and M. Lucarelli. The Integrated Survey for Excavated Architectures: the Complex of Casalnuovo District with the World Heritage Site "Sassi" (Matera, Italy). In *Proceedings of 3D Virtual Reconstruction and Visualization of Complex Architectures*, pages 403–409, 2015.
20 May Cassar. Climate Change and the Historic Environment. Technical report, University College London Centre for Sustainable Heritage, 2005.
21 Roberto Cecchi and Paolo Gasparoli. *Preventive and Planned Maintenance of Protected Buildings*. Alinea International, Italy, 2012.
22 Maria Mercede Cerimele and Rossella Cossu. Decay Regions Segmentation from Color Images of Ancient Monuments Using Fast Marching Method. *Journal of Cultural Heritage*, 8:170–175, 2007.
23 Maria Mercede Cerimele and Rossella Cossu. A Numerical Modelling for the Extraction of Decay Regions from Color Images of Monuments. *Mathematics and Computers in Simulation*, 79:2334–2344, 2009.
24 CIB Working Commission W86. Building Pathology: A State of the Art Report, CIB Report, Publication 155. Technical report, International Council for Research and Innovation in Building and Construction, Rotterdam, the Netherlands, 1993.
25 Rossella Cossu and Laura Chiappini. A Color Image Segmentation Method as Used in the Study of Ancient Monument Decay. *Journal of Cultural Heritage*, 5:385–391, 2004.
26 AECOM. *Spon's Architects' and Builders' Price Book*. London: CRC Press, 2014.
27 Department of Trade and Industry. Annual Construction Statistic. Office for National Statistics, Government Buildings, Cardiff Road, Newport, United Kingdom, 2002.
28 C. Dore and M. Murphy. Semi-Automatic Modelling of Building Facades with Shape Grammars Using Historic Building Information Modelling. In *International Archives of the Photogrammetry, Remote Sensing and Spatial Information Sciences, Volume XL-5/W2, 2013 XXIV International CIPA Symposium*, pages 57–64, Strasbourg, France, 2013.
29 P. Drap, D. Merad, J. Boi, J. Seinturier, D. Peloso, C. Reidinger, G. Vannini, M. Nucciotti, and E. Pruno. Photogrammetry for Medieval Archaeology: A Way to Represent and Analyse Stratigraphy. In *Proceedings of 18th International Conference on Virtual Systems and Multimedia (VSMM)*, pages 157–164, September 2012.
30 P. Drap, J. Seinturier, J.C. Chambelland, and E. Pruno. Going to Shawbak (Jordan) and Getting the Data Back: Toward a 3D GIS Dedicated to Medieval Archaeology. In *Proceedings of 3D Arch 2009, 3D Virtual Reconstruction and Visualization of Complex Architectures*, Trento, Italy, 2009.

31. Pierre Drap, Gilles Gaillard, Pierre Grussenmeyer, and Andreas Hartmann-Virnich. A Stone-by-Stone Photogrammetric Survey Using Architectural Knowledge Formalised on the Arpenteur Photogrammetric Workstation. In *Proceedings of XIXth Congress of the International Society for Photogrammetry and Remote Sensing (ISPRS)*, pages 187–194, Amsterdam, 2000.
32. M.F. Durka, W. Morgan, and D.T. Williams. *Structural Mechanics*. Prentice Hall, Edinburgh, 1996.
33. S. Fai and M. Sydor. Building Information Modelling and the Documentation of Architectural Heritage: Between the 'Typical' and the 'Specific'. In *Proceedings of Digital Heritage International Congress 2013*, volume 1, pages 731–734, Marseille, France, 2013.
34. F. Fassi, L. Fregonese, S. Ackermann, and V. De Troia. Comparison between Laser Scanning and Automated 3D Modelling Techniques to Reconstruct Complex and Extensive Cultural Heritage Areas. In *Proceedings of 3D-ARCH 2013–3D Virtual Reconstruction and Visualization of Complex Architectures*, pages 73–80, Trento, Italy, 2013.
35. A. M. Forster, P. F. G. Banfill, K. Carter, and B. Kayan. Green Maintenance for Historic Masonry Buildings: An Emerging Concept. *Building Research and Information*, 39(6):656–664, 2011.
36. Alan Forster. Building Conservation Philosophy for Masonry Repair: Part 1 – "Ethics". *Structural Survey*, 28(2):91–107, 2010.
37. Alan Forster and K. Carter. A Framework for Specifying Lime Mortars for Rubble Masonry Construction. *Structural Survey*, 29:373–96, 2011.
38. Alan Forster and Brit Kayan. Maintenance for Historic Buildings: A Current Perspective. *Structural Survey*, 27(3):210–229, 2009.
39. Alan Mark Forster and James Douglas. Condition Survey Objectivity and Philosophy Driven Masonry Repair: An Increased Probability for Project Divergence? *Structural Survey*, 28(5):384–407, 2010.
40. Franklin & Andrews. *Griffiths Complete Building Price Book*. London: Franklin & Andrews, 2015.
41. S. Garagnani and A. M. Manferdini. Parametric Accuracy: Building Information Modeling Processs Applied to the Cultural Heritage Preservation. In *Proceedings of 3D-ARCH 2013–3D Virtual Reconstruction and Visualization of Complex Architectures*, pages 87–92, Trento, Italy, 2013.
42. G Garau, E Dal Zio, R Paparella, and M Caini. Deterioration in Buildings because of Design Mistakes. Teaching Prevention in a Civil Engineering Course. In *2nd Symposium on Building Pathology, Durability and Rehabilitation*, 1996.
43. Glasgow. Glasgow West Conservation Trust. Stonework – Decay of Stone and Recommended Repair Techniques. Technical report, Glasgow West Conservation Trust, Glasgow, UK, 1993.
44. P. Grussenmeyer, E. Alby, T. Landes, M. Koehl, S. Guillemin, J.-F. Hullo, P. Assali, and E. Smigiel. Recording Approach of Heritage Sites Based on Merging Point Clouds from High Resolution Photogrammetry and Terrestrial Laser Scanning. In *Proceedings of XXII ISPRS Congress*, pages 553–558, Melbourne, Australia, 2012.
45. Health and Safety Executive. Health and Safety in Construction in Great Britain, 2014, 2015.
46. N. Hichri, C. Stefani, L. De Luca, P. Veron, and G. Hamon. From Point Cloud to BIM: A Survey of Existing Approaches. In *International Archives of the Photogrammetry, Remote Sensing and Spatial Information Sciences, Volume XL-5/W2, 2013 XXIV International CIPA Symposium*, pages 343–348, Strasbourg, France, 2013.
47. Malcolm Hollis. *Surveying Buildings*. London: RICS Books, 2005.
48. ICOMOS. *ICOMOS-ISCS: Illustrated Glossary on Stone Deterioration Patterns*. Champigny/Marne, France: ICOMOS, 2008.
49. P. Kapsalas, M. Zervakis, and P. Maravelaki-Kalaitzaki. Evaluation of Image Segmentation Approaches for Non-Destructive Detection and Quantification of Corrosion Damage on Stonework. *Corrosion Science*, 49:4415–4442, 2007.
50. Anestis Koutsoudis, Blaz Vidmar, George Ioannakis, Fotis Arnaoutoglou, George Pavlidis, and Christodoulos Chamzas. Multi-Image 3D Reconstruction Data Evaluation. *Journal of Cultural Heritage*, 15:73–79, 2014.
51. W.L. Lai, T. Kind, and H. Wiggenhauser. Using Ground Penetrating Radar and Time–Frequency Analysis to Characterize Construction Materials. *NDT & E International*, 44(1):111–120, 2011.
52. Jürg Leckebusch. Two- and Three-Dimensional Ground-Penetrating Radar Surveys Across a Medieval Choir: A Case Study in Archaeology. *Archaeological Prospection*, 7(3):189–200, 2000.
53. J.L. Lerma and C. Muir. Evaluating the 3D Documentation of an Early Christian Upright Stone with Carvings from Scotland with Multiples Images. *Journal of Archaeological Science*, 46:311–318, 2014.

54 David G. Lowe. Distinctive Image Features from Scale-Invariant Keypoints. *International Journal of Computer Vision*, 60(2):91–110, 2004.

55 Hélène Macher, Tania Landes, Pierre Grussenmeyer, and Emmanuel Alby. Semi-automatic Segmentation and Modelling from Point Clouds towards Historical Building Information Modelling. In Marinos Ioannides, Nadia Magnenat-Thalmann, Eleanor Fink, Roko Žarnic, Alex-Yianing Yen, and Ewald Quak, editors, *Digital Heritage. Progress in Cultural Heritage: Documentation, Preservation, and Protection*, volume 8740 of *Lecture Notes in Computer Science*, pages 111–120. Limassol, Cyprus: Springer International Publishing, 2014.

56 Maintain Our Heritage. Putting It Off: How Lack Maintenance Fails Our Heritage. Technical report, Maintain Our Heritage, Bath, UK, 2004.

57 J. De Matías, F. Berenguer, J.P. Cortés, J.J. De Sanjosé, and A. Atkinson. Laser Scanning for the Geometric Study of the Alcantara Bridge and Coria Cathedral. In *Proceedings of 3D-ARCH 2013–3D Virtual Reconstruction and Visualization of Complex Architectures*, pages 51–56, Trento, Italy, 2013.

58 Travis McDonald. Understanding Old Buildings: The Process of Architectural Investigation. In *Preservation Briefs*. Technical Preservation Services, National Park Service, US Department of the Interior, 1994.

59 J.D. Meneely, B.J. Smith, J. Curran, and A. Ruffell. Developing a Non-Destructive Scientific Toolkit to Monitor Monuments and Sites. In *Proceedings of ICOMOS Scientific Symposium*, 2009.

60 Najib Metni and Tarek Hamel. A UAV for Bridge Inspection: Visual Servoing Control Law with Orientation Limits. *Automation in Construction*, 17:3–10, 2007.

61 Laura Moltedo, Giuseppe Mortelliti, Ovidio Salvetti, and Domenico Vitulano. Computer Aided Analysis of the Buildings. *Journal of Cultural Heritage*, 1:59–67, 2000.

62 Maurice Murphy, Eugene McGovern, and Sara Pavia. Historic Building Information Modelling – Adding Intelligence to Laser and Image Based Surveys of European Classical Architecture. *ISPRS Journal of Photogrammetry and Remote Sensing*, 76:89–102, 2013.

63 A. Nettley, K. Anderson, C. DeSilvey, and C. Caseldine. Using Terrestrial Laser Scanning and LIDAR Data for Photo-Realistic Visualization of Climate Impacts at Heritage Sites. In *Proceedings of the International Symposium on Photogrammetry, Remote Sensing and Spatial Information Sciences*, pages 223–229, 2013.

64 Clifford L. Ogleby. Advances in the Digital Recording of Cultural Monuments. *{ISPRS} Journal of Photogrammetry and Remote Sensing*, 50(3):8–19, 1995.

65 Daniela Oreni, Raffaella Brumana, Fabrizio Banfi, Luca Bertola, Luigi Barazzetti, Branka Cuca, Mattia Previtali, and Fabio Roncoroni. Beyond Crude 3D Models: From Point Clouds to Historical Building Information Modeling via NURBS. In Marinos Ioannides, Nadia Magnenat-Thalmann, Eleanor Fink, Roko Žarnic, Alex-Yianing Yen, and Ewald Quak, editors, *Digital Heritage. Progress in Cultural Heritage: Documentation, Preservation, and Protection*, volume 8740 of *Lecture Notes in Computer Science*, pages 166–175. Limassol, Cyprus: Springer International Publishing, 2014.

66 Noelia Oses and Fadi Dornaika. Image-Based Delineation of Built Heritage Masonry for Automatic Classification. In Mohamed Kamel and Aurélio Campilho, editors, *Image Analysis and Recognition*, volume 7950 of *Lecture Notes in Computer Science*, pages 782–789. Springer Berlin Heidelberg, 2013.

67 Pieter Pauwels, Rens Bod, Danilo Di Mascio, and Ronald De Meyer. Integrating Building Information Modelling and Semantic Web Technologies for the Management of Built Heritage Information. In *Proceedings of Digital Heritage International Congress 2013*, volume 1, pages 481–488, Marseille, France, 2013.

68 Hannes Püschel, Martin Sauerbier, and Henri Eisenbeiss. A 3D Model of Castle Landenberg (CH) from Combined Photogrammetric Processing of Terrestrial and UAV-Based Images. In *The International Archives of the Photogrammetry, Remote Sensing and Spatial Information Sciences*, volume XXXVII, pages 93–98, Beijing, 2008.

69 Danilo Ranalli, Marco Scozzafava, and Marco Tallini. Ground Penetrating Radar Investigations for the Restoration of Historic Buildings: The Case Study of the Collemaggio Basilica (L'Aquila, Italy). *Journal of Cultural Heritage*, 5(1):91–99, 2004.

70 F. Remondino, L. Barazzetti, F. Nex, M. Scaioni, and D. Sarazzi. UAV Photogrammetry for Mapping and 3D Modeling – Current Status and Future Perspectives. In *International Archives of the Photogrammetry, Remote Sensing and Spatial Information Sciences*, volume XXXVIII-1/C22, pages 25–31, 2011.

71 RICS. *SMM7- Standard Method of Measurement*. London: RICS Books, 1998.

72 RICS. Guide to Building Surveys and Inspections of Commercial and Industrial Property, 2005.
73 RICS. *RICS New Rules of Measurement – NRM 2: Detailed Measurement for Building Works*. London: RICS Books, 2012.
74 Mike Riley and Alison Cotgrave. *Construction Technology 3. The Technology of Refurbishment and Maintenance*. Palgrave MacMillan, New York, 2nd edition, 2011.
75 Pablo Rodriguez-Gonzalvez, Diego Gonzalez-Aguilera, Gemma Lopez-Jimenez, and Inmaculada Picon-Cabrera. Image-Basedased Modeling of Built Environment from an Unmanned Aerial System. *Automation in Construction*, 48:44–52, 2014.
76 T. Rushton. Understanding Why Buildings Fail and Using 'HEIR' Methodology in Practice. In *Latest Thinking on Building Defects in Commercial Buildings: Their Diagnosis, Causes and Consequences*, 1992.
77 P. Salonia and A. Negri. Historical Buildings and Their Decay: Data Recording, Analysis and Transferring in an ITC Environment. In *Proceedings of the Congress of the International Society for Photogrammetry and Remote Sensing (ISPRS)*, 2003.
78 C. Santagati, L. Inzerillo, and F. Di Paola. Image-Based Modeling Techniques for Architectural Heritage 3D Digitalization: Limits and Potentialities. In *International Archives of the Photogrammetry, Remote Sensing and Spatial Information Sciences, Volume XL-5/W2, 2013 XXIV International CIPA Symposium*, pages 555–560, Strasbourg, France, 2013.
79 Pierre Smars. Software Tools for In-Situ Documentation of Built Heritage. In *International Archives of the Photogrammetry, Remote Sensing and Spatial Information Sciences, Volume XL-5/W2, 2013 XXIV International CIPA Symposium*, pages 589–94, Strasbourg, France, 2013.
80 L. J. Sánchez-Aparicio, A. Villarino, J. García-Gago, and Diego González-Aguilera. Non-Contact Photogrammetric Methodology to Evaluate the Structural Health of Historical Constructions. In *Proceedings of 3D Virtual Reconstruction and Visualization of Complex Architectures*, pages 331–338, 2015.
81 SPAB. *The Manifesto*. London: Society for the Protection of Ancient Buildings, 1877.
82 Chiara Stefani, Xavier Brunetaud, Sarah Janvier-Badosa, Kévin Beck, Livio De Luca, and Muzahim Al-Mukhtar. Developing a Toolkit for Mapping and Displaying Stone Alteration on a Web-Based Documentation Platform. *Journal of Cultural Heritage*, 15(1):1–9, 2014.
83 Ad Straub. Dutch Standard for Condition Assessment of Buildings. *Structural Survey*, 27(1):23–35, 2009.
84 C. Strecha, T. Tuytelaars, and L. Van Gool. Dense Matching of Multiple Wide-baseline Views. In *Proceedings of International Conference on Computer Vision, ICCV 2003*, Nice, France, 2003.
85 T. Taylor, J. Counsell, and S. Gill. Energy Efficiency is More than Skin Deep: Improving Construction Quality Control in New-Build Housing Using Thermography. *Energy and Buildings*, 66:222–231, 2013.
86 T. Temizer, G. Nemli, E. Ekizce, A. Ekizce, S. Demir, B. Bayram, Askin F.H, A.V. Cobanoglu, and H. F. Yilmaz. 3D Documentation of a Historic Monument Using Terrestrial Laser Scanning Case Study: Byzantine Water Cistern, Istanbul. In *International Archives of the Photogrammetry, Remote Sensing and Spatial Information Sciences, Volume XL-5/W2, 2013 XXIV International CIPA Symposium*, pages 623–628, Strasbourg, France, 2013.
87 Mary J. Thornbush and Heather A. Viles. Photographic Monitoring of Soiling and Decay of Roadside Walls in Central Oxford, England. *Environmental Geology*, 56(3–4):777–787, 2008.
88 Darren Turner, Arko Lucieer, and Christopher Watson. An Automated Technique for Generating Georectified Mosaics from Ultra-High Resolution Unmanned Aerial Vehicle (UAV) Imagery, Based on Structure from Motion (SfM) Point Clouds. *Remote Sensing*, 4:1392–1410, 2012.
89 Dennis Urquhart. *Safeguarding Glasgow's Stone-Built Heritage: Skills and Materials Requirements: Executive Summary*. Edinburgh: Scottish Stone Liaison Group, 2006.
90 R Webster. *Stone Cleaning and the Nature, Soiling and Decay Mechanisms of Stone*. Donhead, London, 1992.
91 Zhihua Xu, Lixin Wu, Yonglin Shen, Fashuai Li, Qiuling Wang, and Ran Wang. Tridimensional Reconstruction Applied to Cultural Heritage with the Use of Camera-Equipped UAV and Terrestrial Laser Scanner. *Remote Sensing*, 6:10413–10434, 2014.
92 David Young, Heritage Council of New South Wales, Heritage Victoria, South Australia. Department for Environment, Heritage, and Adelaide (S.A.). Corporation. *Salt Attack and Rising Damp: A Guide to Salt Damp in Historic and Older Buildings*. [Sydney?] : Heritage Council of NSW, Heritage Victoria, South Australian Department for Environment and Heritage, Adelaide City Council, 2nd edition, 2008. Cover title.

93 M. Young. Thermal Imaging in the Investigation of Solid Masonry Structures, 2014. Accessed 20/11/15.
94 C.W. Yu and J.W. Bull (eds.) *Durability of Materials and Structures in Building and Civil Engineering*. Edinburgh: Whittle Publishing, 2006.
95 P.J. Zarco-Tejada, R. Diaz-Varela, V. Angileri, and P. Loudjani. Tree Height Quantification Using Very High Resolution Imagery Acquired from an Unmanned Aerial Vehicle (UAV) and Automatic 3D Photo-Reconstruction Methods. *European Journal of Agronomy*, 55:89–99, 2014.
96 L. Wilson, A. Rawlinson, D. Mitchell, H. McGregor, and R. Parsons, The Scottish Ten Project: Collaborative Heritage Documentation. In *XXIV International CIPA Symposium, Strasbourg*, France, 685–690, 2013.

9 HBIM applications in Egyptian heritage sites

Yasmine Sabry Hegazi

9.1 Introduction

Egypt is like a story formed from thousands of years spent in civilizations; creating a character that was built through ages by adding one layer of history after another. Ever since the construction of the High Dam, the world discovered what huge threats could hurt the Egyptian heritage, with the High Dam causing a rising in the water level of the River Nile, leading to a flooding over the Nubian land. After the Second World War, the United Nations came to light and made an organization called United Nations Educational, Scientific and Cultural Organization (UNESCO), which announced its responsibility of protecting both natural and cultural heritage through its convention in 1972. In 1978, UNESCO made a huge campaign to save Nubian heritage, including the relocation of Egypt's ancient temples "Ramses II" and "Nefertari" away from the water level of Lake Nasser. Then, in 1979, many sites were inscribed in UNESCO's list of world heritage. Historic Cairo not only refers to the heritage of the Islamic era but that which extended into the eighteenth, nineteenth and twentieth centuries as well.

9.2 Explaining HBIM and BIM

To know the meaning of HBIM (Historic Building Information Modeling), we should know what BIM (Building Information Modeling) refers to. The idea behind BIM started before adding the "historic" term; where BIM was defined as "a simulation of construction in a virtual environment by constructing a model that contains all building information".[1] BIM doesn't merely present a 3D model but rather has many more dimensions added to the original model; as follows:[2]

Before BIM

- First dimension: drafting (scratch stage)
- Second dimension: vector drawing the views and plans

With BIM

- Third dimension: forming the model with tools such as laser scanner
- Fourth dimension: time scheduling; i.e. the connection between timing information and the construction progress
- Fifth dimension: cost estimation and value engineering
- Sixth dimension: applying sustainability measures
- Seventh dimension: facility management and the building life cycle
- Eight dimension: accident prevention

By using the previous dimensions, an integrated management approach can be used to control the construction process before it starts, during its operation, and even after it's implemented through dimension seven: the facility management. But what if it's an existing building with a heritage value? What kind of building information modeling could be useful when it comes to a historic building? Even why did we choose to apply HBIM to our significant and valuable heritage?

9.3 Why HBIM?

Early modeling of historic buildings utilized three-dimensional engineering models to analyze the dead and live load risks, in order to decide conservation priorities. In time, more advanced software programs – such as Revit – produced a complete package of architectural and structural drawings known as documentation drawings, which provided data on all elements in the building. In the case of rehabilitating a monument to prepare it for a certain use, HBIM was used to review rehabilitation codes in order to guarantee the rehabilitation process was achieved without conflicting with the monument authenticity. Modeling here assists in coordinating, for example, between the electromechanical works on the permitted level of fixing and all the decorated items in the ceilings. Most of the monuments generally undergo a preventive conservation program; this program is known in BIM as facilities management. Both systems serve the maintenance operation. In valuable buildings, the cost estimate shouldn't just be calculated with value engineering principles. Recommendations of international conventions and charters for conservation and rehabilitation must be taken into consideration when deciding the required cash flow and applying the fifth dimension (cost estimation and value engineering).[3]

9.4 Color coding of model elements

The following color system should be utilized for all federated models – Coordination, As-Built and Record Models – and it has to be applied in HBIM as well, in order to keep to the international language of information modeling:[4]

- Architectural Elements: White (255,255,255)
- Envelope (Curtain wall, Precast, Other): Light gray (211,211,211)

Structural models:

- Steel: Maroon (128,0,0)
- Concrete: Gray (128,128,128)
- Masonry: Gray (128,128,128)

Electromechanical models:

- Mechanical Ductwork Supply: Blue (0,0,255)
- Mechanical Ductwork Return: Blue (153,204,255)
- Mechanical Ductwork Exhaust: Blue (105,105,255)
- Mechanical Piping Supply: Cyan (0,160,155)
- Mechanical Piping Return: Cyan (0,255,255)
- Electrical Conduit/Cable Tray: Yellow (255,255,0)
- Electrical Lighting: Gold (218,165,32)
- Plumbing Domestic Water: Green (124,252,0)

- Plumbing Storm/Roof Drain: Green (0,100,0)
- Plumbing Waste/Vent: Green (154,205,50)
- Medical Gas: Green (0,155,0)
- Fire Protection: Red (255,0,0)
- Fire Alarm/Data/IT/Controls: Coral (255,127,80)
- Pneumatic Tubing: Magenta (139,0,139)

9.5 HBIM coordination

Coordination is necessary in HBIM in order to detect clashes. This coordination has two parts. The first part is between the activities that need to be done, and the second is between the software modeling program – like Revit – and the cost estimate software, scheduling software and property management software, among others. With 70 different software programs listed as support programs for BIM, we need a similar platform to be used in the case of HBIM to enable coordination between all the used data in every dimension. Here are some of the most important coordination software programs and tools to consider using with HBIM:[5]

- Navisworks: An Autodesk software program that can be used to connect any model of almost any file type with navigate models and project schedules (4D), as well as perform clash detection.
- Revit: An Autodesk software tool for designers. The software allows users to produce Building Information Models and corresponding 2D drawings, which are "snapshots" of the model, as well as access the building information from a model database.
- Revit Family: A set of parametric Revit objects used to create building elements such as walls, doors, windows, etc. . . .
- Solibri: A rules-based software tool that analyzes Building Information Models for aspects like code compliance, data integrity, material quantities and clashes.

It is important in HBIM to also consider the level of development (LOD), which is determined according to actual work progress or how close to a finalized design the featured model is. Although people often refer to LOD as "level of detail", a level of development coordinates the required level of details in a model with the maturity of the preservation.

9.6 Using HBIM in documentation and monitoring according to objects scale

The usage of the 3D technology in monuments will make the documentation, monitoring, and even management of monuments much easier. Cairo's cultural heritage can be classified according to size into groups of buildings and sites on the one hand and monuments on the other.[6] In both cases, HBIM is strongly recommended, but different approaches are required when we deal with either via the 3D technology.

9.6.1 Groups of buildings and sites

Modeling of historic buildings depends on existing condition records which might be connected to a geographical information system (GIS) with the goal of coordinating with the environment where the monument is located. If the monument was already in the form of a heritage district or a group of buildings, this coordination becomes very important in providing the built heritage information. The model should combine certain parameters that can be edited according to actual work development, so that updated plans can be provided during the conservation process. Every element should be identified separately on the model to produce the table of elements, with a code given to each material and acting as an important entry to quantities calculation and the listing of restoration materials. The table of elements should also include flooring and roof layers, such as insulation and drainage, in addition to vertical circulation elements.[7]

9.6.1.1 Collecting data for HBIM in groups of buildings and sites

When adopting HBIM with a group of buildings using a geographic information system (GIS), data is collected and analyzed through hardware and software which in turn transform this data into geographical output that can be presented in three dimensions.[8] HBIM also integrates with the Global Positioning System (GPS) which is managed by the American Defense Intelligence Agency through 24 satellites located in different positions around the earth. This system works through a transmitter that sends signals to satellites, and by calculating the time and place of each signal from at least three satellites, the position of the site can be specified in its original place. Accordingly, this position can be connected to the HBIM model; and if a fourth satellite exists, then a façade survey can also be produced.[9]

Another data collector and analyzer used to support HBIM is the remote sensing device, which is mainly classified into aerial and photographic systems. The first is a position finder used in locating sites and groups of buildings where linear drawings can be manually produced. In the second system, sunlight can be reflected by a reflector and received through satellite sensors to create a complete model in 3D that is produced after treating the geometrical errors coming from the earth sphere form. This system is preferred in huge cities, such as historic Cairo, in order to coordinate the massive data and monitor historic districts' buffer zones.[10]

All those previously mentioned systems work with a digital information system,[11] a technology that depends on devices and a bundle of tools such as maps and aerial satellite photos, to collect data that can be stored and updated through four stages:

- Data entry
- Data storage
- Data analysis
- Data presentation

There is also the system used in conservation stages which classifies buildings by typology and levels of danger and hence assists in the design of a preventive conservation plan that can be used through the seventh dimension of HBIM (facility management) for maintenance and operation. HBIM can be an important tool in managing information for a group of buildings and sites, as it provides these benefits:

- urban documentation in order to produce accurate land use
- demographic maps creation as output to monitor large-scale sites and heritage cities
- monitoring of urban growth by comparing satellite images
- coverage of historic sites documentation by an infinite number of accurate photos that can be produced by different satellites at the same time
- utilizing a variation of ultrasonic waves and frequencies to document invisible sites and reveal undiscovered desert heritage districts
- saving documentation process time and money as well as human resources
- enabling compatible documentation data for large sites in good quality

9.6.2 Collecting data for HBIM in the case of monuments

Material basic data can be inscribed in an HBIM system in order to calculate required resources for conservation and maintenance. This data includes a diagnosis of the materials used in monuments, mostly done by using a polarizing microscope, in addition to a petrographic analysis created by X-ray diffraction.[12] A mapping of groundwater can also be achieved by utilizing piezometers and survey data collectors, such as a total station device. Such survey data collectors are not only used to connect information in the process of historic buildings modeling, but also to monitor the stability of the monuments by analyzing the engineering modeling[13] and making a conservation plan with an order of priorities.

When performing stability analysis through engineering modeling using SAP (Systems Applications & Products), which is a software program used for analyzing structural loads, a software clash appeared because the load analysis model done by SAP does not have a connection point with the HBIM software. Yet, it is very important to use this analysis in historic buildings because it can calculate loads in a color output with a key that shows the level of risk; hence a certain order of priority in the process of conservation and maintenance can be followed. If the clash in connection between HBIM and SAP software was solved, it would be a great progress that enables the application of the eighth dimension to HBIM, which is known in BIM as accident prevention.

A photogrammetry tool is also used for historic building three-dimensional modeling. Photogrammetry is terrestrial, aerial, or spatial,[14] based on the position of the camera and the object of documentation. The most common in documenting Egyptian heritage is the terrestrial photogrammetry which is either done by the stereo method, where two cameras are placed in a condition so that each façade would be totally covered, or by the bundle method, which is used if the monument is complicated. In the bundle method, many cameras are placed in a bundle of positions and the model is produced by knowing the camera positions[15] as well as the distances and angles between each camera and the monument. This operation depends on four known points and is supported by mathematical analysis to straighten the lines.

Modeling by 3D laser scanning is another tool used to collect and analyze data in order to produce a model by sending laser rays to a monument surface and reflecting it back to the device again. By defining the scanner positions and coordinating points on the monument, the model can be produced. 3D laser scanning can be fixed terrestrial, kinematic terrestrial or airborne. The scanning idea depends on generating a cloud with millions of points that can produce a multi-points model which is compatible with many software programs where the data can be classified and analyzed to produce a very accurate 3D model.[16] If this model is connected to GPS, a realistic model in situ can be produced; this technology is the chosen data collector for the case study in this chapter.

On the level of information management of monuments, HBIM can help in conservation and maintenance management as follows:

- designing the mortar through a technology of material analysis
- measuring the groundwater level to control its effect on monuments
- documenting deterioration aspects through photogrammetry and 3D laser scanning
- obtaining a high accuracy in documentation output
- insuring construction stability by engineering modeling and total stations
- saving time and cost by using the technologies of HBIM data collectors

9.7 Case study – Baron Empain, Heliopolis, Cairo, Egypt

The district of Heliopolis was founded in the first decade of the twentieth century by Baron Edward Empain, a Belgian banker; with the word "Heliopolis", meaning the city of the sun. Baron Empain lived in Heliopolis at a unique Indian-style palace with traditional architectural design created by

Alexander Marcel. The palace was built between the years 1907 and 1910,[17] with rooms designed to expose sunrays throughout indoor spaces. Mainly with an Indian character, the palace still had Chinese, Arab and Roman features.[18] (See Figure 9.1.)

9.7.1 Using HBIM in Baron Palace

The objective in using the HBIM application in sites of Egyptian heritage is to reach the qualified data analysis that can only be achieved through HBIM with the goals of managing the monument

Figure 9.1 Baron Empain Palace.

conservation process while also using the model as an educational tool. The study of using HBIM in Baron Palace can be divided into two main approaches: experimental and theoretical. The first approach is based on using a 3D laser scanner as a data collector to produce the palace model, create a schematic time schedule for conservation and develop an idea for adaptive reuse. In the second approach, theoretically, we will address how we can apply all the HBIM dimensions to Baron Empain Palace as a future target study. Eventually, we will be able to formulate an integrated approach for conservation and come up with an adaptive reuse plan for the palace.

9.7.2 The experimental approach

Since Baron Palace has a complicated design with many architectural details, a 3D laser scanner was used to produce point clouds in order to make the model and coordinate it with information on the historic building. This scanner belongs to the generation that can make a full dome view with 360 degrees in all directions, and can target up to 270 meters of distance. It has a mixed pixel filtering system, producing very accurate point clouds, and is supported with an external high definition camera that can export a survey with flexible resolution and which can be adjusted according to

Figure 9.2 Baron Palace experimental output point cloud.

building details. The scanner is also compatible with the GPS and so can coordinate the building with the surrounding environmental setting.

The point cloud creates X,Y and Z engineering measurements, which can be digitized via Auto-CAD and Revit using intermediate software that exports to different drawing software programs. The point cloud image can be displayed through TruView software so that the building dimensions can be measured, while also displaying the physical materials so that a map of deterioration can be easily produced. Point clouds captured by the scanner into Revit enable the production of the model, which can provide a virtual visit to the site and thus can be a very useful tool in tourism marketing (see Figure 9.2).

To produce the model, the scanner was used in Baron Palace for eight hours of fieldwork with many positions to cover most of the details. For example, to produce the main façade, three positions were taken as shown (see Figure 9.3). Every position of the scanner was adjusted through survey coordinate points to define X,Y and Z coordinates, and then all positions were merged together. Any clashes resulting from the merging process were automatically corrected by the software.

9.7.3 Steps of implementation

1 Choose scanner positions
2 Make coordinate points in every position (see Figure 9.4)

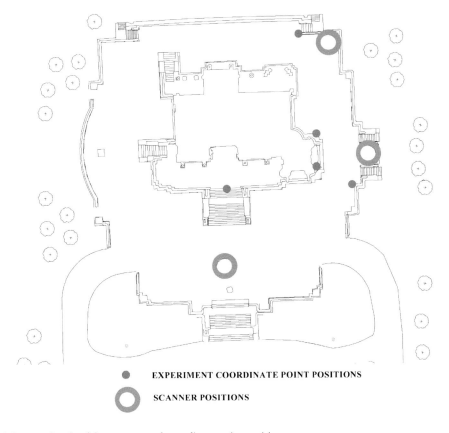

Figure 9.3 Layout sketch of the scanner and coordinate point positions.

3 Control the amount of data by adjusting camera resolution
4 Start the scanning process
5 Join the scanning positions to create a complete point cloud
6 Import model to computer (see Figure 9.5)

Figure 9.4 Scanner positions and coordinate points.

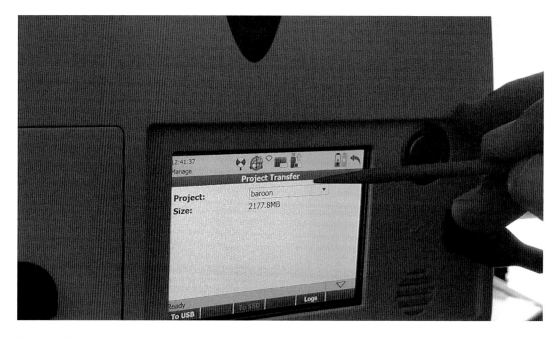

Figure 9.5 Transfer of the point clouds to computer for the model production.

7   Refinish the surfaces (smoothen, close holes, decimate, etc.)
8   Import the polygon surfaces by the intermediate software to the Revit family in order to apply the full set of HBIM dimensions

9.7.4 Theoretical approach

While the experiment is still in the stage of collecting data to formulate the HBIM for Baron Empain Palace, the theoretical approach mentioned below acts as an idea of how to apply HBIM dimensions to the project in a full process that can be developed through the project stages.

9.7.4.1 Applying fourth and fifth dimensions

The fourth and fifth dimensions (time scheduling and cost evaluation/value engineering) can be applied to Baron Palace as a project of conservation and adaptive reuse through project management software. The conservation bill of quantity for implementation should be inscribed into the program to keep the time and financial resources monitored all the time. The most common program in dealing with time scheduling and cost evaluation is Primavera, which can be connected to the monument model in Revit for all of the stored building information.

9.7.4.2 Applying 6D (sustainability)

Sustainability is a well-known concept in heritage conservation and a much-appreciated aspect in most of the old monuments. With the emergence of the LEED certification (Leadership in Energy and Environmental Design), the approach of saving energy by using natural materials and solar power was demonstrated in almost all features of rehabilitation projects. For example, in the proposed idea of adaptive reuse for Baron Palace, the project signs are now already operating using solar power (see Figure 9.6). Although sustainability measures can be connected to HBIM through compatible intermediate software that coordinates sustainability measures with the palace model in Revit, the idea of

Figure 9.6 Project sign.

applying sustainability inside the palace is still in discussion with the Ministry of Antiquities. As the owning entity responsible for managing the palace, the ministry has so far shown signs of willingness to apply sustainability procedures in the conservation and adaptive reuse processes.

9.7.4.3 Applying 7D (facility management)

From a future perspective, after the palace is conserved and adapted for its new use, a complete preventive conservation plan should follow, and it has to combine short-, medium- and long-term sub-plans. It is important to keep the monument in a good state of conservation – to enhance users' and visitors' experience by saving and restoring the values of the palace for the coming generations. The role of preventive conservation comes here, which is identified in Building Information Modeling (BIM) as facility management and maintenance, and which can equally formulate the seventh dimension in HBIM.

9.7.4.4 Is the eighth dimension applicable in Baron Palace while using HBIM?

The application of the eighth dimension (accident prevention) in HBIM is not that widespread in Egypt, as much as it is in Building Information Modeling. But let's explain the eighth dimension first. We can simply say that it represents the prevention of accident through design – a process that mostly consists of three tasks. The first is defining hazard profiling elements, which could be easily achieved in historic buildings and even districts by identifying expected hazards; a very close approach to the idea of monuments risk analysis. The second task in accident prevention is providing a safe design suggestion, which is known in HBIM as the hazard procedures program, where every element is connected with its history of hazards and the actions that were taken. The third task is specifying the emergency procedures to be taken in case of accident occurrence, which is possible by connecting every element with short- and medium-term conservation procedures. All this information is easily shared with Revit applications through intermediate software, enabling the application of the eighth dimension in Baron Palace and any other monument.[19]

9.8 Conclusion

HBIM can solve the general problem that most Egyptian heritage sites have, which is information management. If the documentation is done by an accurate data collector showing full deterioration information, a complete bill of quantities (BOQ) can be created for the conservation process, which will include all required conservation and adaptive reuse elements. This step can then be followed by time scheduling and cost evaluation, depending on the provided BOQ. As recommended by the Ministry of Antiquities consultants, a friendly environmental approach of sustainability should be adopted when dealing with monuments. Hence, LEED measures can be applied to Egyptian monuments, representing the sixth dimension (sustainability). As for the seventh and eighth dimensions, they are already considered in the form of preventive conservation. We thus find how a compatible system of managing information can write a new page in saving Egypt's groups of buildings, sites and monuments.

Notes

1 Eastman, Chuck and others, BIM handbook "A Guide to Building Information Modeling for Owners, Managers, Designers, Engineers and Contractors", second edition, John Wiley & Sons, 2011, Hoboken, New Jersey, US, page 1.

2 Conover, Dave and others, "An Introduction to Building Information Modeling (BIM)," 2009, American Society of Heating, Refrigerating and Air-Conditioning Engineers, Inc. 1791 Tullie Circle, N.E., Atlanta, Georgia 30329, www.ashrae.org, page 5.
3 Hung-Ming Cheng, Wun-Bin Yang, Ya-Ning Yen, BIM applied in historical building documentation and refurbishing, The International Archives of the Photogrammetry, Remote Sensing and Spatial Information Sciences, Volume XL-5/W7, 2015. In: 25th International CIPA Symposium 2015, 31 August–04 September 2015, Taipei, Taiwan, page 86.
4 Gresham-Barlow School District, 2013.
5 David McNell, http://community.infocomm.org, last accessed 2015.
6 UNESCO, "The Convention Concerning the Protection of the World Cultural and Natural Heritage," Whc.unesco.org/convention.
7 C. Dore and M. Murphy, Integration of Historic Building Information Modeling and 3D GIS for Recording and Managing Cultural Heritage Sites. In: 18th International Conference on Virtual Systems and Multimedia: "Virtual Systems in the Information Society", 2–5 September, 2012, Milan, Italy, Pages 369–376, 2012.
8 Cultnat, "Experience in the Documentation Using GIS in Egypt," Bibliotheca Alexandrina, 2009, Alexandria.
9 Suki Dixon, Editor, "GIS and Cultural Resource Management: A Manual for Heritage Managers," UNESCO, 1999, Bangkok.
10 English Heritage, "Measured and Drawn: Techniques and Practice for the Metric Survey of Historic Buildings," second edition, Historic England Publishing, 2007, Swindon, UK.
11 Mario Santana, Bill Blake and Tand Eppich, "Conservation of Architectural Heritage: The Role of Digital Documentation Tools: The Need for Appropriate Teaching Material," *International Journal of Architectural Computing* 5(2), June 2007.
12 Sabry Hegazi, Yasmine, Splendor of Islamic Architecture "A Mission of Conservation 2002: 2006", SCA Publishing, 2006.
13 Ministry of Culture, "Historic Cairo," SCA Publishing, 2002.
14 D. Skarlatos and S. Kiparissi, "Comparison of Laser Scanning, Photogrammetry and SFM-MVS Pipeline Applied in Structures and Artificial Surfaces," University of Technology, Cyprus, 2012.
15 Fabio Remondino, Heritage recording and 3D modeling with photogrammetry and 3D scanning, *The Journal of Remote Sensing* 3(6), 2011.
16 Skarlatos and Kiparissi, Comparison of Laser Scanning.
17 A. Dobrowolska and J. Dobrowolski, Heliopolis "Rebirth the City of the Sun", 2006, Cairo, Egypt, American University in Cairo Press.
18 Sh. K. Amin, M. M. Maarouf and S. S. Ali, *Australian Journal of Basics and Applied Sciences*, 616, 227: 236, 2012, ISSN 1991–8178.
19 I. Kamardeen 8D BIM Modelling Tool for Accident Prevention through Design. In: C. Egbu (Ed), Procs 26th Annual ARCOM Conference, 6–8 September 2010, Leeds, UK, Association of Researchers in Construction Management, 281–289, 2010.

10 Planning of sustainable bridges using building information modeling

Mohamed Marzouk and Mohamed Hisham

Building Information Modeling (BIM) is considered an innovative transition from conventional separate engineering techniques to model-based integrated technique. BIM has been used for creating construction shop drawings, 4D modeling, cost estimation, clash detection, and documentation of construction processes. Applying BIM in bridge projects is named Bridge Information Modeling (BrIM). BrIM is widely used for construction of new bridges. This chapter presents a framework for using BrIM as an effective tool in the management of existing and heritage bridges. The proposed framework utilizes BrIM features to manage the knowledge related to the existing and heritage bridges by developing a database module, inspection and condition assessment module. The created database is used to retrieve the information related to each element of the bridge, and it is linked to BrIM elements using Structured Query Language (SQL) in order to visualize each element with its related information in a road network. The structural condition of a bridge is calculated using ANSYS finite element simulation software, which is linked to the proposed BrIM via C# programming language in order to extract the geometric and material properties that are fed to the ANSYS software, which calculates the structural condition of the existing bridge, and thus, suitable maintenance and rehabilitation techniques could be defined and applied. The proposed BrIM can be integrated with different systems including geographic information systems (GIS) and Google Earth to ease the process of data exchange. A numerical example is presented to demonstrate the use of the proposed system.

10.1 Introduction

Building Information Modeling (BIM) is a digital representation of physical and functional characteristics of a facility. It serves as a shared knowledge resource for facility information, forming a reliable basis for decisions during the life cycle of this facility. The basic premise of BIM is collaboration by different stakeholders at different phases of the life cycle of a facility to insert, extract, update, or modify information. BIM is a shared digital representation founded on open standards for interoperability (NIBS 2007). BIM is the development and use of a computer software model to simulate the construction and operation of a facility (AGC 2006). The main concern of BIM is the development of an integrated model that can be used in all stages of a project's life cycle. A BIM model can be connected to schedules of construction activities. It can be used to generate accurate shop drawings, perform quantity surveys and cost estimates, and for facility management. BIM provides great benefits to all project parties. For designers, BIM eases the creation of different design alternatives, where designers can modify their design in a very short time with minimal effort. For contractors, BIM reduces site errors because of early coordination between different models, and better project visualization. For owners, an integrated Building Information Model is created at the end of the project construction phase. The developed model contains all information about the executed project. Building Information Modeling allows teams in different regions and countries to work together to create designs, solve problems, and complete projects faster (Marzouk et al. 2010).

BIM supports parametric modeling, which is an important feature that enables objects and components within a model to be parametrically related. BIM enables the collaboration of team members in the early phases of a project through the use of consistent and more complete information; thus, it is more effective than traditional approaches (ASHRAE 2010). The BIM model can be a 4D model by integrating the schedule of construction activities to the model elements. The fifth dimension (5D) uses the 3D model data to quantify materials and apply cost information (McCuen 2008). BIM has been utilized in different applications, such as: cost estimation (Shen and Issa 2010); lean construction (Sacks et al. 2009; Sacks et al. 2010); facility management (Liu and Issa 2012); sustainable development (Barnes and Castro-Lacouture 2009); and construction process documentation (Goedert and Meadati 2008).

BIM has been considered in modeling existing and heritage buildings. Arayici (2008) described the research towards BIM for existing structures via the point cloud data captured by 3D laser scanner technology. Murphy et al. (2011) have proposed an application of Historic Building Information Modeling by using geometric descriptive language (GDL). Liu et al. (2012) presented a case study in which a laser scanner and a camera were used to capture the construction history and develop a more complete as-built BIM. Fai et al. (2013) presented a study that considers challenges, limits, and advantages of using a BIM for the documentation of architectural heritage; they have demonstrated that it is possible to take advantage of the parametric components that are inherent in the software in order to create highly detailed and unique iterations of typical construction details.

Applying BIM technology on bridges is named Bridge Information Modeling (BrIM). Bridge Information Modeling has widely become an effective tool in the bridge engineering and construction industry. Bridge Information Modeling goes beyond traditional bridge design by fostering data reuse in different processes. The 3D model of the bridge can serve as a window into the vast information assets, and organizations can begin to optimize business processes that cross the bridge life cycle by more flexible access to information about the bridge (Peters 2009). The 3D bridge model can be used for up-to-date shop drawings, quantity takeoffs and bills of materials, CNC (computer numerically controlled) input files to drive automated shop equipment, material estimating and shop material management, and so forth. (Chen and Shirole 2007). Marzouk and Hisham (2013) integrated BrIM with genetic algorithms to optimize the locations of mobile cranes during the construction phase of bridges, taking into consideration existing conditions of site, surrounding areas, safety, and schedule constraints.

Bridges have great impact on the economy and the society because the transportation system would not exist without bridges. Bridges are usually subjected to dead loads; live loads; traffic; horizontal forces due to braking, wind, earthquakes; and other environmental conditions. These conditions affect the bridges through their service lives and lead to deterioration. Any cracks, deformations, or deflections must be within the acceptable limits. In order to achieve that, bridges need periodical maintenance.

A bridge management system is a rational and systematic approach for organizing and carrying out all the activities related to managing a network of bridges; it includes optimizing the selection of maintenance and improvement actions to maximize the benefits while minimizing the costs (Hudson et al. 1993).

Bridge management systems are software systems that depend on different types of data to perform their functions. They have a great influence on the decisions related to maintenance, repair, rehabilitation, improvements, and replacement of bridges on a network while using the available resources in the most efficient and cost-effective manner (Abu-Hamd 2006).

Bridge management systems are composed of several modules that integrate together to perform the system activities and achieve the defined goals. The main modules are: database module; inspection and condition assessment module; cost module; and optimization module. Bridge management systems can be classified into two types, which are project level and network level. The project-level system focuses on an individual bridge and suggests suitable maintenance and rehabilitation strategies for that bridge. The network-level system focuses on prioritization and selection of bridges in a network for maintenance and rehabilitation. Several bridge management systems have been developed. A management system for the bridges located in rural environments was presented (Gralund and Puckett 1996). Reliability theory and expert system techniques were applied in the development of a bridge management system (Christensen 1995). A comprehensive framework for a bridge-deck management system that aims at integrating project-level and network-level decisions was presented (Hegazy et al. 2004). A geographic information system (GIS) module and an object-oriented database module were integrated in a bridge life-cycle management system (Itoh et al. 1997). A smart, client-based bridge management system was presented (Shan and Li 2009).

This chapter presents a proposed bridge management system for existing and heritage bridges by utilizing Bridge Information Modeling (BrIM). The proposed BrIM management system comprises several modules, such as database module and condition assessment module. A visualization feature is applied with the system modules to allow better understanding of bridge status.

10.2 BrIM management system for existing and heritage bridges

10.2.1 Proposed system features

Utilizing Bridge Information Modeling as an effective tool in the management of existing and heritage bridges is proposed in this research. The proposed BrIM management system depends on developing a 3D BrIM model of the heritage or existing bridge then integrating it with the other system modules, which are the database module and the condition assessment module. The 3D BrIM model of the heritage or existing bridge could be developed using several commercial software packages; integrating laser scanning technique could assist and ease the process of developing an accurate BrIM model.

The database module contains information relative to each bridge in the network. Information systematized in the database must include general data about each bridge, like the description of the structure and its environment (geometry, materials, etc.) (De Sousa et al. 2009). All functions and decisions of bridge management systems are based on the information in the database module (Hudson et al. 1993), so it is very important to maintain accurate information in the database, and this can be achieved by effective utilization of Bridge Information Modeling. The database module contains data related to bridge stakeholders, bridge history, service data, geometric data, material data, and navigation data. It also includes inspection data, which comprises the inspection checklist and inspection findings. Related bridge photos and documents are also integrated with the database module in addition to GIS data (such as data related to soil, roads, etc.). Bridge Information Modeling is proposed to be used as a tool for developing an effective and integrated database for existing and heritage bridges.

Checking bridges' structural health is a main concern of the condition assessment module. This module uses the data obtained from the site measurements and visual inspection and analyzes them based on rating codes to obtain a value that represents the structural integrity of a bridge. Bridge Information Modeling is proposed to assist in the condition assessment process by integrating Bridge

Information Modeling with finite element simulation software to obtain a value that can be used to compare the structural integrity of different bridges to assist in decision making related to maintenance and rehabilitation.

The visualization feature is integrated with the proposed management system modules to achieve better understanding that assists decision makers to make optimum decisions regarding existing or heritage bridges. Figure 10.1 shows the different modules and components of the proposed BrIM management system.

The requirements to achieve connectivity among the proposed BrIM management system components include commercial software packages (Tekla Structures, Navisworks Manage, MS Excel, ANSYS, Google Earth, and Google SketchUp); customized programs that were developed using C# programming language to extract and utilize information; and other requirements (site inspection results, bridge documents and photos, and GIS data). Figure 10.2 shows the connectivity amongst the different components and their respective methodologies and software packages. For example, Tekla Structures is integrated with MS Excel and ANSYS software packages via C# programming language; while DWG format is used to integrate Tekla Structures with Google SketchUp software, and IFC format is used to integrate Tekla Structures with Navisworks Manage software. Navisworks Manage software is integrated with MS Excel via Structured Query Language (SQL), while Navisworks Manage software is integrated with Google Earth and bridge photos via direct hyperlinks.

10.2.2 Developing 3D bridge information model

The 3D Bridge Information Model can be developed using several commercial software packages. In the proposed framework, Tekla Structures software is used. The 3D Bridge Information Model is not only a 3D geometrical representation of the bridge, but it also contains attributes such as material type, area of cross-section, and weights.

The 3D Bridge Information Model of a previously built bridge can be developed using different methods such as the as-built CAD drawings, or survey techniques. The survey techniques can be

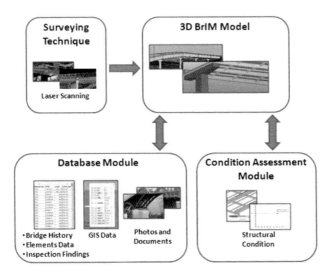

Figure 10.1 Proposed BrIM management system components.

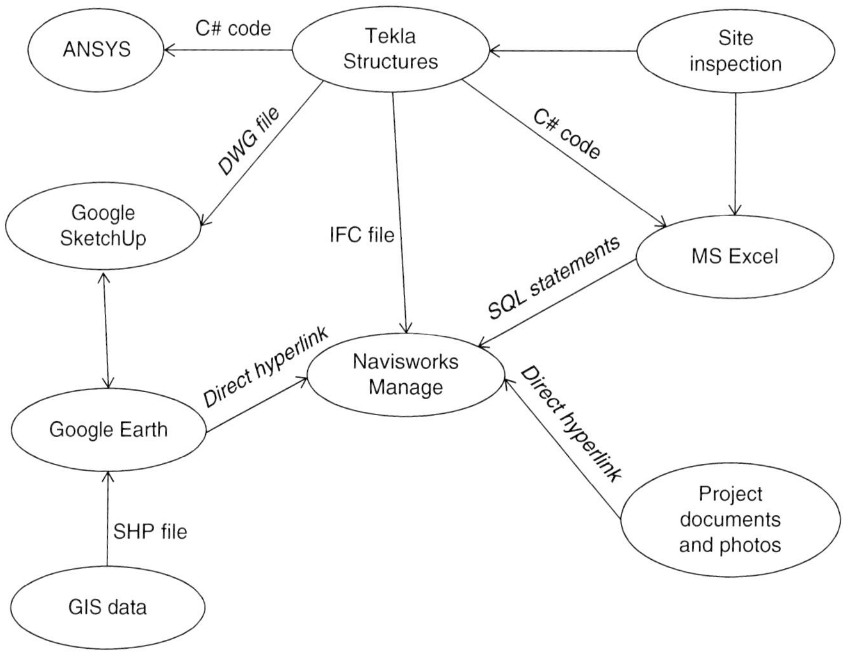

Figure 10.2 Connectivity amongst Bridge Information Modeling components.

integrated with the Bridge Information Modeling software such as laser scanning, as presented by Giel and Issa (2011) and by Arayici (2008).

10.2.3 Database module

10.2.3.1 Extracting bridge elements data

Most Bridge Information Modeling software packages allow an application programming interface (API). The API allows extending the uses of the software packages by creating new features and applications related to the 3D Bridge Information Model by using several programming languages such as C# and Visual Basic. A program was created using C# programming language to extract information of each component in a bridge to create a database. This program also creates inspection spreadsheets. The developed program captures BrIM intelligent attributes, such as:

- element ID, which is considered a unique property of each element
- coordinates (X, Y, and Z) of start and end points of each element
- material type (concrete, steel, etc.)
- profile (dimensions of cross section)
- element length
- casting type (precast or cast in place)

The developed program then creates a database of the bridge components by writing the extracted information to an MS Excel file. This program can also be modified easily to extract any user-defined attribute within the 3D Bridge Information Model. The developed program performs its function for each BrIM element (i); where (i) is a positive integer with a minimum value of 1 and a maximum value of (n), where (n) is the total number of selected BrIM elements. The extracted information acts as a database for bridge components, as shown in Figure 10.3. The developed program also creates inspection sheets as MS Excel files. These sheets contain component ID, component position, and the required inspection checklist to be filled by the site inspector directly using a laptop or ultra-mobile personal computer (PC), as shown in Figure 10.4.

10.2.3.2 Visualizing bridge data

Visualization is one of the effective features in Bridge Information Modeling. It allows better planning and helps in the decision-making process. In order to visualize each element with its related attributes and its inspection results, the 3D Bridge Information Model is exported as an IFC file to be opened with Navisworks Manage software.

Using Navisworks Manage API, Structured Query Language (SQL) statements were written to link the previously exported database to the bridge model components in order to visualize the attributes of each component by pointing to the component, as depicted in Figure 10.5. The inspection results in the inspection sheets are linked to the bridge model components by using the same method of linking the components attributes to the model components. Before linking any Microsoft Excel file to components in Navisworks Manage, the CONCATENATE function is applied in the MS Excel files to add a string, which is "TS_", before the ID value of each component. This string refers to "Tekla Structures", since any component ID in Navisworks Manage must begin with an abbreviation of the software that developed the 3D Bridge Information Model. The potential of linking the MS Excel files to the model components is that by modifying or adding any results or attributes to the Microsoft Excel files, this change is directly visualized when pointing to the component in Navisworks Manage.

Figure 10.3 Extracted bridge components data.

	A	B	C	D	E	F	G	H	I	J	K
1	ID	Start X	Start Y	Start Z	End X	End Y	End Z	Cracks	Deformation	corossion	Next maintenance
2	2113	-1.10E-11	30000	6090	4000	35000	6090				
3	2097	-1.72E-11	35000	6090	4000	30000	6090				
4	2067	-9.78E-12	35000	6090	4000	35000	6090				
5	2051	-7.97E-12	30000	6090	-7.96E-12	35000	6090				
6	2035	4000	30000	6090	4000	35000	6090				
7	2003	-7.96E-12	30000	6090	4000	30000	6090				
8	1987	-6.15E-12	25000	6090	-6.14E-12	30000	6090				
9	1971	4000	25000	6090	4000	30000	6090				
10	1941	-7.96E-12	25000	6090	4000	25000	6090				
11	1925	-6.15E-12	20000	6090	-6.14E-12	25000	6090				
12	1909	4000	20000	6090	4000	25000	6090				
13	1879	-7.96E-12	20000	6090	4000	20000	6090				
14	1863	-6.15E-12	15000	6090	-6.14E-12	20000	6090				
15	1847	4000	15000	6090	4000	20000	6090				
16	1817	-7.96E-12	15000	6090	4000	15000	6090				
17	1801	-6.15E-12	10000	6090	-6.14E-12	15000	6090				
18	1785	4000	10000	6090	4000	15000	6090				
19	1753	-7.96E-12	10000	6090	4000	10000	6090				
20	1721	4000	5000	6090	4000	10000	6090				
21	1145	4000	-6.82E-13	6090	4000	5000	6090				

Figure 10.4 Generated inspection spreadsheets.

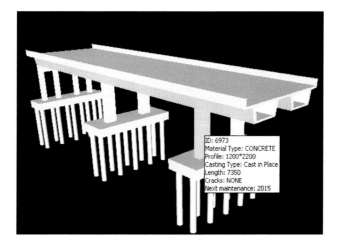

Figure 10.5 Visualizing component properties.

The link between the components of the 3D Bridge Information Model and the MS Excel files was referred to as *Maintenance*. Consequently, Maintenance is considered a category in Navisworks Manage. Every category in Navisworks Manage contains attributes and their values. Examples of these attributes are component ID, component material type, and component condition. The values of these attributes can be numerical values (e.g. length, weight, and condition) or string values (e.g. material type). Maintenance and rehabilitation are always done based on policies which are set based on the required conditions to be achieved and the amount of available funds. A policy may recommend to perform maintenance for components that have a condition less than or equal a specific value (for example, 4). For any category, Navisworks Manage allows highlighting components that have an attribute value less than, greater than, or equal to a specific value. By setting the category equals to *Maintenance*, the attribute property equals to *Condition*, and attribute value less than or equals 4, it is easy to visualize all bridge components that have a condition less than or equal to 4, as shown in Figure 10.6. As such, a rehabilitation plan is prepared for those elements.

10.2.3.3 Integrating BrIM with GIS data

The use of GIS is becoming common in the development of applications in various fields of civil engineering. In previous research studies, GIS was applied to bridge planning (Hammad et al. 1993). The concepts, features, and strategy for a GIS-based bridge management system in which GIS was proposed to be treated as a tool to enhance an existing bridge management system were outlined (Ng and She 1993). A GIS system was used as a platform to evaluate the impact of an earthquake on a system of bridges in a region (Kim 1993). GIS was used to integrate the bridge data and other data related to roads, rivers, soils, and so forth (Itoh et al. 1997). GIS data have great importance in visualizing the geographical data that may influence bridge management and in clarifying the spatial relationships between a road and the bridge (Itoh et al. 1997).

The proposed methodology depends on integrating the 3D Bridge Information Model with the GIS data related to the location of the bridge such as the administrative zone, earthquakes, waterways, roads, soil type, and traffic flow. This integration could be achieved using Google Earth software to allow the visualization of 3D bridge models of a network with the related GIS data. A Google Earth file for the location of each bridge is then created and linked to its related 3D bridge model in Navisworks Manage using a direct hyperlink.

To achieve this integration, three main steps are required. The first step is to import the 3D bridge model to Google Earth; the second step is to import the GIS data to Google Earth; the third step is to link the Google Earth file to the 3D bridge model in Navisworks Manage software. Detailed

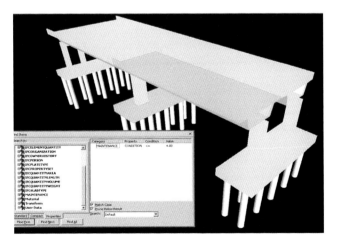

Figure 10.6 Highlighting components at specific threshold value.

workflow of the required steps is shown in Figure 10.7. In order to import the 3D bridge model to Google Earth, the model is exported from Tekla Structures as DWG file. This DWG file is imported to Google SketchUp software. Google SketchUp allows the user to select and capture the location of the bridge from Google Earth and to import the exported bridge DWG file to this location then export the bridge model to Google Earth.

Figure 10.8 shows importing the bridge model to its location using Google SketchUp; Figure 10.9 shows the bridge model after being exported to Google Earth. In order to import GIS data to Google Earth, GIS shapefiles were used. Shapefiles are a common geospatial vector data format for geographic information systems software. Shapefiles were developed and regulated by ESRI. They spatially describe geometries such as points, polylines, and polygons. These, for example, could represent rivers and lakes. Each item may also have attributes that describe the item, such as the name or the length. The GIS data is imported to Google Earth as a shapefile (.shp). Figure 10.10 shows the visualization of the 3D bridge model with the imported GIS data. The Google Earth file that integrates the GIS data and 3D bridge model is saved in KMZ format, which is then linked to the BrIM model (IFC file) in Navisworks Manage software by performing a hyperlink that directly opens the Google Earth KMZ file.

Also, photos and documents related to the bridge can be linked to the Bridge Information Model with the same method of linking. Figure 10.11 shows linking photos and documents to the 3D bridge model components. As such, an integrated 3D Bridge Information Model containing all information related to the bridge is created, allowing visualization of bridge attributes and GIS data. The results allow better management, planning, and more effective decisions.

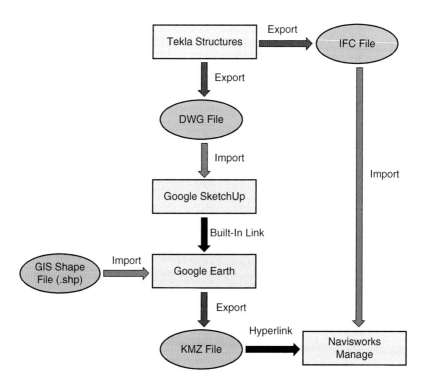

Figure 10.7 Workflow of integrating BrIM with GIS.

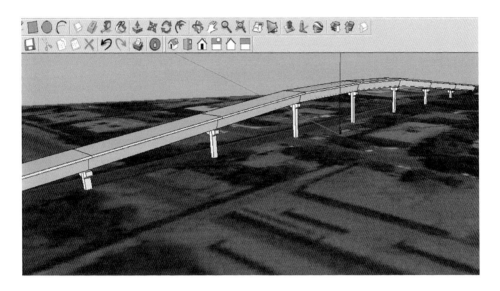

Figure 10.8 Importing bridge model to its location using Google SketchUp.

Figure 10.9 Exporting bridge model to Google Earth.

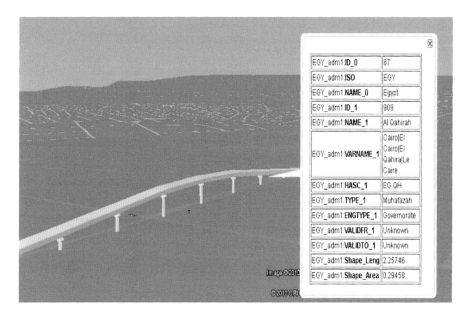

Figure 10.10 Visualizing 3D bridge model with GIS data.

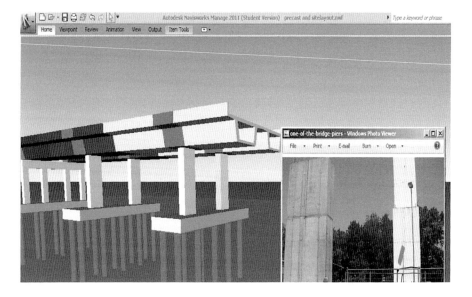

Figure 10.11 Linking bridge photos to 3D Bridge Information Model.

10.3 Structural condition assessment

10.3.1 Structural integrity of bridges

Structural integrity of bridges is the main concern of bridges' stakeholders. Previous research integrated structural condition, hydraulic vulnerability, and strategic importance of bridges in prioritization of maintenance (Valenzuela et al. 2010). A three-dimensional (3D) visualization system was developed for rating reinforced concrete deck-girder bridges (Gregory et al. 2009). Bridges are usually deteriorated because they exist in severe conditions, and they are subjected to dead loads, live loads, and horizontal forces generated by braking, winds, and earthquakes. In order to keep bridges' deterioration within acceptable limits, bridges have to be inspected and rehabilitated periodically.

The site inspections are carried out to determine the physical and the functional conditions of the bridge. They are used to initiate maintenance actions and to provide a continuous record of bridges' conditions to establish priorities for repair and rehabilitation programs. Structural advanced analysis technique can be used to obtain a numerical value that represents the structural integrity of bridges and to determine priorities for maintenance and rehabilitation.

10.3.2 Structural advanced analysis

Structural advanced analysis indicates any method that can sufficiently capture the limit state strength and stability of a structural system and its individual members so that separate member capacity checks encompassed by the specification equations are not required (El Samman 2010). Advanced analysis has the advantage that it predicts the strength of the entire structure because it explicitly considers both material and geometric nonlinearity. In addition, advanced analysis does not need separate member capacity checks and calculations. Therefore, for regular structures, advanced analysis has great benefits in accuracy and time saving. Moreover, for irregular structures, code requirements are not clear, so advanced analysis becomes very essential.

The output of the advanced structural analysis is a numerical value named "load factor". The load factor determines the ability of a structure to afford the applied loads. If the load factor equals 1, this indicates that there is no factor of safety. If the load factor is less than 1, this indicates that the structure could not afford the applied loads. If the load factor is greater than 1, this indicates that there is a safety factor that allows the structure to afford more than the applied loads.

10.3.3 Integrating BrIM with structural advanced analysis

In order to integrate Bridge Information Modeling with structural advanced analysis software, such as ANSYS, a program was developed using C# programming language. The C# program extracts the intelligent attributes of elements from BrIM and creates a text file that can be used by the ANSYS software package. The extracted attributes include:

- material type and grade
- element position in the form of X, Y, and Z coordinates of start and end points
- element profile attributes, such as profile type (I, C, etc.) and profile dimensions (such as web and flanges dimensions of steel elements)

The developed program then utilizes these attributes to create an ANSYS input file. This file is a text file that includes positions, profiles, and materials, while loads are inserted from the ANSYS software. Figure 10.12 shows an example of the created text file presenting the extracted BrIM attributes. The C# code used is presented in Appendix A.

10.3.4 Numerical example

This section presents an example of assessing the structural condition of an existing steel bridge using Bridge Information Modeling and ANSYS software. The bridge main girder data is:

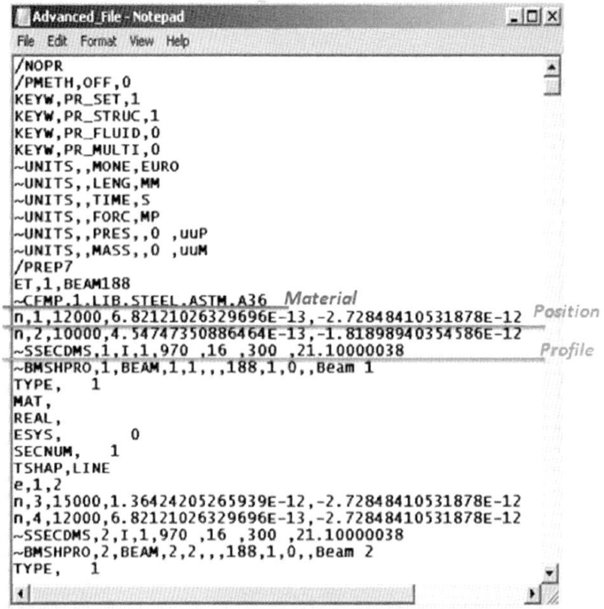

Figure 10.12 Sample of the created ANSYS input file.

- Material type: Steel
- Profile type: I beam
- Flanges dimensions: 300 × 31 mm
- Web dimensions: 990 × 16.5 mm
- Deterioration: Corrosion of 9 mm in the bottom flange

The bridge was modeled using Tekla Structures software. In order to achieve the required integration between BrIM and ANSYS, the handles (control points of modeling) of all beams must be modeled on the same horizontal plane; any difference in the planes of beams has to be considered by changing the End Offset while keeping handles on the same horizontal plane. The corrosion is modeled by splitting the location of corrosion identified during inspection and modifying the beam profile to match the real status of the beam. Figure 10.13 shows the 3D Bridge Information Model indicating corrosion position.

The bridge model was then exported to ANSYS software, where the model was loaded by double the value of the real loads applied to the bridge. By performing the analysis, and as shown in Figure 10.14, it was found that the bridge afforded 60% of the applied load, which indicates that the bridge could afford 120% of the real loads (load factor equals 1.2). It was also found that the stress in the corroded flange is the highest stress as it reached a value equal to 0.034 t/mm^2, which equals to the yielding stress approximately. A snapshot of the analysis results is shown in Figure 10.15.

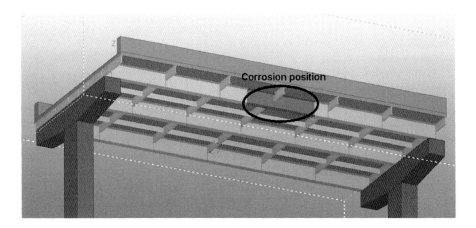

Figure 10.13 3D Bridge Information Model.

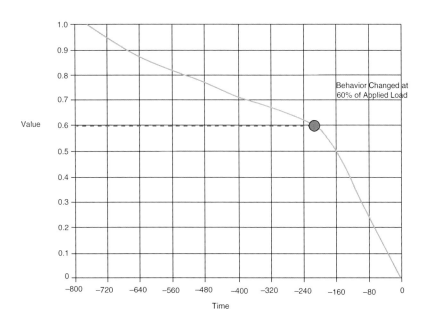

Figure 10.14 Identifying load factor.

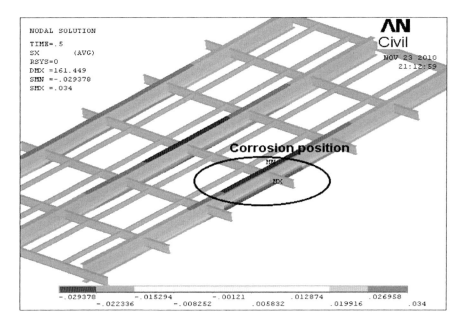

Figure 10.15 Structural advanced analysis results.

10.4 Summary

The chapter presented a proposed Bridge Information Modeling (BrIM) management system for existing and heritage bridges. The chapter presented the creation of database of bridge components extracted from the Bridge Information Model. The database included information related to the components position, material type, length, profile, and casting type. Inspection sheets were created to include inspection information such as components' ID, position, and an inspection checklist to be filled in at the inspection site. The link between the database and inspection sheets with the Bridge Information Model was illustrated. The visualization feature facilitated identifying the defective components and led to more accurate decisions and planning. The chapter also showed the link of photos to its related bridge components using direct hyperlinks in Navisworks software. Visualizing and integrating the Bridge Information Model with related GIS data were demonstrated. Different software packages were utilized, including Tekla Structures, Navisworks Manage, Google SketchUp, and Google Earth. Finally, the chapter presented a methodology of assessing the structural condition of bridges using structural advanced analysis technique. It also illustrated how to integrate Bridge Information Modeling with ANSYS advanced analysis software. This integration is achieved by a developed program that creates a text file which acts as an input file for the ANSYS software, where loads are inserted and analysis is performed. A numerical example is presented to assess the structural condition of a steel bridge using Bridge Information Modeling.

Acknowledgement

The authors acknowledge financial support from ITIDA (Information Technology Industry Development Agency) for funding of this research, which is carried out at Nile University.

Appendix A

C# Code of Integrating BrIM with ANSYS

```csharp
using System;
usingSystem.Collections.Generic;
usingSystem.ComponentModel;
usingSystem.Data;
usingSystem.Drawing;
usingSystem.Linq;
usingSystem.Text;
usingSystem.Windows.Forms;
usingTekla.Structures.Model;
using TSM = Tekla.Structures.Model;
using TSG = Tekla.Structures.Geometry3d;
namespace ADVANCED
{
public partial class Advanced : Form
{
TSM.Model model1 = null;
public Advanced()
{
InitializeComponent();
}
private void Form1_Load(object sender, EventArgs e)
{
model1 = new Model();
if (!model1.GetConnectionStatus())
{
MessageBox.Show("couldn't connect to tekla model");
Application.Exit();
}
}
private void button2_Click(object sender, EventArgs e)
{
Application.Exit();
}
private void button1_Click(object sender, EventArgs e)
{
if (model1.GetConnectionStatus())
{
string DOC1 = "/NOPR\r\n";
DOC1 += "/PMETH,OFF,0\r\n";
DOC1 += "KEYW,PR_SET,1\r\n";
DOC1 += "KEYW,PR_STRUC,1\r\n";
DOC1 += "KEYW,PR_FLUID,0\r\n";
DOC1 += "KEYW,PR_MULTI,0\r\n";
DOC1 += "~UNITS,,MONE,EURO\r\n";
```

```
DOC1 += "~UNITS,,LENG,MM\r\n";
DOC1 += "~UNITS,,TIME,S\r\n";
DOC1 += "~UNITS,,FORC,MP\r\n";
DOC1 += "~UNITS,,PRES,,0, uuP\r\n";
DOC1 += "~UNITS,,MASS,,0, uuM\r\n";
DOC1 += "/PREP7\r\n";
DOC1 += "ET,1,BEAM188\r\n";
DOC1 += "~CFMP,1,LIB,STEEL,ASTM,A36\r\n";
TSM.ModelObjectEnumerator parts = model1.
GetModelObjectSelector().GetSelectedObjects();
if (parts.GetSize() > 0)
{
inti = 1;
int j = 2;
int k = 1;
while (parts.MoveNext())
{
TSM.Beam beam1 = parts.Current as Beam;
if (beam1 != null)
{
DOC1 += string.Format("{0},{1},{2},{3},{4}\r\n", "n", i, beam1.
StartPoint.X, beam1.StartPoint.Z, beam1.StartPoint.Y);
DOC1 += string.Format("{0},{1},{2},{3},{4}\r\n", "n", j, beam1.
EndPoint.X, beam1.EndPoint.Z, beam1.EndPoint.Y);
stringptype = "";
beam1.GetReportProperty("PROFILE_TYPE", ref ptype);
double a =0.0;
beam1.GetReportProperty("PROFILE.HEIGHT",ref a);
double b = 0.0;
beam1.GetReportProperty("PROFILE.WEB_THICKNESS", ref b);
double c = 0.0;
beam1.GetReportProperty("PROFILE.WIDTH", ref c);
double d = 0.0;
beam1.GetReportProperty("PROFILE.FLANGE_THICKNESS", ref d);
DOC1 += string.Format("{0},{1},{2},{3},{4}, {5}, {6}, {7}\r\n",
"~SSECDMS", k, ptype, 1, a,b,c,d);
DOC1 += string.
Format("{0},{1},{2},{3},{4},{5},{6},{7},{8},{9},{10} {11}\r\n","~
BMSHPRO",k,"BEAM",k,k,","",188,1,0,"","Beam",k);
DOC1 += string.Format("{0}\r\n", "TYPE, 1");
DOC1 += string.Format("{0}\r\n", "MAT,");
DOC1 += string.Format("{0}\r\n", "REAL,");
DOC1 += string.Format("{0}\r\n", "ESYS, 0");
DOC1 += string.Format("{0}, {1}\r\n", "SECNUM",k);
DOC1 += string.Format("{0}\r\n", "TSHAP,LINE");
DOC1 += string.Format("{0},{1},{2}\r\n", "e", i, j);
i = i + 2;
j = j + 2;
```

```
k = k + 1;
}
}
MessageBox.Show("File created");
System.IO.StreamWriter s1 = new System.IO.StreamWriter(model1.
GetInfo().ModelPath + @"\Advanced_File.txt");
s1.Write(DOC1);
s1.Close();
System.Diagnostics.Process.Start(model1.GetInfo().ModelPath + @"\
Advanced_File.txt");
}
else
{
MessageBox.Show("No parts are selected");
}
}
else
{
MessageBox.Show("Model is not open and file couldn't be created");
}
}
}
}
```

References

Abu-Hamd, I. (2006). *A Proposed Bridge Management System for Egypt: Development of Database Structural Analysis, and Rating Modules.* MSc thesis, Faculty of Engineering, Cairo University.

AGC (2006). The Contractors' Guide to BIM. Ed. 1. The Associated General Contractors of America. http://www.agc.org/ (Accessed April 10, 2010).

Arayici, Y. (2008) "Towards Building Information Modelling for Existing Structures", *Structural Survey*, 26(3), 210–222.

ASHRAE (2010). An Introduction to Building Information Modeling (BIM): A Guide for ASHRAE Members. http://cms.ashrae.biz/bim/ (Accessed May 23, 2011).

Barnes, S. and Castro-Lacouture, D. (2009). "BIM-Enabled Integrated Optimization Tool for LEED Decisions." *Computing in Civil Engineering*, 258–268. doi: 10.1061/41052(346)26.

Chen, S.S. and Shirole, A.M. (2007). Parametric 3D-Centric design and Construction of Steel Bridges. In *Proceedings of 2007 World Steel Bridge Symposium, National Steel Bridge Alliance*, New Orleans, LA.

Christensen, P. (1995). "Advanced Bridge Management Systems." *Structural Engineering Review*, 7(3), 151–163.

De Sousa, C., Almeida, J. and Delgado, R. (2009). "Bridge Management System as an Instrument of Risk Mitigation." In *3rd International Conference on Integrity, Reliability and Failure*, Porto/Portugal.

El Samman, B. (2010). *Advanced Design of Steel Structures*. MSc. thesis, Faculty of Engineering, Cairo University.

Fai, S., Filippi, M. and Paliaga, S. (2013). Parametric Modelling (BIM) for the Documentation of Vernacular Construction Methods: A BIM Model for the Commissariat Building, Ottawa, Canada. *ISPRS Annals of the Photogrammetry, Remote Sensing and Spatial Information Sciences*, Volume II-5/W1.

Giel, B. and Issa, R. (2011). "Using Laser Scanning to Access the Accuracy of As-Built BIM." *Computing in Civil Engineering*, 665–672. doi: 10.1061/41182(416)82.

Goedert, J. and Meadati, P. (2008). "Integrating Construction Process Documentation into Building Information Modeling." *Journal of Construction Engineering and Management*, ASCE, 134(7): 509–516.

Gralund, M. and Puckett, J. (1996). "System for Bridge Management in a Rural Environment." *Journal of Computing in Civil Engineering*, ASCE, 10(2), 97–105.

Gregory, E., Michael, H. and Eugene, Z. (2009). "Graphical 3D Visualization of Highway Bridge Ratings." *Journal of Computing in Civil Engineering*, ASCE, 23(6), 355–362.

Hammad, A., Itoh, Y. and Nishido, T. (1993). "Bridge Planning Using GIS and Expert System Approach." *Journal of Computing in Civil Engineering*, ASCE, 7(4), 278–295.

Hegazy, T., Elbeltagi, E. and Elbehairy, H. (2004). "Bridge Deck Management System with Integrated Life Cycle Cost Optimization." *Transportation Research Record: Journal of the Transportation Research Board*, No. 1866, TRB, National Research Council, Washington, DC, 1866(1), 44–50. doi: 10.3141/1866-06.

Hudson, R., Carmichael, R., Hudson, S., Diaz, M. and Moser, L. (1993). "Microcomputer Bridge Management System." *Journal of Transportation Engineering*, ASCE, 119(1), 59–76.

Itoh, Y., Hammad, A., Liu, C. and Shintoku, Y. (1997). "Network-Level Bridge Life-Cycle Management System". *Journal of Infrastructure Systems*, 3(1), 31–39.

Kim, S. (1993). *A GIS-Based Regional Risk Analysis Approach for Bridges against Natural Hazards*. PhD thesis, State University of New York at Buffalo, Buffalo, NY.

Liu, X., Eybpoosh, M. and Akinci, B. (2012). "Developing As-Built Building Information Model Using Construction Process History Captured by a Laser Scanner and a Camera". In *Construction Research Congress 2012*, pp. 1232–1241.

Liu, R. and Issa, R. (2012). "Automatically Updating Maintenance Information from a BIM Database." *Computing in Civil Engineering*, 373–380. doi: 10.1061/9780784412343.0047.

Marzouk, M. and Hisham, M. (2013). "A Hybrid Model for Selecting Location of Mobile Cranes in Bridge Construction Projects." *The Baltic Journal of Road and Bridge Engineering*, 8(3), 184–189.

Marzouk, M., Hisham, M., Ismail, S., Youssef, M. and Seif, O. (2010). "On the Use of Building Information Modeling in Infrastructure Bridges", In *Proceedings of 27th International Conference – Applications of IT in the AEC Industry (CIB W78)*, Cairo, Egypt, 135, pp. 1–11.

McCuen, M. (2008). "Scheduling, Estimating, and BIM: A Profitable Combination", In *Proceedings of 2008 AACE International Transactions*, BIM.01.

Murphy, M., McGovern, E., and Pavia, S. (2011). "Historic Building Information Modelling – Adding Intelligence to Laser and Image Based Surveys", *International Archives of the Photogrammetry, Remote Sensing and Spatial Information Science*, XXXVIII-5/W16.

Ng, S.K. and She, T.K. (1993). "Towards the Development of a GIS Based Bridge Management System." *Bridge Management 2*, Thomas Telford, London, England, pp. 998–1007.

NIBS (2007). *United States National Building Information Modeling Standard. Version 1 – Part 1: Overview, Principles, and Mythologies*. National Institute of Building Sciences, Washington, DC.

Peters, D. (2009). *Bridge Information Modeling to Cover a Complete Set of Processes*. Bentley systems, http://ftp2.bentley.com/dist/collateral/docs/press/brim-cover-complete-set-processes_cecr.pdf (accessed Sept. 15, 2010).

Sacks, R., Koskela, L., Dave, B., and Owen, R. (2010). "Interaction of Lean and Building Information Modeling in Construction," *Journal of Construction Engineering and Management*, ASCE, 136(9), 968–980.

Sacks, R., Treckmann, M. and Rozenfeld, O. (2009). Visualization of Work Flow to Support Lean Construction. *Journal of Construction Engineering and Management*, ASCE, 135(12): 1307–1315.

Shan, D. and Li, Q. (2009). "Development of a Smart, Client-Based Bridge Management and Maintenance System for Existing Highway Bridges." In *Proceedings of the Second International Conference on Transportation Engineering*.

Shen, Z. and Issa R.R.A. (2010). Quantitative Evaluation of the BIM-Assisted Construction Detailed Cost Estimates. *Journal of Information Technology in Construction (ITcon)*, 15, 234–257.

Valenzuela, S., De Solminihac, H. and Echaveguren, T. (2010). "Proposal of an Integrated Index for Prioritization of Bridge Maintenance." *Journal of Bridge Engineering*, ASCE, 15(3), 337–343.

11 Jeddah Heritage Building Information Modelling (JHBIM)

Ahmad Baik and Jan Boehm

The main issues that face the historic district of Jeddah city today are in which way the Jeddah municipality can manage, preserve and record these heritage buildings, which were built in the sixteenth century, as historical documentations and protect these monuments from any chance of collapse and erosion through human factors, and also natural disasters such as floods and fires. Furthermore, the concept of employing Building Information Modelling (BIM) in the historical field has been employed in many worldwide historical sites through documenting, managing, conserving, and generating complete engineering information and drawings. In the last 20 years, the historic Jeddah municipality has employed a number of traditional surveying techniques to manage and document these buildings. On the other hand, these techniques can sometimes provide unreliable information, take a long time and often lack completeness. Through this chapter, a new method will be examined for historical documentation via Jeddah Heritage Building Information Modelling (JHBIM) and further via employing the Hijazi Architectural Objects Library.

11.1 Introduction

11.1.1 Historic Jeddah: Background

The majority of cultural heritage experts are in agreement that the significant provenance of civilization and culture is our heritage and past, which is usually evinced through language, music, habits and buildings. Indeed, monuments such as historical buildings represent knowledge of the past, which were built in consideration of such things as the economic conditions, the social status, lifestyle and even the climate. Since the United Nations Educational, Scientific and Cultural Organization (UNESCO) convention in Paris in 1972, the conservation and protection of cultural heritage has constituted a significant pillar of sustainable development efforts.

As such, several heritage locations across the world were recorded. Unluckily some were not, mainly in poor and developing nations. Historic Jeddah city is an outstanding cultural heritage site which was inducted into the world heritage list of UNESCO in 2012 (TECTURAE, 2012).

The historic district of Jeddah city is one of the most important and valuable parts of the Hijaz region. The district is situated on the western side of the Kingdom of Saudi Arabia, with a coastline on the Red Sea. Today, Jeddah city is known as the capital of business and tourism of the country after Riyadh city (Worldatlas.com, 2013). According to Al-Fakahani (2005), the city has "a long history as it dates back to 3000 years ago when groups of fishermen settled in the region after their fishing trips around the sea". In this historical district of Jeddah there are many buildings which were built 100 years ago, such as Nasif Historical House and Jamjoum Historical House (Figure 11.1).

134 *Ahmad Baik and Jan Boehm*

Figure 11.1 Examples of the buildings in historic Jeddah.

The architectural characters of these historic houses are very unique, exhibiting the very cultures of the nation (Telmesani et al., 2009). According to Feilden (2012), the local residents "have interacted with people of different background and cultures"; moreover, "this knowledge is reflected in their way of life and how they built their buildings."

11.1.2 Historic district issues

The major issue that faces the historical district of Jeddah city today, as the director of the historic area, Sami Nawwar, pointed out, is "how the local authorities preserve such buildings from the risk of collapse and erosion by aging and human factors, as well as disasters such as fires and floods". Indeed, in the last 20 years, huge numbers of historical houses and buildings in the historical district of Jeddah city have been destroyed, erasing hundreds of years of culture and history. For instance, in March 2010, the historic district in Jeddah city was hit by the worst fire in years and more than six buildings were burnt down (Alawi, 2013).

According to engineer Sami Nawwar, "these Houses were important Historical Landmarks", and unluckily, "the number of historic houses has declined from 557 to 350" in recent years. Most of these houses were "A Class" historical buildings. The historic district of Jeddah city requires a quick and effective solution for resolving these matters. In 2012, after historic Jeddah was added to the UNESCO World Heritage Site's list, the numbers of the historical buildings increased to 1,447 buildings, and a large number of these buildings were in a dangerous situation according to Mohammad Yosof al-Aidaroos, the supervisor of archaeological and historical sites at the Supreme Commission for Tourism, Saudi Arabia.

Many heritage experts suggest that these historical monuments must be conserved, preserved, maintained and reused in a better and more organized way, or else they will lose their essence of history and culture over time. Additionally, the main element for heritage preservation is via documenting these heritage sites. Presently, there is no official Hijazi architectural database or adequate records in existence, this being due to the absence of specialists and experts to carry out this job. Hence, Alitany et al. (2013) pointed out that "there is an important need for trained professionals and infrastructure to preserve the city and its buildings."

Ten years ago, according to Nawwar (2013), "the municipality of Historic Jeddah City decided to preserve and develop this area by using independent engineering survey offices which often take a long time and can sometimes provide unreliable information." These engineering survey offices used traditional surveying techniques in order to perform the task, which took around 10 months to completion and incurred high costs. Since 2011, the historic Jeddah municipality has taken action to develop and preserve the historic district by defining historical paths for tourists. These tourist paths need to be promoted and developed in order to be attractive to the visitors and, in addition, also to the investors (TECTURAE, 2012). However, the existing preservation methods lack several key aspects, such as remote management, shared databases and integration with any future building's condition updates.

11.2 Heritage building information modelling (HBIM)

Employing BIM into the historical field has emerged as a new method over the past few years. Already this method has been used for a small number of projects worldwide, such as in the research of Fai et al. (2011), Murphy (2012), Oreni (2013) and Penttilä et al. (2007). The main aim of these research projects was to provide intelligent data (Fai et al. 2011), in order for it to be used for several purposes, such as documentation, conservation and management. Moreover, Murphy (2012) has described HBIM as "a novel prototype library of parametric objects, based on historic architectural data", as well as "a mapping system for modelling the objects library", based on the survey data of terrestrial laser scanning and image survey data.

Penttilä et al. (2007) employed BIM to "demonstrate how modern digital methods can be used in reconstruction design" and in addition "in renovation projects, with special emphasis on BIM". In this study, the idea of an "inventory model" has been applied on an existing building (which was the building of the Architectural Department in Finland). This represents an important base for the inventory data being well structured, presenting both the present and past situations of the building. Furthermore, it can be noted in Penttilä et al.'s (2007) study that the inventory model idea covers the definition of historic data and how it can be implemented. However, using Building Information Modelling as a database for documentation has not been fully investigated yet.

Fai et al. (2011) employed BIM for the documentation and conservation of architectural heritage based on terrestrial laser scan survey data in Toronto, Canada. Fai et al. (2011) focused on the issues related to linking the laser scanning data within BIM. In addition, they investigated how to model generic objects like library objects based on laser scanning data within a BIM environment. Moreover, the hybrid documentation model, which was offered by Sabry El-Hakim (El-Hakim et al., 2005) and Fabio Remondino (Remondino et al., 2009), was used to develop the case study model of Fai et al. (2011) in Toronto. This model contained different types of data sets, such as building type, performance, construction and material, in "a digital object" model, which permits access into and ongoing authentication of the separate assets that make up the whole (Fai et al., 2011). Furthermore, according to Fai et al. (2011), the case model "is not wholly parametric, it points to the potential of parametric relationships between all data types for heritage documentation". Moreover, Fai et al. (2011) developed a BIM database, which is the CDMICA (Cultural Diversity and Material Imagination in Canadian Architecture); and this database can be used "as a tool for heritage documentation, conservation, and dissemination".

Murphy (2012) has applied BIM for recording and documenting the classical architecture in Dublin during the period of 1700 to 1830 by using "historic architectural data", as well as "a mapping system for creating the library objects onto laser scan survey data". This method started with remote collection by using the terrestrial laser scanner survey data, followed by point cloud processing and generating an ortho-image model. The next step of Murphy's method was to create a parametric library based on the laser scanning survey data using the architectural shape rules from the "18th century architectural pattern books". Murphy's library was created via employing the geometric descriptive language (GDL) of the ArchiCAD software. As the final step, Murphy stated that the "final HBIM" outcome was a "product consisting of 3D models of the building including the detail behind the object's surface, relating to its methods of construction and material makeup", and HBIM could also "automatically generate conservation documentation in the form of survey data, orthographic drawings, schedules and 3D CAD models for both the analysis and conservation of historic objects, structures and environments". In relation to the Venice Charter in 1964, it is very common in the conservation field to use the heritage information to rebuild the past of such historical monuments or conserve them (Charter, 1964).

Recently, a number of research projects have focused on using the 3D parametric model to present heritage sites based on AEC rules (architecture, engineering and construction). On the other hand, using BIM in the historical field, as explained by Murphy (2012), "differs in its approach to the analysis of historic data and parametric design".

11.2.1 Using BIM in historical conservation

Building Information Modelling can offer several benefits to the field of heritage conservation in different ways. For example, a complete study of planned restorations and developing can be allowed before any decisions are made, providing assistance and guidance in the case of any maintenance for the heritage building, assisting in the costing for maintenance and repairs and permitting a wider public building knowledge as models have the ability to be viewed remotely and via a free professional browser. Moreover, according to Fai et al. (2011), "HBIM can provide automated conservation documentation and differs from the sophisticated 3D models produced from procedural and other parametric modelling approaches, whereby the main product is a visualisation tool."

11.2.2 HBIM challenges

Evidently there are also a lot of challenges to be faced by those using BIM in the historical field. These challenges can include organization, technical issues and problems with the site. The organization issues can be related to the cost to the employers of utilizing new methods, as well as the cost of the new technology and training. The technical issues can be things such as the high cost of BIM applications, errors in the system and training time. Besides that, compatibility issues can arise with files and the computer platform. Regarding historic site issues, one of the important things to consider is how complicated they are, which can affect how long it takes to undertake the survey, and also the accuracy of the survey can be problematic.

Moreover, a lot of the historical sites have unique architectural elements, and each of these unique elements and parts need to be modelled up, such as walls, structures, windows, doors and so forth. On the other hand, using BIM for a new project has the advantage of providing available libraries of the modern building elements, which are available with the BIM applications or even through online 3D libraries. Moreover, another important obstacle is the access and permission issues which can come up when attempting to undertake recordings in these historical sites. Furthermore, all these issues can possibly increase the costs of the project.

11.3 Data acquisition in HBIM

11.3.1 *The concept of using terrestrial laser scanning (TLS) in HBIM*

There are several reasons to use terrestrial laser scanning technology in HBIM, and these relate to the high level of accuracy that the laser scanning can offer, in addition to the time spent doing the scans on the cultural sites. Moreover, this technology can be described as an automatic measurement process in a 3D coordinate environment. Figure 11.2 shows the Leica laser scanner C10.

Figure 11.2 Leica laser scanner C10.

The scanner output data is represented as a point cloud, and each of the cloud points has x, y and z coordinates. Furthermore, as Murphy (2012) pointed out, the "laser ranger is directed towards an object by reflective surfaces that are encoded so that their angular orientation can be determined for each range measurement". There are a number of laser scanner vendors present in the geo-engineering market recently. Conversely, as Murphy (2013) stated, "There are three types of scanners suitable for metric surveys for cultural heritage," which are "time of flight scanners, triangulation and phase comparison".

The main difference between these scanner systems is regarding the method by which the 3D coordinate measurements will be analyzed by the scanner. For instance, in the case of the triangulation method, the scanner uses the spot of the laser ray on the surface of the object which was captured via one or more cameras (Murphy, 2013). However, according to Boehler et al. (2003), the "time of flight scanners calculate the range, or distance, from the time taken for a laser pulse to travel from its source to an object and be reflected back to a receiving detector."

11.3.2 Combining digital images and laser scanning

The built-in camera feature for the image survey data is very common in the modern laser scanning systems. Additionally, the feature of applying colour to the laser scanning data "3D point cloud" can be done via linking the multi-image batch to the laser scanning data. As Abmayr et al. (2005) stated, "The RGB colour data from the images can be mapped onto range data by taking account of point translation, instrument rotation and perspective projection." For this, according to Murphy (2013), both the camera and the laser scanning "must be correctly geometrically calibrated" and "the correction of the camera is presented to correct the distortion of camera lenses, and by mapping onto the point cloud any perspective contained in the images is removed."

Additionally, the high dynamic range (HDR) colour images, according to Beraldin (2004), "can be precisely mapped onto a geometric model represented by a point cloud, provided that the camera position and orientation are known in the coordinate system of the geometric model".

11.3.3 Terrestrial laser scanning data processing

A massive range of data is captured by laser scanning and represented in 3D coordinates, which is known as "point cloud data". In order to work with this massive amount of data, professional software is required. However, the scanning can take time in order to capture millions of accurate 3D points, and further, there is the huge job of transporting the laser data into a 3D model containing useable information.

Moreover, as Murphy (2013) explains, "Dedicated software programs such as Leica CloudWorx, Polyworks, AutoCAD and RiSCAN Pro have highly improved the processing, manipulation and analysis of vector and image data from the point cloud" and "all of these software platforms have combined algorithms for triangulation and surfacing of the point cloud."

11.3.3.1 Data cleaning

After transferring the point cloud data from the laser scanning, there are many suitable programs that have the ability to remove the distortions and noise. Besides, many of the laser scanners companies have their own special application. For example, Leica laser scanning systems are compatible with the Cyclone program to deal with the data from the scanner, and FARO laser scanners are compatible with FARO SCENE and Autodesk ReCap.

Generally, the modelling process in these programs includes creating the best fit for the geometric objects from the point cloud. Furthermore, Leica Cyclone has a number of object-suitable utilities for the specified purpose, from which the user can choose, depending on the topology of the scanned point cloud. On the other hand, as Ikeuchi (2001) pointed out, "Other applications are used instead of polygonal 3D models: NURBS surface models, or editable feature-based CAD models."

11.3.3.2 Point cloud registration

The concept of the registration step is combining a range of scan stations with different views of the scanned object. According to Geosystems (2006), "Integration is derived by a system of constraints, which are pairs of equivalent tie-points or overlapping point clouds that exist in two Scan-Worlds." The registration method "computes the optimal overall alignment transformations for each component Scan-World in the Registration such that the constraints are matched as closely as possible".

There are several methods which can be used to register these scan stations, by using one of the two methods, or a combination of them, for example, target based or point cloud based (Mills and Barber, 2004; Rajendra et al., 2014). Furthermore, by using the Global Positioning System (GPS), the coordinates of the laser scanner location can be determined, which can then, according to Cheok et al. (2000), "allow for the scans from each position to be brought into a common frame of reference in a global or project co-ordinate system".

The first method, target-based or target-to-target registration, is a registration between multiple point cloud scans into one single point cloud model and is done via using the control targets in the point clouds. Moreover, to achieve perfection, or at least minimum errors, three corresponding points or more at each point cloud registration must be in common between them. These points can be either natural or artificial targets.

Although natural targets are assigned manually, the artificial targets can be assigned automatically by employing certain algorithms. The automatic assigning process is based on detecting certain shapes of targets, like spheres, or black and white targets. Some advanced algorithms, which can detect correspondence to enable a fully automatic registration process, are implemented in commercial software such as Leica Cyclone (Abdelhafiz, 2009).

The second method is the overlap or the cloud-to-cloud registration method, which involves aligning overlapping scans into a single point cloud by using a number of constraints that are selected within the point cloud software such as Autodesk ReCap or Leica Cyclone model space. It is very important in this method that the selected features match in both registered model spaces (Darie, 2014). The registration result can be used as a base frame for the geometry of the historic structure (Murphy, 2012), which can be modelled in different software such as Autodesk Revit, 3D MAX, Rhinoceros, SketchUp and AutoCAD.

11.4 Jeddah heritage building information modelling (JHBIM)

The purpose of employing JHBIM in the heritage sector, according to Baik et al. (2014), is "to provide an interactive solution in order to move from the zero level of BIM (CAD and 2D drawings) to more advanced levels of BIM (level 2 and level 3)", in the situation of assisting the conservation and the sharing of information about such heritage monuments with experts and societies involved in the decision-making process (Eastman et al., 2011; Fussel et al., 2009). Furthermore, Saygi et al. (2013) state, "BIM will provide the possibility to represent all views (3D models, plans, sections, elevations, and details) automatically."

Furthermore, according to Murphy et al. (2009), the purpose of using Historic Building Information Modelling (HBIM) is to offer interactive parametric objects that represent architectural elements, in order to model historical monuments for the TLS point cloud, for example. These parametric objects represent the components of the building; they are characterized by their descriptive data and their relationships with other components of the building. Figure 11.3 shows the JHBIM method and its workflow.

The heritage preservation of Jeddah historical city may profit from the development of JHBIM because it helps to document the context, knowledge of materials, construction techniques and the building pathologies, and JHBIM supports a wide range of materials and assemblies that are not available from stock libraries of 3D model parts. Besides, JHBIM will provide several advantages to the historical conservation of historic Jeddah, such as offering full study of proposed renovations and changes before final decisions are made, assisting in maintenance efforts, damage detection on the building's surfaces, planning for maintenance and repairs, and allowing wider public building experience as models can be viewed with free viewer software from remote locations.

Moreover, JHBIM will offer virtual visits for the exterior and interior of the heritage building, thus improving the management performance and creating a better understanding for the decision-making procedure of the conservation.

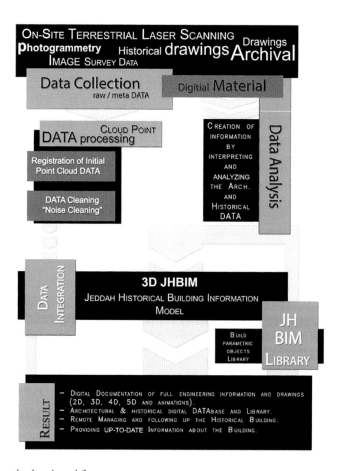

Figure 11.3 JHBIM method and workflow.

Jeddah HBIM 141

The JHBIM model will be remotely reviewed for both interior and exterior in view of offering better management and understanding prior to the decision-making process of any conservation plans. In the case of decision makers, the decision can be more effective and efficient. Many advantages can be provided from employing JHBIM to the heritage field, for example, providing scheduling (4D), estimating or costing (5D), facility management (FM) application or life cycle management (6D), sustainability (7D), maintainability (8D), acoustics (9D) and security (10D).

During the process of establishing the JHBIM, a plug-in called Hijazi Architectural Objects Library (HAOL) was developed, which was mainly born out of the Hijazi pattern book (Greenlaw, 1995).

Furthermore, one of the most important houses in historic Jeddah, Nasif Historical House, was selected as a case study to examine the JHBIM method (Figure 11.4). During the latter part of the nineteenth century, the Nasif family built the house. The work started in 1872 and the building was finished in 1881 (Saban, n.d.).

The house is located in the centre of the historical district of Jeddah city. At the present time, the municipality of historic Jeddah has restored the house and turned it into a museum and cultural centre (SCTA, 2013).

11.4.1 The Islamic Hijazi Architectural Objects Library (HAOL)

The Hijazi Architectural Objects Library (HAOL), according to Baik et al. (2014), "will be dependent on the image survey and point cloud data, and will recreate some of the past, in order to conserve

Figure 11.4 Nasif Historical House.

or to restore these unique parts of these Historical Buildings". Besides, these Hijazi features, such as Manjur, Mashrabiyah and Roshan, have "become the vocabulary of the Historic Jeddah building".

These architectural vocabularies are the key reason for the beauty of the Jeddah historical building. However, there is a huge gap in the Hijazi architectural library in order to provide these unique elements. Additionally, the HAOL of JHBIM will be linked to the data collected in a database, where each single modifier of a parameter bases a change in the shape of the object; as a result, considering the level of detail is very important, as well as considering how can we simplify these object models in order to be suitable for preservation plan, linked to the real opportunity in order to modify the parameters of the shape of the architectonic elements, in particular, of Old Jeddah historical building objects that are at all times unrivalled and irregular. As Dore and Murphy (2013) pointed out, "due to the individuation of the form, grammar and stylistic rules can be used to create a library of historical elements" of the historical buildings in historic Jeddah.

The result of HAOL will allow building an abacus of local constructive objects in order to match the objects' real dimensions with the information derived from any previous architectural drawings. This will make the JHBIM models as detailed as the real objects. The concept of the Hijazi library is to use it as plug-in for existing BIM software platforms such as Autodesk Revit. Also, the HAOL will be introduced and documented and will support any future projects in the historic district of Jeddah city.

11.5 Toward the JHBIM "method"

11.5.1 Image survey data and terrestrial laser scanning

The on-site work started with the data capture step, which includes an image survey in order to highlight and document the architectural features of the historical house, taking around 12 working

Figure 11.5 Nasif Historical House elevations.

days to complete the task. Nasif Historical House is characterized by different Islamic architectural characteristics, which were borrowed from different cultures and ages; for instance, the Mashrabiyahs and Roshans were borrowed from the days of the Ottoman Empire. Moreover, in this step, the architectural photogrammetry technique has been tested and, by employing professional software such as Autodesk ReCap, the 3D image modelling built for some parts of the house. Figures 11.5, 11.6 and 11.7 show the case study buildings for image survey and terrestrial laser scanning.

Before starting to use the Leica ScanStation C10 laser scanner in order to start the geometry capturing for Nasif Historical House, it was very important to prepare the site for this process through planning and defining the targets and ScanStations locations. Moreover, the HDS White and Black targets technique has been used for the purpose of capturing around 400 HDS targets. This step took about 10 working days to be completed.

Figure 11.6 Al-Tyramah at Nasif Historical House.

Figure 11.7 Some of the architectural features of Nasif house.

After preparing the site, the laser scanning was conducted in order to capture the geometry of the house, starting from inside to outside. The laser scanning step took around 20 working days in order to complete, with more than 150 scan stations and more than 70 GBs of point cloud data.

11.5.2 Laser scanning data processing

After exporting the point cloud data from the laser scanner, different processing steps could be used in order to create the required 3D models of the point cloud data and by using a number of programs, for example, Leica Cyclone and Autodesk ReCap. In the case study of Nasif Historical House, Leica Cyclone software was used to remove the noise. The data processing step took around 10 working days to complete.

Then, a number of Scan-Worlds corresponding points in overlapping sections were linked for registration purposes. Figure 11.8 shows the Scan-Worlds registration steps for Nasif Historical House. In Figure 11.9 the result of the scan registrations is shown.

The average point density of the 3D point cloud model of Nasif Historical House was 7 cm on the object surface.

11.5.3 The method of creating the JHBIM object library

The concept of inserting the Hijazi Architectural Objects Library (HAOL) into the JHBIM model started after creating these Hijazi objects in BIM programs such as Autodesk Revit. The outline for building the HAOL, in the case of the Nasif Historical House project, began by understanding and analyzing the architectural manufacturer rules of the house objects. The second step started with understanding the purposes and the level of details (LoD). This was the main key in order to develop the best quality for the JHBIM model that would be needed. The third step started with classifying the Hijazi Architectural Objects based on numbers of standards such as the amount of details, the style, the shape, and the similarity between these objects. At the beginning, the library was divided into three main types: Roshan, Doors and Windows (see Figure 11.10).

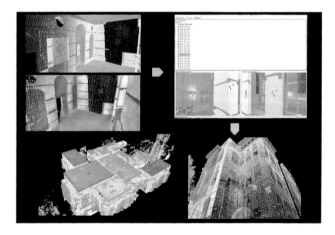

Figure 11.8 The Scan-Worlds registration steps for Nasif Historical House.

Jeddah HBIM 145

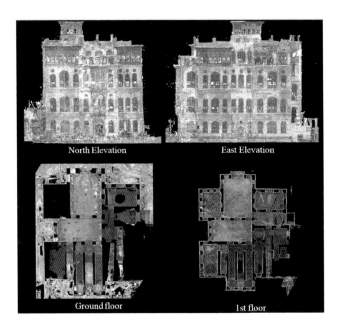

Figure 11.9 The results of the registration for the 3D point cloud model.

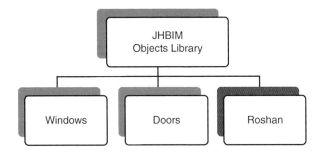

Figure 11.10 JHBIM objects library layout.

In order to create the HAOL objects, a Roshan object is selected to be modelled. After understanding the architectural roles of the Roshan manufacturing, the Roshan has been divided into three parts (i.e. head, body and base). Each of these parts was divided into three subtypes. This method reduced the complexity and was simple to be understood for the modelling. Creating the structure frame of the Roshan was the next step, followed by creating each single part (the detailed parts) as close to the real detail as possible (i.e. as-built in level of detail). Finally, this object was inserted into the JHBIM model of the Nasif Historical House. Figure 11.11 shows the modelling process for the structural parts of the Roshan case study.

In order to model these libraries, there are a number of programs which can be used for this purpose, such as Autodesk Revit, ArchiCAD and Rhinoceros. In the case of creating the HAOL, Autodesk Revit "Revit Families" have been used. These Revit Families are built in an environment

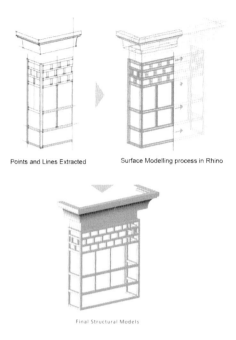

Figure 11.11 Modelling processes of structural parts of the Roshan.

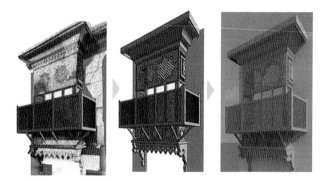

Figure 11.12 Steps of the Roshan modelling.

described as the "Family Editor", which allows for working directly inside of the family file. All these families are saved as "RFA" extensions. The first method is by using the "Traditional Family Editor". In this method, the object must be sketched on a 2D work plan. The second method is by using the "Massing Family". The common thing between the two methods is saving the file in the "RFA" extension; however, the main difference is that the Massing environment is a 3D work environment, which allows us to work directly in a 3D view. However, according to Paul (2013), "The mass category is only useful as a clay model or study model in the project." Figure 11.12 shows the steps for the Roshan modelling, whereas Figures 11.13 and 11.14 show examples of the case study structures' modelling and library development and use.

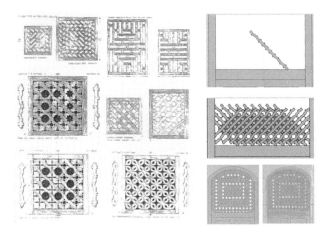

Figure 11.13 Example of the Manjur pattern in old Jeddah (Greenlaw, 1995), and modelling the Manjur.

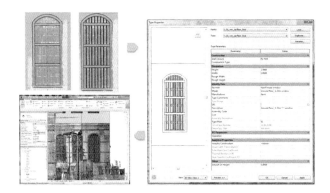

Figure 11.14 Example of the Hijazi Architectural Object in the library and inserting in the JHBIM model.

11.5.4 The JHBIM model (the modelling)

The modelling step started once the 3D point cloud had been established and the HAOL created (Figure 11.15). Models such as the JHBIM model can be created in several BIM programs such as ArchiCAD and Autodesk Revit.

In the case of Nasif Historical House, Autodesk Revit was used as it provided quick modelling and allowed changes to the 3D model, as well as a high quality of construction documents and a high level of flexibility.

In the case of the Nasif Historical House project, the modelling step started with building the house façades as simple as they could be (Figure 11.16). Next, starting with the first level, each floor and wall was built (Figure 11.17).

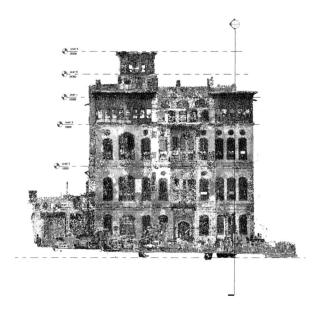

Figure 11.15 The 3D point cloud inserted in BIM software.

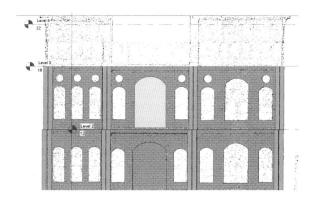

Figure 11.16 Modelling the main façade of Nasif house.

In this step, it was very important to build the house by using traditional methods of construction, to allow us to calculate the loads on the walls and the floors for any future use. The second step, which involved using the HAOL as a plug-in for the model, saved a lot of time and work. However, some objects of the HAOL, such as the windows, Rowshan and doors, were modified to fit exactly in the model. This is because most of these elements had been thought over and designed on-site in order to be fitted into specific locations in the house. The third step was to add any available BIM

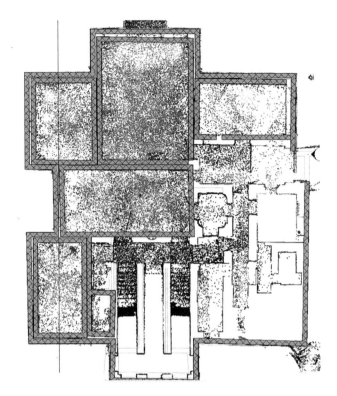

Figure 11.17 Modelling the first-floor walls.

layers such as structures, ventilation, electricity, water, sewerage, and air conditioning. The fourth step can be described as the "make-up" step for the JHBIM model, which requires adding the details, which can give a realistic sense to the model.

As a result of the complexity of reverse engineering a current structure, the creation of the JHBIM model of the Nasif Historical House has been treated as an iterative modelling application. The first challenge to overcome was the incredible amount of information, which was collected via the terrestrial laser scanning, image survey and on-site inspections, besides the huge amount of Mashrabiyah and Roshan details. Figure 11.18 shows the HAOL being inserted into the #d model..

Moreover, the first version of the JHBIM model provided a great environment, which we could use to compare and reconcile the complexities of the many existing data sets. On the other hand, it can be found that the JHBIM approach for the difficult task of modelling a historical building needs to be reviewed, especially in the case of the level of details and the structures of such buildings. However, JHBIM offers a powerful tool for the assembly of the components that can be placed and repeated in different places. Figure 11.19 shows the 3D point cloud model of Nasif Historical House, Figure 11.20 illustrates the 3D section on the JHBIM model, while Figure 11.21 shows the rendered view of the model.

Figure 11.18 Preparing the 3D model for inserting the HAOL.

Figure 11.19 3D LiDAR point cloud of Nasif Historical House and 3D model of Nasif Historical House based on JHBIM.

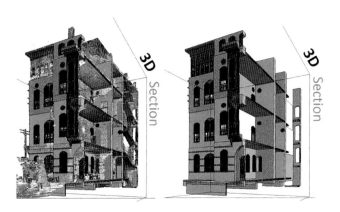

Figure 11.20 3D section on the JHBIM model and the point cloud.

Figure 11.21 Nasif Historical House JHBIM model after the rendering in Autodesk Revit.

11.6 Conclusion

Building Information Modelling is perfectly suitable for providing the kind of data and information that can be used to improve designs and building performance. On the other hand, applying BIM in the heritage field is still a new area to investigate (Wong and Fan, 2013). There are several programs which can serve the BIM concept, such as Autodesk Revit and Graphisoft ArchiCAD; however, there are some limitations in terms of exchanging the information between different vendors' programs.

In the case of applying the BIM method to Jeddah's historical buildings, it will provide many benefits, such as a better understanding and context of heritage buildings, knowledge of materials and also construction techniques. On the other hand, it seems there are a lot of challenges facing those applying BIM to the historical buildings in Jeddah, including organization, technical obstacles and site issues. Another important point regarding applying BIM to the historic field is determining suitable levels of detail or the levels of development (LoD). In the case of Jeddah's historical buildings, one of the most significant points is to rebuild the past, which requires a very high level of detail (i.e. as-built level). This level of detail can be reached via implementing advanced technologies such as laser scanning and photogrammetry. These technologies provide a very rich point cloud model, which can be used as a base for the HBIM model.

The project of JHBIM will be expanded and developed in the future to cover the other buildings and houses in the historic district of Jeddah city in order to form a process of complete documentation and a management system. Moreover, all the extracted information from the constructed 3D JHBIM models, such as the house's structural condition and maintenance activities, can be stored in an interactive database for spatial modelling and follow-up purposes. Furthermore, the system could be linked with fire systems and security systems to protect these buildings from other dangers. This system, upon completion, will be a necessary tool for all individuals and organizations in the field of heritage management and urban planning, in terms of assessment, maintenance and monitoring for each house.

References

Abdelhafiz, A., 2009. Integrating digital photogrammetry and terrestrial laser scanning. Techn. Univ., Inst. für Geodäsie und Photogrammetrie.

Abmayr, T., Härtl, F., Reinköster, M., Fröhlich, C., 2005. Terrestrial laser scanning–applications in cultural heritage conservation and civil engineering, in: Proceedings of the ISPRS Working Group V/4 Workshop 3D-ARCH 2005, Virtual Reconstruction and Visualization of Complex Architectures, International Archives of Photogrammetry, Remote Sensing and Spatial Information Sciences, Mestre-Venice.

Alawi, I., 2013. Saudi Gazette – Old Jeddah hit by worst fire in years [WWW Document]. URL http://www.saudigazette.com.sa/index.cfm?method=home.regcon&contentID=2010030465281 (accessed 10.21.13).

Al-Fakahani, H., 2005. *Jeddah: The bridge of the red sea: progress and development*. The Arab Publishing House for Encyclopedias, Jeddah, KSA.

Alitany, A., Redondo, E., Fonseca, D., Riera, A.S., 2013. Hybrid-ICT. Integrated methodologies for heritage documentation: evaluation of the combined use of digital photogrammetry, 3D modeling and augmented reality in the documentation of architectural heritage elements, in: Information Systems and Technologies (CISTI), 2013 8th Iberian Conference on. IEEE, pp. 1–7.

Baik, A., Alitany, A., Boehm, J., Robson, S., 2014. Jeddah historical building information modelling "JHBIM"–object library, in: International Society for Photogrammetry and Remote Sensing.

Beraldin, J.-A., 2004. Integration of laser scanning and close-range photogrammetry-the last decade and beyond, in: International Society for Photogrammetry and Remote Sensing.

Boehler, W., Vicent, M.B., Marbs, A., 2003. Investigating laser scanner accuracy. Presented at the XIXth CIPA Symposium.

Charter, V., 1964. International charter for the conservation and restoration of monuments and sites, in: Second International Congress of Architects and Technicians of Historic Monuments.

Cheok, G.S., Stone, W.C., Lipman, R.R., Witzgall, C., 2000. Ladars for construction assessment and update. *Autom. Constr.* 9, 463–477.

Darie, D., 2014. Building Information Modelling for Museums (MSC in Surveying). University College London, London, UK.

Dore, C., Murphy, M., 2013. Semi-automatic modelling of building facades with shape grammars using historic building information modelling. *ISPRS Int. Arch. Photogramm. Remote Sens. Spat. Inf. Sci.* 40, 5.

Eastman, C., Teicholz, P., Sacks, R., Liston, K., 2011. *BIM handbook: A guide to building information modeling for owners, managers, designers, engineers and contractors*. Wiley.com.

El-Hakim, S., Beraldin, J.-A., Gonzo, L., Whiting, E., Jemtrud, M., Valzano, V., 2005. A hierarchical 3D reconstruction approach for documenting complex heritage sites.

Fai, S., Graham, K., Duckworth, T., Wood, N., Attar, R., 2011. Building information modelling and heritage documentation, paper presented to XXIII CIPA International Symposium, Prague, Czech Republic, 12th–16th September.

Feilden, B., 2012. *Conservation of historic buildings*. Butterworths, London.

Fussel, T., Beazley, S., Aranda-Mena, G., Chevez, A., Crawford, J., Succar, B., Drogemuller, R., Gard, S., Nielsen, D., 2009. National guidelines for digital modelling.

Geosystems, L., 2006. Inc. HDS Training Manual: Scanning & Cyclone 5.4. 1. Leica Geosystems Inc San Ramon CA.

Greenlaw, J., 1995. *The coral buildings of Suakin, Islamic architecture, planning, design and domestic arrangements in a Red Sea port*, 2nd ed. Kegan Paul International limited, London, New York.

Ikeuchi, K., 2001. Modeling from reality, in: 3-D Digital Imaging and Modeling, 2001, in: Proceedings of the Third International Conference on. IEEE, pp. 117–124.

Mills, J., Barber, D., 2004. Geomatics techniques for structural surveying. *J. Surv. Eng.* 130, 56–64.

Murphy, M., 2013. Historic building information modelling – Adding intelligence to laser and image based surveys of European classical architecture. *ISPRS J. Photogramm. Remote Sens.* 76, 89–102.

Murphy, M., 2012. Historic Building Information Modelling (HBIM) For Recording and Documenting Classical Architecture in Dublin 1700 to 1830 (Doctor of Philosophy thesis). Trinity College Dublin, Dublin.

Murphy, M., McGovern, E., Pavia, S., 2009. Historic building information modelling (HBIM). *Struct. Surv.* 27, 311–327. doi:10.1108/02630800910985108

Nawwar, S., 2013. Jeddah Historic Preservation Department, Jeddah Municipality.

Oreni, D., 2013. From 3D content models to HBIM for conservation and management of built heritage, in: Computational Science and Its Applications – ICCSA 2013, 0302–9743. Springer, Berlin Heidelberg, pp. 344–357.

Paul, A., 2013. *Renaissance revit: creating classical architecture with modern software, 1B ed.* CreateSpace Independent Publishing Platform, Oak Lawn, IL.

Penttilä, H., Rajala, M., Freese, S., 2007. Building information modelling of modern historic buildings, in: Predict. Future 25th ECAADe Konf. Frankf. Am Main Ger., pp. 607–613.

Rajendra, M.Y., Mehrotra, S.C., Kale, K.V., Manza, R.R., Dhumal, R.K., Nagne, A.D., Vibhute, A.D., 2014. Evaluation of partially overlapping 3D point cloud's registration by using ICP variant and cloudcompare. *ISPRS-Int. Arch. Photogramm. Remote Sens. Spat. Inf. Sci.* 1, 891–897.

Remondino, F., El-Hakim, S., Girardi, S., Rizzi, A., Benedetti, S., Gonzo, L., 2009. 3D virtual reconstruction and visualization of complex architectures-the "3D-ARCH" project, in: Proceedings of the ISPRS Working Group V/4 Workshop 3D-ARCH "Virtual Reconstruction and Visualization of Complex Architectures".

Saban, A., n.d. The Nasif House Jeddah, 1st ed. ed. Printing CO. LTD, Jeddah, Saudi Arabia.

Saygi, gamze, Agugiaro, G., Hamamcıoğlu – Turan, M., Remondino, F., 2013. Evaluation of GIS and BIM roles for the information management of historical buildings. ISPRS Ann. Photogramm. Remote Sens. Spat. Inf. Sci. Vol. II-5W1 2013 XXIV Int. CIPA Symp. 2–6 Sept. 2013 Strasbg. Fr. 283–288.

SCTA, 2013. Historic Jeddah, the gate to Makkah. Saudi commission for tourism and antiquities, Saudi Arabia, Jeddah.

TECTURAE, I. design group, 2012. Al Balad Historic Dstrict Survey (Eng).

Telmesani, A., Sarouji, F., Adas, A., 2009. *Old Jeddah a traditional Arab Muslim city in Saudi Arabia*, 1st ed. King Fahad national library, Jeddah.

Wong, K., Fan, Q., 2013. Building information modelling (BIM) for sustainable building design. *Facilities* 31, 138–157.

Worldatlas.com, 2013. Saudi Arabia Map / Geography of Saudi Arabia/Map of Saudi Arabia – Worldatlas.com [WWW Document]. URL http://www.worldatlas.com/webimage/countrys/asia/sa.htm (accessed 10.21.13).

12 Algorithmic approaches to BIM modelling from reality

Ebenhaeser Joubert and Yusuf Arayici

As part of the design of new buildings, a parametric model is usually developed. This model has a high level of semantic information that has been created or implanted by the designer. For an as-built BIM, the modeller does not have that level of knowledge of the existing building. Much of the necessary semantic information is not available to imprint the model with, and the time necessary to model to a good level of detail is not economically viable. Despite these drawbacks, it is sometimes necessary to create as-built models. This is usually done by using a point cloud as trace reference and modelling the building features manually.

To speed up the production of as-built models and increase the accuracy of the results, algorithmic developments are attempting to automate the translation of point data to parametric model. The methodology used to achieve this is mostly based on planarity testing in three-dimensional space, but linear regression in 2D and clustering of rasters are also employed. Geometry is created and then further analysed to subdivide the new vector model into semantic elements that resemble building elements in nesting and function.

This chapter explores the principles and features of algorithmic development to create parametric building models from 3D data. Case studies are employed to assist in this, and their different approaches are used to highlight strengths and weaknesses.

12.1 Introduction

Fast as-built BIM creation from raw geometry would be a very powerful tool. The translation of scanned data into effective parametric models will have a dramatic effect on the efficiency of design, maintenance and preservation of buildings within their context. If building elements are recognisable from scans, the manufacturer specifications and building regulations associated with the model can be embedded as parametric information. With the parametric data updated, the scanned data will provide maintenance and preventative information that cannot be captured with manpower alone. In construction of new buildings, the environment or building site has characteristics that need consideration before construction can begin. Existing buildings need to be preserved or demolished. Rock strata may need to be excavated and trees may need to be protected. These are examples of data that can be recorded and understood by an as-built algorithm, and consequently the necessary design or management functions can be assisted. Parametric information generation from point data is a complicated problem that is likely to endure for the foreseeable future, unless a machine learning method can be found to streamline it.

12.1.1 Why as-built BIM from scanned data?

Parametric as-built building models consisting of millimeter-accurate vector models are very rare. The time required to model an object to this level of detail renders it unfeasible for all but the most critical of elements in a project. Mechanical parts design and industrial design are modelled to this level of detail, but existing buildings are mostly modelled to a higher-order detail and the parametric information included is limited. When reverse engineering an object, it is very convenient to be able to scan and then use dimensions captured in a manual modelling process. But the manual modelling process is flawed and inaccurate. Having an accurate and fast way of capturing detail in a parametric environment would enable analysis of specific attributes of materiality, geometry and interrelationship with other building elements. By association, other parametric values can be implanted in these models. This will in turn enable classifiers to be searched for in models, similar to the machine learning algorithms that identify features in images.

12.1.2 What is parametric data?

A Building Information Model is classified by Isikdag and Zlatanova (2009) and Lee, Sacks and Eastman (2006) as object based, with building elements that have parametric relationships that are object oriented. Furthermore, it has a data-rich representation of physical and functional characteristics. The topologies of the building elements are maintained hierarchically. Its semantic classification lets objects inherit properties and behaviour. 2D and 3D views are available, with annotation that can be added. Finally, elements are definable from the properties of other objects, so it is parametric. To turn a point cloud into a parametric model, the data needs to become intelligent.

Our understanding of intelligent data exists in a digital context with a binary foundation. The data needs to be structured in a machine-interpretable context. Point data does not conform to this. Although it is recorded in coordinates that are coded in machine language, it performs as analogue data would. The data is rich and dense, but it speaks a different language than our interpreter – the machine. "In order to obtain BIM outputs from 3D point cloud data of huge size, a series of processes including data editing, registration, data structuring and modelling are required" (Hajian et al., 2009). The visual processing capacity of the human brain is powerful enough to translate neuron pulses from the eye into parametric information. The shape, materiality and origin of what we see are often known to the observer because of its prior experience.

What we are trying to accomplish in creating an algorithm that interprets point data into a parametric model is to give the machine the ability to interpret the data as we do. For that the machine needs a frame of reference and the encoding capability to compare the data with it – not only to be proficient in reading, classifying and interpret the context of the data, but also to retain information and apply it on tasks data in future tasks. "Modelling of surface shapes is an especially labour-intensive and error prone operation; even well trained modellers sometimes produce significantly different results" (Adan et al., 2011).

12.1.3 Implicit versus explicit data

The ability to discern vectors and planes directly in a mathematical sense from the model data qualifies it as a vector model. The data occupies more than one dimension, whereas a point cloud is essentially one-dimensional data on a large scale. The multidimensionality of a BIM model includes information beyond geometry. This is referred to as parametric information. The data in a parametric model is directly readable to analytical software. For instance, thermal conductivity is expressed as a U-value. A parametric element can have the U-value embedded in the model. This has a specified format and can be read by software simulating the energy consumption of a building. The data is thus explicit. Point clouds do not contain multidimensional data. The Cartesian system used to describe a single point needs three fields for the data to be of use. More fields are employed in certain formats, where a colour or intensity designation is made. These fields add to the usefulness of the data but do not add further geometric dimensionality to the model. The second and third geometric dimensions are defined by the way point data is clustered. Information deduced from the point data is thus implicit.

12.1.4 Compiling data in parametric form

Level of detail for parametric information in a BIM is described by Pătrăucean et al. (2015) as follows. IFC provides the platform for an object-based inheritance hierarchy, defining three abstract concepts: the specialised subclasses of object definitions, relationships and property. Where a completed as-built BIM contains the geometric data as captured by the scanning process, it needs to at least aspire to also contain visible objects coded in IFC as "IfcBuildingElement" class. Without creation of these classes, the model will not be able to show parametric information, whether assumed or sensed. This could be shown for instance as beam, door, window, wall and so forth. Another level of semantics would be "IfcSpatialStructureElement", of which examples would be building storey, site and building. This will place the model within context and further divide the model into manageable zones. The further breakdown of subclasses such as "IfcElementComponent", such as latch, fastener and mesh, would not qualify as an acceptable level of detail to include. Parametric information also consists of material properties which are difficult to infer. Relationships between elements could be important. "IfcRelDecomposes" and "IfcRelConnects" should be included, as proximity and fit of building elements can be captured from as-built view.

12.1.5 Reasons for fully automated vectorising of point data

Jung et al. (2014) describe vectorising of point data as having to deal with complexity of design attributed to indoor structures, often leading to errors in the modelling process. Shadowing by household items such as chairs, desks, tables, flowerpots and decorations represents occluded areas. Lastly, the size of 3D data files uses more memory than the hardware can handle, and the system slows down or crashes. Pătrăucean et al. (2015) have a similar opinion on the reasons for automated as-built creation having not yet achieved its potential. There is also the general strategy when designing the algorithm that needs consideration.

As-designed models are semantically rich due to the designer's high level of involvement and knowledge of the specific building. This information cannot be inferred from the as-built state alone. Practicalities limit the levels of detail that can be adopted in the as-built model. Where the data collection method does allow for fine detail collection, the usefulness of the model does not increase linearly with the amount of time and cost associated with high-level detailed modelling. Obscured or partially occluded elements are missed during data collection. It cannot be shown in the completed as-built model if not captured as data. Methods for detecting obscured elements like ground-penetrating radar are in use, but are not relevant to the discussion in this chapter.

In addition to the above-mentioned concepts of constraints encountered, there are other issues when point data is used to construct as-built models with semantically rich elements. The technique used as standard when grouping point clusters is to identify planar surfaces. These surfaces are interpreted in vector form and then further refined to acquire edge boundaries. The interrelationships found in the model as defined by the planar surfaces are investigated, and very basic assumptions can be made about the position and function of the surface identified. An example would be the identification of a floor element by virtue of its position in relation to the rest of the data.

A floor element would typically have a horizontal orientation with vertical elements at a 90-degree angle butting up to it. This method does not allow for very high resolution when buildings are examined. The tolerances built into the algorithm performing gradient descent prevent the algorithm from examining small point clusters. This is necessary for the process to complete within a reasonable time and also to prevent the memory use to spiral out of control. A different perspective of looking at detail extraction will be examined in the following text.

Geometric drawing is defined as: "The process of defining identical planes and representing them with simple lines rather than point clouds . . . this reduces the data size and guides the manual modelling process" (Jung et al., 2014).

12.2 Recent commercial development into BIM modelling from scanned data

Commercial software that has been in use in the recent past makes use of various degrees of automation. Following below are some of the packages available.

Leica CloudWorx: Automated pipe centreline identification from preselected elements. Diameter and recognisable features are entered as a learning set. The parametric element can then be manually modelled from the centreline and path provided.

Intergraph Smart 3D for Plants: Automatic modelling of pipes from user-defined centreline in the point data.

AutoCAD Plant 3D, Kubit PointSense Plant: Manually selected coordinates at extremities of a pipe enable automated modelling.

AVEVA Laser Model Interface, Trimble RealWorks, ClearEdge3D: Semi-automated creation of 3D geometry by manually segmenting point clouds and assigning categories for each segment.

Trimble RealWorks, ClearEdge3D: Semiautomatic creation of a 3D model by manually segmenting the point cloud and selecting the corresponding catalogues for each segment.

For a model to be considered parametric, the key building elements need to be defined and modelled with their relationships amongst one another. Further steps to achieve the above includes geometric primitive detection, point cloud clustering, shape fitting and classification.

12.3 Point clouds and their manipulation

In recent years, the hardware used to produce 3D data sets or point clouds has evolved sufficiently to support time-efficient takeoff of geometric information in higher definition. Computing power also increased sufficiently to carry higher-order algorithms analysing these data sets. The method of scanning can vary, but in essence, the result is a point cloud with coordinates, colour (RBG), intensity of the propagated signal's reflection and a direction of gravity. The signal does not always need to be propagated, and passive collection like Structure Through Movement (STM) or photogrammetry is also used. The data set is either registered when captured – that is, it is given a specific orientation in space – or it is associated with another data set which has an orientation and then registered accordingly by calculating area comparative features.

From this it is possible to derive elevation for all points, and another parameter is established thus. The difference between parametric models and point clouds is that pure geometry doesn't have explicit data available on the characteristics of the subject. The fields in the IFC format are not identified and no indication is given for materiality, supply, cost and so forth. Therefore, the implicit data carried by the position and reflective value of points needs to be evaluated and manipulated to establish its significance when compared and summed with the values of the rest of the data set. By evaluating and comparing relative available values of points (x-coordinate, y-coordinate, z-coordinate, red, green, blue, intensity and orientation), the human brain can distinguish between points and what the points as a collective represent on the monitor.

The scanned scene could depict a room for instance with a window, door and pipe-work against the ceiling. The trained eye would recognise the type of window or perhaps the diameter of pipe used. From such knowledge, the experienced individual could cite where the window was bought or what type of liquid the pipes are likely to carry. The research described here is developing an efficient method of transplanting the ability of a human to recognise and parameterise features in a point cloud onto a machine. This is already being done in broad terms and in niche environments, but how to relate this to Building Information Models is still in its infancy. The process of interpreting point clouds to produce parametric models is mostly done by manual means and constitutes an enormously expensive and time-consuming process.

12.3.1 Data capture

Laser scanning, photogrammetry and LIDAR are popular methods to produce point clouds. Understanding the basic workings of the methods utilised to produce 3D data is important. This allows the reader to put the new approach towards data processing within the context of current physical technology. Different methods of data collection offer a wide range of strengths and weaknesses. Laser scanning, for instance, is relatively fast and accurate in comparison to photogrammetry but is affected by shadowing and leaves gaps in the data consequently. Using microwave scanning affords much freedom in the wavelength propagated.

This adds the dimension of comparing different materials against their reflective qualities when scanned by varying wavelength categories. Also, techniques in compiling photogrammetrical imagery have a big influence on the quality of point cloud produced. Point cloud data is, in its simplest form, a database of coordinates in three axes. These points in space can have attributes assigned to them such as reflection strength and colour. Collection methods are diverse and consist of two basic philosophies: propagated and passive scanning. Figure 12.1 shows different spatial data captured and organised in different techniques.

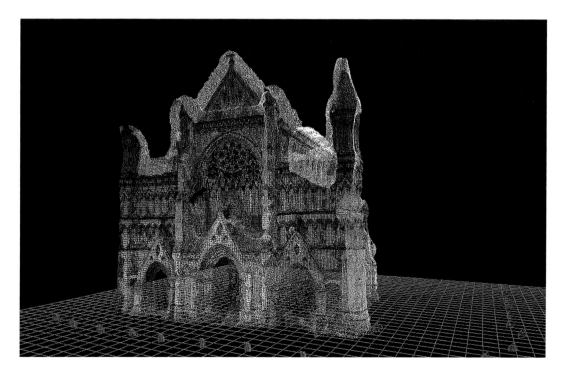

Figure 12.1 Photogrammetry, meshed and unmeshed, and an organised laser scan.

Figure 12.1 (Continued)

Propagated signal scanning consists of laser scanning, ultra-high frequency electromagnetic scanning (radar), sonar and so forth. Passive scanning makes use of existing or ambient energy reflecting off the geometry to be recorded (Ulaby et al., 1982). An attribute consistent to propagated scanning is a measurable reflection intensity that can be an indication of material properties of the reflecting object. Propagated scanning has a set orientation which provides a scope of view or a known set of coordinates within which the scan is situated. The direction of gravity is also a known factor in most propagated scans.

Passive scanning has attributes consistent with the range of frequency of the energy collected. Such attributes are qualities like colour in photogrammetry. The RGB (red, green and blue) assigned data fields in some point cloud data formats are a good example of utilising passive scanning attributes. During data collection, the imaging system (SFM – Structure From Motion) loses the third dimension. The scene is translated into a 2D image. Corresponding points in multiple images taken from different viewpoints are triangulated to infer depth and angle (Hartley and Zisserman, 2003). With laser scanning the registration process is dependent on geometric information and doesn't use appearance cues (Musialski et al., 2013). The 3D reconstruction process, regardless of the collection method, consists of a registration procedure that allows different data sets to be read under the same coordinate system. There are deformations associated with both propagated scanning systems and SFM. Rigid deformation (translation and rotation) is present due to the need of moving the observation point in order to ensure sufficient coverage (Pătrăucean et al., 2015).

12.3.2 *Data formatting*

Simply finding another algorithm to derive vectors and surfaces from scanned data is not the aim. The ability to associate shapes, colours and textures with underlying truths about the points viewed as a whole is what is of interest. In setting up a learning database, an algorithm can reference previous attempts to identify data sets and classify them as elements known to the user. This process is by no means simple, and many different disciplines are involved. Dealing with large data sets where billions of points are present can slow down a computer enough to crash the system. In order to store and utilise these large environments in the data set, innovative ways are used to "zoom" into detail and use only the necessary detail when looking at a scene from a distance.

Techniques like voxelisation are performed to generate an even distribution of point data throughout the scene. In the same process, the data is organised into a regular format with known coordinates for each point. The size of each voxel determines the density of the point data, and any voxel space can only be occupied or not. Data regarding colour can also be associated with each voxel. This is called an organised point cloud and is easier to manipulate due to its known structure. Pathways to each voxel are partially shared, thus saving computational cost.

The economy of working with organised data is well known, but it doesn't achieve the higher priority of providing higher-resolution data where needed and discarding points where detail is not required. The normal segmentation process requires data to be culled in this way in order to allow for the high cost of gradient descent on many points in a randomly scattered pattern. Data manipulation in the latter proposed method of segmentation requires data to be abundant in areas where the geometry scanned is nonplanar. Planar areas need very little data if the boundary edges are established. One way of achieving a data set with the above-mentioned qualities would be to process it in real time and influence the scanning process as it takes place. Another alternative is to voxelise the data set but don't discard the data contained in voxels. The data can then be manipulated in isolated regions. This enables the algorithm to identify high-value coordinates, like those on curved areas. Once boundary edges have been established around planar surfaces, the data between can be culled according to the estimated accuracy the plane achieved.

12.3.3 Registration

The term "registration" refers to the alignment or fitting of a point cloud or data set. This can be aligned to a local grid, other point cloud or global grid. In reality, point cloud data often consists of many scans that need to be aligned with each other to form a more complete image of the subject. The aim with registration then is to find a rigid transformation that optimises the data position with respect to the model (Gelfand et al., 2005). Terrestrial laser scanning necessitates the need to move the scanner position in order to acquire a complete view of large subjects. A common coordinate system is established, and pair-wise registration is usual in this practice (Mitra et al., 2004). Where larger areas are to be scanned and objects to be compared to other structures, a global coordinate system is necessary.

12.3.3.1 Sensor-driven registration

Registration from a sensor-based perspective makes use of additional equipment to determine the position and orientation of the scanner. Global Positioning Systems (GPS), digital compasses or inertial measurement unit (IMU) sensors are used to achieve this (Wilkonson et al., 2010). By establishing position and orientation of a scanner, the points produced can be registered in a global coordinate system (Bohm and Haala, 2005). This enables the real-time acquisition of scans from different positions into the same coordinate system without the need for further registration. The drawback of this method is the greater cost of high-end GPS equipment (Schuhmacher and Bohm, 2005). Additional control stations and clear line of sight to approximately five satellites is required for high-accuracy GPS (Monteiro and Moore, 2005). In built-up areas, shadowing of the scanned position can cause intermittent GPS sensor performance (Laefer and Ryan, 2007).

12.3.3.2 Data-driven registration

The basis of data-driven registration is matching features in two data sets overlapping partially. The features shown in both data sets are identified, aligned and brought into a common coordinate system (Eggert, Lorusso and Fisher, 1997). High degrees of accuracy are possible where many geometric features are recognisable (Pottmann et al., 2006). No additional sensory equipment is required for a data-driven approach. To ease feature detection, targets can be added to a scene before scanning. These targets are easily recognisable to registration software and can clarify data sets with bland features. It is often difficult to place targets, though, and the amount of time added to the acquisition process necessitated by planning and placement makes feature detection from the directly acquired data more appealing.

12.3.3.3 ICP (iterative closest point) method

This iterative refinement step fits two or more point clouds so the corresponding planes or points fit each other in the best way. The reference point cloud is kept stationary while the source (secondary) point cloud's transformation is iteratively revised. The position of the secondary point cloud and rotational accuracy is alternated iteratively until the best fit is found. The two point clouds do need to be roughly aligned for this process to commence (Mitra et al., 2004). Between the point-to-point and point-to-plane methods, the point-to-plane method works best.

12.3.3.4 Georeferencing

Where point data is concerned, georeferencing refers to inserting a scan, usually taken in a local coordinate system, into a global coordinate system. This is achieved by using a primary data set with features that overlap secondary data sets and using a data-driven approach to orientate the subsequent data (Bohm and Haala, 2005). Manual methods can also be used where coordinates of known features are inserted in the scene during preprocessing or the scanner is placed and oriented before scanning starts.

12.3.3.5 Segmentation and challenges

Point density in areas varies as the distance from the collection device changes. Shadows and exclusions of unwanted data also cause areas with lacking data. Occluded areas and noise are the main drawbacks of conventional laser scanning techniques, and the roughness of local surfaces varies from smooth and reflective to very rough. For the algorithm to distinguish a rough surface from several small objects is problematic. Similarly, smooth-surfaced areas can display reflection, confusing triangulation and time of flight.

The surface texture contributes greatly to visual cues on materiality, but it causes many problems when constants are to be selected for segmentation algorithms. Curvature variations of surfaces that can be flat or multi-curved with undulations at multiple scales present a challenge for segmentation algorithms. This makes boundaries hard to identify. Segmentation by means of primarily identifying surfaces, curved or planar, relies mostly on gradient descent to identify the points that fit, to a predetermined accuracy, a function describing a curve or more commonly a plane. Due to the incomplete nature of the data, the boundaries of such a plane as algorithmically identified are uncertain.

The method of intersecting planes and using the intersections as edge boundaries allows for a greater degree of accuracy but, as seen from the case studies, still doesn't achieve an acceptable edge tolerance from which to discern building element characteristics. Being able to accurately separate data from regions made up of surfaces forms a first-tier operation for the conventional vectorising algorithm; edge boundary detection forms the second tier. Segmentation is also used on a third tier of complexity. This occurs where clusters of edges and boundaries are recognised as a discernible object. The function and shape of an object can be stored when its boundaries and surfaces are captured in vectors that conform to simple mathematical functions. Conversion from point data to vector model is the crucial step, though, in which the data is made translatable to the recognition algorithm.

12.3.4 State-of-the-art segmentation algorithms

12.3.4.1 Region growing

The algorithm is conceptually simple and allows application in a wide range of settings. The basic operation merges the points that are close enough to each other in terms of the smoothness constraint into one plane cluster. It sorts the points by their curvature value "P" which added to the "seed points" set. For each seed point chosen, the algorithm finds its neighbour points $\{PN\}$ and tests each neighbour point $N \in \{PN\}$ for the angle between its normal and the normal of the current seed point. The current seed point is added to the current region if the angle is less than the threshold value θ_{th}. As shown in Figure 12.2, the curvature value of the neighbour point is compared with the value of the seed point. If the curvature value is less than the threshold value θ_{th}, this

Figure 12.2 The normal function in viewing software allows a visual aid in identifying planarity.

neighbour point is added to the set of seed points and the current tested seed point is removed from the set. The algorithm repeats this process until the set of seed points is empty. The output of this algorithm is a set of segmented point cloud clusters with the points in each cluster considered to be in the same plane.

12.3.4.2 Bundle adjustment

Bundle adjustment is the minimization of the re-projection error between the locations of observed image points, which is then equated as the sum of squares of a large number of nonlinear, real-valued functions. Nonlinear least-squares algorithms are used to perform the minimization. Due to the ease of implementation and the use of a dampening strategy, the

Levenberg–Marquardt algorithm is one of the most successful. Normal equations are solved to iteratively linearize the function to be minimized in the neighbourhood of the current estimate. A sparse block structure is obtained by the normal equations due to the lack of interaction between parameters for different 3D points. Employing a sparse variant of the Levenberg–Marquardt algorithm, the above statement can be used to gain computational benefits which explicitly take advantage of the normal equations zeros pattern, avoiding storing and operating on zero elements (Triggs et al., 1999).

12.3.4.3 Hough transform

The 2D Hough transform is used in recognising lines in imagery. The 3D Hough transform is an extension of this. Non-vertical planes can be described by the following equation:

$$Z = s_x X + s_y + d$$

sy and **sx** represent the slope of the plane along the x- and y-axis.
d is the height of the plane at (0; 0).

The three above-mentioned parameters define the parameter space. In the parameter space, every point corresponds with a plane in the object space. Every point (x, y, z) defines a plane in the parameter space, because of the duality of these two spaces (Maas and Vosselman, 1999). To detect planar surfaces in a point cloud, these object points are mapped in the parameter space.

12.3.4.4 RANSAC

The learning technique called RANSAC estimates parameters in a model by randomly sampling data (Fischler and Bolles, 1981). With inliers and outliers in a data set, a voting scheme is used to find the best fit. RANSAC is composed of two steps that are iteratively repeated. In the first step, minimal data containing a sample subset is randomly selected from the input data set. This sample subset's elements are used to compute corresponding model parameters to a fitting model. To determine the model parameters, the smallest sufficient cardinality for the sample subset is used. The entire data set's elements are checked by the algorithm for consistency with the model represented by the parameters from the first step. If a data element doesn't fit the model parameters within a set threshold, it will be considered an outlier.

The RANSAC process follows five steps:

- A random subset of the original data is selected, called *hypothetical inliers*.
- The hypothetical inliers lead a model to be fitted to it.
- Other data is tested against the fitted model. The points that fit the estimated model well are considered as part of the *consensus set*.
- If sufficient points have been classified as part of the consensus set, the estimated model is considered as reasonably good.
- All members of the consensus set can be used afterwards, to reestimate the model in order to improve it.

The above procedure is iterated a fixed number of times, producing either a model which is rejected because too few points are part of the consensus set, or a refined model together with a corresponding consensus set size.

12.3.4.5 Total least squares

Calculating the mean for a data set: $(x_1 + \ldots + x_n)/N$ $\qquad \bar{x} = \dfrac{1}{N}\sum_{n=1}^{N} x_n$

The variance in a data set:

For $\{x_1 + \ldots + x_n\}$ the variance is denoted $s_x^2 \; s_x^2 = \dfrac{1}{N}\sum_{n=1}^{N}(x_i - \bar{x})^2$

The standard deviation σx is the square root of the variance, denoted as:

$$\sigma_x = \sqrt{\dfrac{1}{N}\sum_{n=1}^{N}(x_i - \bar{x})^2}$$

Finally, the best fit calculation is written out as follows:

If $y = ax + b$ then $y - (ax + b)$ could be zero $\{(x_1, y_1), \ldots, (x_n, y_n)\}$
$\{y_1 - (axx_1 + b), \ldots, y_n - (ax_n + b)\}$

The mean should be small if a good fit is found. The variance will be a measure of the how good the fit is.

The variance for this data set is: $\sigma^2_{x(ax+b)} = \dfrac{1}{N}\sum_{n=1}^{N}(y_n - (ax_n + b))^2$

Least squares method in 2D is helpful to locate the best fit for lines formed by multiple points. It is akin to linear regression when the linear method is used. Matrix calculations are used to perform least squares in a nonlinear environment.

12.3.5 The point cloud classification

12.3.5.1 Performance metrics for predictive modelling

To solve classification problems, the most commonly used measurement tool is a coincidence matrix. Numbers along the diagonal from top left to bottom right indicate correct outcomes. The numbers in other cells represent errors. The true positive rate or recall is obtained by dividing the correctly classified positives by the total number of positives. The false positive rate is calculated by dividing the incorrectly classified negatives by the total negatives. The overall accuracy of a classifier is calculated by dividing the total number of correctly classified positives and negatives by the total number of samples (Olson and Delen, 2008). Table 12.1 shows a coincidence matrix.

12.3.5.2 Estimation methodology for classification models

It is important to estimate the accuracy of a classifier, because it provides a prediction for its future accuracy. This influences the level of confidence the classifier's output invokes when chosen for a prediction system. The classifiers can be rated by its accuracy, and the classifying system can be optimised by choosing the best model from two or more classifying models. Lastly, when combining classifiers, the outcome can be rated with a confidence level. When the final accuracy of a classifier is estimated when using combined classifiers, the estimation method is best when the bias and variance are low (Olson and Delen, 2008). The variance and bias are also relevant indicators when working with linear regression in algorithms and specifically important when least squares are performed.

12.3.5.3 Simple split (holdout)

The data is partitioned into two subsets which are mutually exclusive. These are called the training and test set. The proportions are often one-third as test set and two-thirds for the training set. Because the data in the two subsets is randomly partitioned, the data is skewed on the classification variable and does not consist of the same properties. This is a considerable point of criticism. To alleviate this problem, the data could be stratified, where the strata becomes the output. This is an improvement, but it still suffers from bias with the single random partitioning (Olson and Delen, 2008).

12.4 Case study examples

The case studies have adopted varied approaches, but some basic assumptions were followed. These assumptions were used to infer a common purpose and comparable methodologies. Even if the steps mentioned in the general methodology were not followed in all four of the studies, it was assumed that such a stage would eventually form part of a complete method. The data input were assumed to be from laser scanning equipment. The data format would consist exclusively of x, y and z coordinates only in the case of the Wang et al. (2015) study. In other studies, the red, green and blue fields are visible but do not feature in any of the results. Beam intensity is a common field and was used in part when semantic feature recognition rules were applied. The direction of gravity is another inferred understanding that was used in semantic feature recognition and needs to be qualified as correct when registration is examined.

The algorithms developed in all four studies concentrated on extracting geometrical shape from the data set as a priority. This was mainly done by segmentation in various forms. The basic shape of the building was then determined by fitting the segments together by either boundary tracing or region growing or a combination of the two. The studies were able to automate this phase in full with different levels of success.

The second major phase labelled building elements such as windows, doors, ceilings and walls. These labels are crucial for a model to be deemed intelligent. In Wang et al. (2015) and Pu and Vosselman (2009), the algorithms have specific stages in which building elements are recognised and labelled. This is also a part of the semantic feature recognition phase in Pu and Vosselman (2009). Jung et al. (2014) and Xiong et al. (2013) do not fully automate the recognition of building elements, with Jung et al. (2014) using the segments produced to be productively labelled by a human modeller and Xiong et al. (2013) labelling window openings and walls successfully. Automation of the as-built model creation process does not stop at labelling building elements. Once the elements have been labelled, it is important to recognise interdependencies between the attributes of these elements. Supplier data, building regulations, specifications and British Standards can be appended to such a label, and the nesting of elements affects how a model makes use of the before-mentioned data. Table 12.2 compares the case studies in terms of their approach in point cloud data modelling techniques.

Case Study A: Obtained data set is preprocessed by noise filtering, and the data is brought to a density that best suits the method. Segmentation is performed by a region-growing algorithm to break the data up into planes of similar flatness and orientation. Then boundary detection algorithms are used to find the edges of planes. These boundaries can be enclosed or on the periphery. Following the boundary detection, building components can be detected and classified by using the

Table 12.1 Coincidence matrix (Olson and Delen, 2008)

		True Class			
		Positive	Negative		
Predicted Class	Positive	True Positive Count (TP)	False Positive Count (FP)	True Positive Rate	$= \dfrac{TP}{TP+FN}$
				True Negative Rate	$= \dfrac{TN}{TN+FP}$
				Accuracy	$= \dfrac{TP+TN}{TP+TN+FP+FN}$
				Precision	$= \dfrac{TP}{TP+FP}$
	Negative	False Negative Count (FN)	True Negative Count (TN)	Recall	$= \dfrac{TP}{TP+FN}$

boundaries found in the previous step by the semantic rules applied. Geometry is created to represent the simple 3D shape of a building. Finally, the simple geometry is assigned labels and given an object category, material properties and topological relationships between objects.

Case Study B: A dense point cloud is recorded by laser scanner. The point data is filtered and registered as a single point cloud in a common coordinate system. Then the 3D building components are reconstructed a simplified 3D shape. Boundary tracing takes place and 3D plane objects are produced and merged into single polygons. No line generalisation is applied at this point in order not to cause information loss. The remaining points not utilised in the boundary trace are imported for detailed manual modelling. Topological relationships between objects are established by a manual modeller, and they get annotated semantically into object categories. Using the automatically extracted boundary lines as guides, this method of manual modelling allows for easy identification of objects, and the reduced data size (down to approximately 5% of the original) eliminates system failure due to memory overload. A low-level surface model is transformed into semantically rich BIM by manually assigning an object category, material properties and topological relationships between objects.

Case Study C: After data is scanned and cleaned, segmentation is used to extract geometry features. Then semantic features are extracted by using defined recognition rules. Each semantic feature is fitted with polygons. General building knowledge is used to manually fill occlusions. Lastly, a polyhedron model is constructed from the hypothesised and directly fitted contour polygons.

Case Study D: This is a context-based modelling algorithm, intended for use on the interior of buildings where components associated with the indoor environment and regular in shape are recognised. Input is a data set consisting of registered point clouds, scanned from multiple positions in the volume. It uses automatic identification and modelling of planar surfaces and openings. The method is applied to multiple volumes in a building, which produces a compact, semantically rich 3D model. Then labelling windows and doors takes place, and the conversion from surface representation to volumetric representation is not included.

In this study, the ray-tracing algorithm reasons throughout the process about the distinction between clutter and non-clutter. The algorithm is taught what clutter looks like and how it differs from walls, ceilings and floors. Areas obscured from every available viewpoint are identified, and differences identified between these occlusions and openings in the surface such as windows and doors.

Context-based modelling is achieved by recognition and modelling of key structural components. A context-based machine learning algorithm extracts planar patches from the point cloud and labels them as wall, ceiling, floor or clutter. The surfaces resulting are patched together, forming a simple planar model. Then, detailed surface modelling is performed by detection and modelling of openings for windows and doors and identifying where occluded data caused an opening in a surface that should be solid. The planar surfaces created in the previous step are analysed individually to identify and model occlusions and openings. By using a learning algorithm, the shape and positions of openings are encoded. This enables the model to act intuitively when confronted with an occluded space.

An implanting algorithm infers the texture and surface for visual purposes in occluded areas. The algorithm processes the data for each room separately to conserve computing time. Several scans can be included in such a separate data set, but it is assumed that they are registered together. For this study, the registration process was conducted manually. The direction of gravity is assumed to be known. Planar patches form the basis of the algorithm's calculations. Nonplanar surfaces are subject to ongoing work.

12.4.1 Data preprocessing

Producing a point cloud to the correct specifications is not always possible due to many and varied factors coming into play when a scene is scanned. The result is more often than not a memory-heavy, cluttered and unwieldy document that represents aspects of reality more closely than what the user would prefer. The preprocessing step consists largely in developing the raw data set into a more usable point cloud that has been cleaned up to only represent the areas of interest and converted into file formats that are accepted by the software of choice.

This step consists of deleting irrelevant data, misleading effects, humans and moving objects that can cause confusing patterns. The registration of individual scans to conform to a common coordinate system is also important. The accuracy with which this is done can vary greatly. The most

Table 12.2 Comparison of case study examples in point cloud data modelling aspects

Methodologies summarised

	Case Study A by Wang et al.	*Case Study B by Jung et al.*	*Case Study C by Pu and Vosselman*	*Case Study D by Xiong et al.*
Input	Unorganised data	Unorganised data	Low density 500–1000 points per m^2	Multi scan registered data
Data preprocessing	Noise removal/decimation	Voxelisation 20% of raw point cloud	N/A	Voxelisation
Shape and surface recognition	Region growing plane segmentation	Root mean squared error	Semantic feature recognition	Patch detection
Surface growth and outline generation	Boundary detection	Boundary tracing/Tracing Grid Cell	Semantic polygon fitting	Patch classification
Shape and surface interdependencies	Categorising into building components	Manual fitting of geometry	Occlusion fitting by assumption	Semantic opening detection
Automated parametric labelling	Yes	No	Yes	Partial

common process followed in practice is the back-sighting of targets which were placed in pre-surveyed positions. These targets are easily recognised by scanning software and render the registration process easy and accurate. Registration from feature-detected attributes in the point data is also becoming common. This automated process is fast and can be very accurate depending on the computing power and time available to the user.

12.4.1.1 Outlier removal

An outlier is defined as "an object that is significantly different from its neighbours" (Papadimitriou et al., 2003). The case studies investigated in this chapter don't address the above-mentioned procedures in great detail as it doesn't impact the emphasis of any of the methodologies described. What is of interest to all of the cases is the density of the point data examined. The study by Pu and Vosselman (2009) was conducted with a maximum data density of 500 points per m^2. This is a very sparse cloud when considering that most scanners commonly used today can easily maintain a point density of 10,000 points per m^2. Wang et al. (2015) do not mention the density of the raw point data, but it is stated that voxelisation has been employed to bring the data down to a size more easily manipulated.

12.4.1.2 Five categories of outlier detection

Distribution-based approach: A standard distribution model is defined, and objects are identified as outliers when they deviate from the model. Most distribution models are univariate (few degrees of freedom) and apply directly to the feature space. This is unsuitable for data sets where multiple factors give rise to a tolerance in accuracy. Also, reality is not directly comparable to mathematically defined shapes and spaces. The distribution of points within a model needs to be established, which is costly in computation time (Barnett and Lewis, 1994).

Depth-based approach: Computational geometry based, this approach determines a series of convex hulls and flags objects in the outer layer as outliers. These algorithms are sensitive to scale, and increased size to a data set will increase the computational cost exponentially (Johnson et al., 1998).

Clustering approach: These algorithms will detect outliers as by-product. The aim of the algorithm is not outlier detection, though, and the criteria are implied (Jain et al., 1999).

Distance-based approach: A distance-based outlier is determined by defining a distance between it and a predetermined fraction of the data set as a whole. All objects outside the diameter of the predetermined distance from the data set's defined density are deemed outliers. Problems occur when the data set has regions of variable density (Knorr and Ng, 1997).

Density-based approach: The local density of a cluster is defined by its neighbourhood points as a predetermined number. The local outlier factor is calculated by defining the distance from the object to the nearest neighbour in the cluster. This method is not sensitive to local density, but surprisingly responsive to the number of neighbours chosen (Papadimitriou et al., 2003).

The first three approaches are not appropriate for large, arbitrary data sets. Approach 4 and 5 are better suited for point cloud analysis.

12.4.1.3 Outlier removal by tensor voting in case study A

The relationships between points are what define the scanned geometry. It follows therefore that a point significantly different from its neighbours does not carry the same relevance and can be ignored. Most data processing methods for 3D point data includes removal of these outliers using local statistics. These local factors could be density (Schall et al., 2005), distance to neighbour (Knorr and Ng, 1997) or eigenvalues of the local covariance matrix (Wand et al., 2008).

There are two phases to the tensor voting algorithm: tensor encoding, where information (curve, surface and intersections) is collected for each 3D point, and tensor voting, where the tensor passes its information to its neighbours via a predefined tensor field. Each neighbour point collects all the votes from its neighbours and encodes them into a new tensor. A surface feature map is created. Noise points can be eliminated by removing those with lower feature values (low scores in the voting phase). "The tensor voting algorithm considers outlier noise explicitly, but may result in serious problems if the inlier data is also noisy" (Wang et al., 2015). The non-parametric tensor voting algorithm can infer local data geometric structure and is used to distinguish and remove isolated points (Kim et al., 2013).

12.4.1.4 Voxelisation

Voxelisation has more benefits than just the downsizing of a raw point cloud. With the points being placed in a regular grid pattern, it is now possible to structure the point cloud in a tree format that allows for much reduced computing time. Logical inference is also possible from a voxel structure, as the known relationship with neighbouring points allows for semantic reasoning.

In Jung et al. (2014)'s study, the preprocessing of data is approached as a segmentation and 2D projection. This in itself can be a time-consuming process. By utilising a brute force search to identify segments or the RANSAC method, the majority of time spent will be in sifting through potentially billions of points to establish relationships that will identify segments. These segments will be partial surfaces that then need further manipulation.

The problem of large data sets is not always as incriminating as it would be presented in most methodologies in feature detection. The study by Xiong et al. (2013) is using voxelisation to produce a point cloud that has an even distribution of points throughout the data set. Patch detection is then used to isolate planar patches for further manipulation. The choice of size for the voxel volume is critical, as it influences directly the tolerances of planarity that can be detected. If the voxel sides are too big, the amount of detail that can be obtained from the data will be less. The size of the plane detected versus the voxel size corresponds to a plane with a radius of roughly 10 times the voxel size. Thus, the decimation of point data is limited.

12.4.1.5 Voxelisation in case study A

The amount of overlay dense data is decreased, thereby increasing the processing speed. Voxels are structured in the model space, and data points fall within its boundaries. The points within a voxel are removed, and a central point representing all the enclosed points takes their place (Moravec, 1996). The larger dimension affects a bigger volume taken up by the voxel and more points captured. Figure 12.3 shows the difference between the centroid of a voxel and a voxel with enclosed points.

12.4.1.6 Voxelisation in case study D

The raw 3D data is arranged into a uniformly spaced grid structure. This is known as a voxel space. This process removes data from areas that have an excessive number of points due to the distance away from the scanner or areas captured several times in different scans. Sparsely sensed areas will not be reduced further in density, but the original position of single points will alter to the centre of the voxel for that volume. The whole data field is volumised into cubes of equal size. The size of the cubes can be adjusted according to the required density of the resultant voxel space. The centre of each voxel will be representative of all the points contained within. The centroid of all the points contained in a voxel can also be taken as representative, but this introduces further computational cost. A small amount of geometric error results from voxelisation, but the amount of cloud data is drastically reduced. It is also possible to control the error amount by varying the voxel size.

12.4.2 Shape and surface recognition

12.4.2.1 Region growing plane segmentation in case study A

Most residential building envelopes have plane-surfaced components due to foundation design and cost. A region growing plane segmentation algorithm is applied to segment the point cloud into a set of disjointed clusters, located in the same plane. Region growing plane segmentation is a conceptually simple algorithm that allows applications in a wide range of settings. It merges the points that are close enough to each other in terms of the smoothness constraint into one plane cluster.

12.4.2.2 RANSAC – robust against outliers in case study B

Three points are selected at random from a point cloud, after which each three-point set forms a plane hypothesis (within *dmax* distance criterion). If the current plane hypothesis has more support than the best plane hypothesis found so far, it becomes the new best plane hypothesis. For best results, this procedure is iterated until (k) reaches $k = \dfrac{\log(1-p)}{\log(1-w^n)}$

p = *probability of best plane being returned after k iterations*
n = *number of samples selected for the hypothesis (in this case 3)*

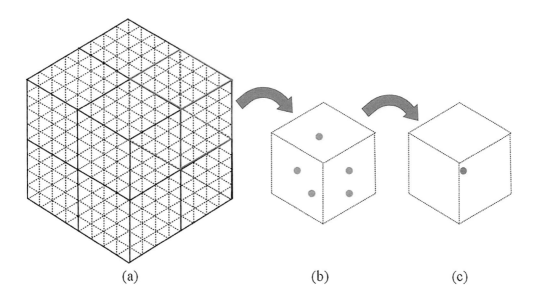

Figure 12.3 (a) The uniform voxel grid; (b) a voxel with enclosed points; (c) centroid of a voxel cell (Wang et al., 2015).

w = the probability that the point belongs to the best plane
[p and w are defined prior]

For determination of best planes, root mean squared error (RMSE) is calculated, where x, y, z and N are Cartesian coordinates and a to d are parameter vectors that define the plane in 3D space:

$$RMSE = \sqrt{\sum_{i=1}^{N}\left(\frac{|ax_i + by_i + cz_i + d|}{\sqrt{a^2 + b^2 + c^2}}\right)^{2} \cdot \frac{1}{N}}$$

See Figure 12.4 for an example of planes segmented by the RANSAC algorithm.

REFINEMENT VIA LABELLING TO FILTER POINTS THAT DON'T BELONG TO PLANES OF INTEREST

Inlier points composing a segmented plane are placed in x-y plane as a 2D binary image. Morphological processing defines which pixels are connected based on the number of connected neighbours. Each connected component is given a unique integer value and its pixels counted. Then the area with the most pixels is retained, and the others are considered noise and removed. A Refinement Grid Cell (RGC) size is manually defined as cell size of the binary image and a Tracing Grid Cell (TGC) size is fixed to a predetermined optimal value.

Typically, a small RGC determines connectivity of points more precisely, but it increases the number of detected planes and thus requires excessive processing time. If the RGC is smaller than the average spacing of the point cloud, it will filter some normal points and lead to serious disconnections in traced boundaries. A large RGC can avoid over-segmentation and reduce processing time but cannot classify small noisy points close to the largest segment. In this study 0.05 m was chosen for RGC, because it is larger than the average spacing of point cloud acquisition (0.01 m) and, in several tests, effectively removed the noisy points within a reasonable time.

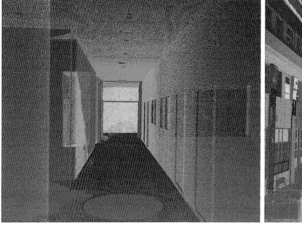

Figure 12.4 An example of planes segmented by the RANSAC algorithm (Jung et al., 2014).

12.4.2.3 Segmentation in case study C

Semantic features usually consist of either planar or smooth-fitting surfaces. Geometric features are extracted using planar geometry. In the ideal situation each of the six semantic features will capture all the segments from the planar segmentation, but this is unlikely because data that doesn't contribute to the intended geometry to capture is present in the data set. Objects like people, cars and what is reflected in windows are examples of this irrelevant data.

The segmented results often also contain the clutter as well. Semantic feature extraction can be used to filter the irrelevant segments from the data. Furthermore, segmentation parameters are difficult to pinpoint. This will change from one data set to another. Even with the ideal parameters selected, there will be over-segmentation, under-segmentation and miss detection – a feature broken up into several segments, several features captured in one segment and features not segmented at all. Lastly, point density affects the ideal segmentation parameters. Areas further away from the scanner will have a lower density and may be missed because of their lower density than the rest of the data set. See Figure 12.5 for examples of semantic feature extraction.

SEGMENTATION PARAMETERS

The ideal segmentation parameter values are represented by the setting producing the fewest number of segments per feature. Conversely, when objects of varying size are found in the same data set, over-segmentation will be a better solution than under-segmentation. Over-segmented areas can be

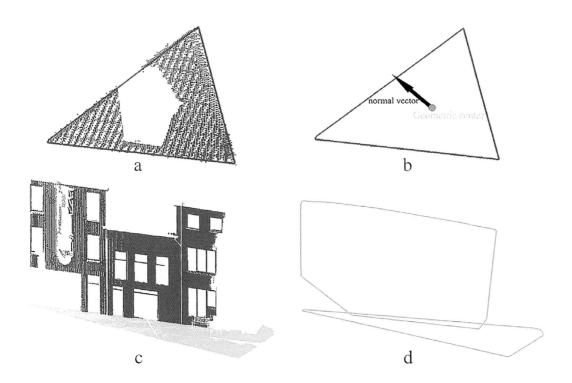

Figure 12.5 (a) Roof section with its convex hull; (b) with convex hull properties derived; (c) wall segment intersecting a ground segment; (d) intersecting convex hulls (Pu and Vosselman, 2009).

consolidated at a later stage dependent on shared properties. Under-segmented parts are problematic, as it is difficult to split. For the paper here discussed, an ideal segmentation parameter setting has been determined for point clouds of 10 cm spacing.

In the segmentation there are good results, as most features have been identified. Some features, like the wall on the right, have been segmented in two pieces. The dormers are segmented into more than two pieces. Colour information contained in each laser scan point can be used as a further guide in identifying features, but shadows and occlusion render this approach fallible and not guaranteed to produce better results than pure geometry (see Figure 12.6).

12.4.2.4 Patch detection in case study D

Planar patches are extracted from the voxelised data. A region-growing algorithm connects neighbouring or nearby points that fall within the planar model's constraints and have similar planar normals. The minimum area bounding rectangle of each planar patch describes its boundary. The total least squares (TLS) algorithm fits a plane to localised points within a predetermined radius (in this case 10 times the voxel size). This provides a starting point for the surface normal. The flatness or planarity of each point is described by the smallest eigenvalue (equal or inverse directional of the vector) for the scatter matrix's centroid or centre. The scatter matrix is the result of the least squares fitting. The planarity of the points are then compared and the most planar point (not already part of a patch) chosen as seed and incorporated to a new patch. This procedure is iterated with points adjacent to points in the patch whose angle between the patch normal and point normal is within the set threshold. (In this study, the threshold was chosen as 2 cm and 45°.)

When all the points that qualified are added to the patch, the patch is used to create a new plane in the data. The new plane is then subjected to the same iteration as above to check points potentially fitting to the patch. As shown in Figure 12.7, this refitting procedure improves the segmentation results significantly. Nonplanar areas create small patches. This is used to filter out patches below a chosen value and keep planes with larger patches and therefore planes that are more planar.

12.4.2.5 Patch classification in case study D

The patches produced in the patch detection step are used to create a graph by connecting each planar patch to its four nearest neighbours. The nearest neighbours are determined by measuring the Euclidean distance between boundaries of patches. Stacked generalization is used to train a sequence of base learners that models the underlying structure of the data being examined. Each base learner is augmented by adding features in the data set – in this case, adding planar patches.

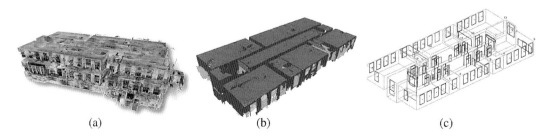

Figure 12.6 (a) Context-based modelling; (b) key structural components; (c) model openings (rectangles) in occluding (Xiong et al., 2013).

Algorithmic approaches to BIM 175

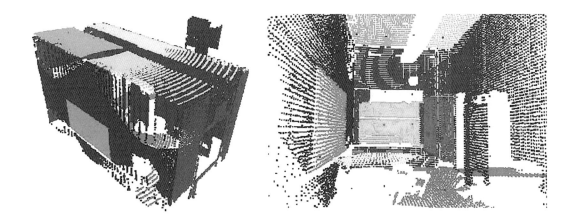

Figure 12.7 Planar patches by region growing for a room from the exterior and interior (Xiong et al., 2013).

12.4.3 Boundary detection

12.4.3.1 Edge and boundary point extraction in case study A

Laser scanning is not suited to low-reflectivity objects. Thus no points show in the window glass area. The edge and boundary detection algorithm is applied to isolate the edge and boundary points from the rest of the point cloud. Edges of objects can be extracted based on the curvature information which is characterised by high changes in curvature. The boundary points on the outer border of the point cloud cannot be found based on curvature data as there is no change for these points (no neighbours).

Since all the data points in each cluster are on the same plane, they can be projected in the 2D plane, where the maximised angle formed by vectors towards the neighbouring points is larger for the boundary points than for the points that are on the inside of the object. Edge points are identified as the edge of openings and boundary points are boundaries of walls and roofs. As shown in Figure 12.8, the 2D concave hull algorithm is used for this.

12.4.3.2 Boundary tracing in case study B

The binary image is being searched from the leftmost occupied pixel in the first row and continues in a clockwise direction to adjacent pixels. An inverse binary image is produced of a plane's empty parts, representing the windows, doors and so forth. These binary images can be boundary traced in the same way as normal planes. When traced boundaries have been produced, they are substituted for the point cloud data of a plane component and inversely rotated to their original coordinates. The points inside the boundaries have been removed, resulting in a much-reduced data set to be manually processed.

PARAMETERS NEEDED FOR BOUNDARY TRACING

A small Tracing Grid Cell (TGC) produces a complex and precise boundary but requires more processing time. As shown in Figure 12.9, if the TGC is smaller than the Refinement Grid Cell (RGC)

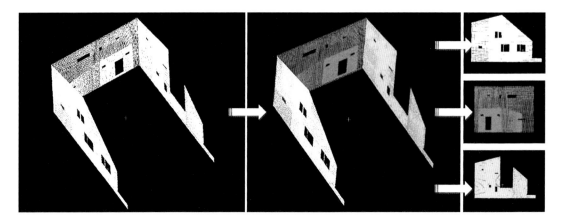

Figure 12.8 Segmented point cloud clusters (Wang et al., 2015).

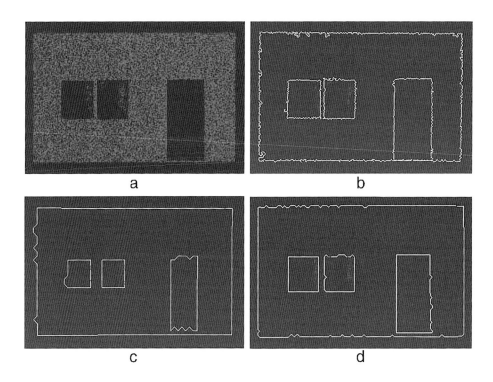

Figure 12.9 TGC's boundary tracing results: (a) point cloud; (b) small TGC; (c) large TGC; (d) optimised TGC (Jung et al., 2014).

size, the pixels being processed can lose their connectivity to each other. This could result in data being missed or excluded from the boundary trace. It is recommended that the TGC is the same size or larger than the RGC. A larger TGC will result in more straight lines within a shorter time, but could produce more distortions. The largest TGC was limited to half the size of the objects observed.

The optimal TGC was determined to be between this minimum setting and the RGC. The TGC for this study was set at 0.05 m.

12.4.3.3 Outline generation in case study C

LEAST SQUARES FITTING

Only the upper contour lines are generated for wall features by the least square algorithm. The outer points of the fitted lines are then projected to the ground level in order to create a surface. This corresponds with the hypothesis that walls are vertical. The lower boundaries of wall outlines are not reliable because of occlusions in this area being frequent. Hence the roundabout procedure here described. The assumption that most parts of a wall intersect with the ground is therefore made. Segmentation errors or small occlusions can cause irregular edges on the outline. More than two nearby edges causes irregularity and needs to be removed, which in turn causes gaps in the outline. If the left and right long edges to the gap belong to the same line, the gap is merely filled in by connecting the two. If the lines are parallel, they are slightly elongated and a perpendicular line segment is inserted to fill the gap. The two edges can be elongated or shortened until they intersect.

CONVEX AND CONCAVE POLYGON FITTING

Roofs differ more in their shapes and orientation than walls. Therefore, it is necessary to use scanned data more intensively than generic model knowledge. The quick hull algorithm (Barber et al., 1996) is used for the generation of convex polygons (Pu, 2008) and the Hough transform for concave polygons. Most roofs visible from the street have a convex shape, so the quick hull method is used.

FOR PROTRUSIONS WITH CONCAVE SHAPES, THE CONCAVE POLYGON FITTING ALGORITHM IS USED

A triangulated irregular network (TIN) for all points in a segment is generated. Long TIN edges are deleted, as the majority of them lie within the concave parts. To infer the contour points, the point with the largest x-coordinate in the segment is chosen. This point is overwhelmingly likely to be a contour point. By using the TIN edges, all the points connected to the starting point are identified. To find the next contour point, the first point in a counterclockwise direction is located amongst the connected points. The procedure is repeated until the contour is closed. Linear features are extracted from the contour points by the Hough transform, which determines the dominant direction of lines, and a perpendicular line segment is inserted to fill gaps between line segments. By connecting the line segments, the concave polygon is combined. The fitted outlines will form a watertight model under ideal circumstances, but due to the fact that parts of the building are usually missed in the scan, the fitted outline polygons are invariably smaller than reality due to occlusions and point density. To compensate for this, the feature outline edges close to another feature with the planes' intersection line.

MINIMUM BOUNDING RECTANGLE

Where windows and doors are identified in the point data, the openings are fitted with minimum bounding rectangles to demarcate their edges. To create a more realistic model, the windows and doors are intruded into the wall surface. This distance is inferred from the window frame or, in the absence of a window frame, the few points that managed to get reflected back from the window surface. To compensate for parts of a building not scanned, certain assumptions regarding the building make-up need to be made in order to achieve a solid polyhedron model.

12.4.4 Geometric modelling and classification

12.4.4.1 Rule-based building envelope component classification in case study A

The boundary points are used to automatically identify the building envelope components. Only building components covered in gbXML schema are considered to be recognised. The process is commenced by determining vertical surfaces and defining them as walls. Separate doors and windows from recognised openings of all wall components are identified. It was assumed that all openings were closed. If a similar surface which is parallel and close to the opening is detected, it is defined as door panel or window blinds. The openings are labelled as doors if they are close to the bottom boundary of the wall, otherwise as windows. The location is recorded and door components are further categorised into normal and glass door according to existence of door panel. Then windows are categorised into clear and blinded based on existence of window blinds. Next, the wall category is divided into exterior walls, which are the areas not below door surfaces, and foundation walls, which are the areas below door surfaces. The roof is recognised next by the fact that it's above walls and adjacent to at least one exterior wall. Unclassified surfaces are classified as raised floor, with surfaces horizontal or below the door.

GEOMETRY SIZE FITTING

The rough semantic model has gaps between the component surfaces. This can be attributed to four reasons. Limited scan resolution of the laser scanner results in gaps being left in the corners. The roof thickness cannot be recorded by the scanner, as it is shaded by ceiling from below and the walls extend above the ceiling. Data downsizing will increase gaps left due to scanner resolution limits (voxelisation). Lastly, the proposed algorithm extends the surfaces of walls, roofs and raised floor and replaces its surface edges with the intersection lines created by two extended surfaces.

12.4.4.2 Manual modelling in BIM software in case study B

The aim of this study was to produce a more productive method for manual modelling of an as-built BIM as opposed to a fully automated approach. It is the opinion of the authors of the study that fully automated approaches are viable under ideal conditions with highly qualitative components.

12.4.4.3 Geometric reconstruction and feature recognition in case study C

The process in this study is based on polygon fitting, with the addition of knowledge-based assumption generation where occlusions are involved. Least squares fitting is directly applied to achieve polygon fitting and the quick hull method or Hough transform for the extraction of feature segments. For a solid polyhedron model to be composed, assumptions about a scanned building need to be made, as it is not possible to scan the building completely. To avoid the final model only looking realistic from scan positions, the assumptions made need to join the polygons together and fill in occluded areas.

SEMANTIC FEATURE RECOGNITION

Six semantic features to be extracted from the terrestrial point cloud are walls, doors, roofs, protrusions, intrusions and windows. Segmentation isolates building features, but it is not clear whether the segments are semantic features. Feature constraints or recognition rules can be formulated to identify each semantic feature. These rules are fed into an algorithm, and automatic recognition achieved thereby. Because a segment is merely a group of points, characteristics of it cannot be inferred before a convex hull is constructed of each segment. The properties of this convex hull are then examined. Different characteristics were found to be efficient in identifying semantic features. Some of these characteristics are:

Size: The relative sizes of segments are in themselves an indication of a segment's semantic feature. Walls for instance are distinguishable from other building elements and noise by comparing their sizes and bringing orientation into the equation. The polygon area of the convex hull is used as a measure.

Position: The position and relative position of building elements can be used to identify them. Some examples of this would be a window or door being dependent on being in a wall and a roof being dependent on walls to carry it. The polygon's geometric centre of the convex hull is used as a measure.

Orientation: By making the assumption that certain building elements have a universal truth to comply with their function, it offers the opportunity to identify the element as a wall, for instance, because of the fact that it is vertical. Another example would be that a roof is never vertical. The normal vector of the plane of the convex hull is used as a measure.

Topology: The way elements interact can provide information. An example is that walls intersect with the ground and roofs intersect walls. The measure is the intersection of two convex hulls' polygons.

Point density: If a laser beam is directed against a highly reflective or transparent object, the reflection is either directed away from the receiving sensor or passes through the transparent surface. Either way, the sensor records a lower point density in such an area. The sudden difference in point density can be used to infer an opening or more specifically windows with glass panes, as illustrated in Figure 12.10 for the feature extraction for protrusion, roof, wall and door.

By running each segment through a constraints check, the semantic features it represents can be isolated. If none of the constraints fit the segment, it is seen as irrelevant or, in other words, clutter. If two attached segments are identified as having the same semantic features, they are merged, thus compensating for over-segmentation.

HOLE-BASED WINDOW EXTRACTION

Feature constraints methods are well suited to identify large building elements, but poor in recognising windows. The reason for this is that the quality of segmentation needs to be good for feature constraints to be recognised. If the point density is very low, such methods fail. A result from the segmentation and feature recognition process is conspicuous gaps in the data or discontinuous segments. Based on the hypothesis of using the omission of sections in a wall panel, a method has been developed, producing a triangulated irregular network (mesh) for walls. Boundary points are extracted by detecting long TIN edges. These long TIN edges are unique to outer boundary (outline) and inner boundary (holes). Points part of the same hole are grouped together.

An interior triangle in a TIN mesh has three neighbours, whilst outer boundary triangles have only two. The triangles that have a boundary point as vertex are selected and checked for the number of neighbouring triangles. The triangles with three neighbours are interior and the boundary point is also interior or a hole. Else, the boundary point is a wall outline or outer boundary. A bounding rectangle is fitted to each hole cluster. Where curved edges are present, a curved fitting method is applied. Holes recorded by tracing boundary points are compared to segments assigned to extrusions and doors. As shown in Figure 12.11, if the hole and segment do not overlap, the hole is considered a window. Holes that have irregular shapes, very long, narrow or small, are also removed.

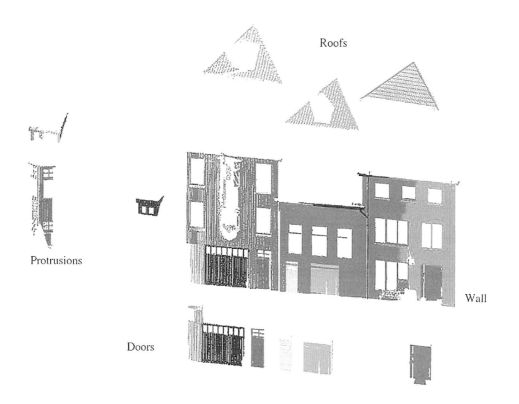

Figure 12.10 Feature extraction for protrusion, roof, wall and door (Pu and Vosselman, 2009).

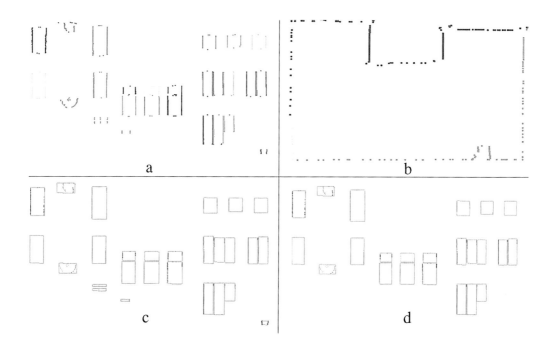

Figure 12.11 Windows extracted from walls: (a) inner hole points; (b) outer hole boundaries; (c) inner hole points; (d) extrusions and door holes filtered out (Pu and Vosselman, 2009).

LIMITATIONS TO THE ABOVE HOLE-BASED EXTRACTION METHOD

When a door, protrusion or intrusion is not detected in error by constraint feature testing, it will then be classed as a window. Only rectangular shape fitting is currently done on windows, and this should be extended to concave hull fitting or curve fitting. The minimum width of holes that can be classified as windows should be set as a constant parameter. Normally this parameter is set to 500 mm to filter out especially people in the scan causing occlusions.

12.4.4.4 Patch intersection, clutter removal and modelling in case study D

The cluttered nature of most indoor spaces is likely to occlude the boundaries of surfaces estimated in the previous step. The patches are assigned the labels wall, floor, ceiling and clutter. By intersecting wall patches with floor patches or ceiling patches, the boundaries are reestimated. The enclosed indoor space as being defined for having a floor, walls and ceiling provides a building structural constraint on which basis the redefinition of boundaries are assumed.

DETAILED SURFACE MODELLING

Each planar patch is being considered by the detailed surface modelling algorithm. It detects occluded areas and areas within openings in the surface. Three steps are followed in the process:

OCCLUSION LABELLING

The axes of the voxel space are aligned with the plane of the patch. The centre of the patch passes through the centre of one layer of voxels. This single layer of voxels is used and each is given one of three labels: F for occupied, E for empty and O for occluded.

The points of a data set representative of a surface are inserted into the voxel space. Each voxel on the surface is labelled as occupied or F if any of these points falls within. Otherwise the voxel is labelled as empty. At this stage it is not possible to distinguish between empty voxels representative of free space and those occluded. A ray-tracing algorithm is used to plot a path between the sensor and surface. If the point being traced lies beyond the surface, the surface voxel can safely be labelled as empty. If the point is observed closer than the surface voxel, the voxel is labelled as occluded.

Large database processing is well served by a voxel-based method, but when detailed modelling of a surface is done a higher resolution of data is required. The voxel surface labelling is converted into a high-resolution image-based format by using the voxel centres as seed for a region-growing algorithm.

This fills in the gaps between voxel centres. This image corresponds to the surface map and corresponding position on the wall, in size and scale. The image is essentially a texture map.

OPENING DETECTION AND MODELLING

Openings could theoretically be extracted directly from the regions labelled as "empty" in the previous step. In real-world conditions, though, this doesn't work. Openings are often occluded by objects, and they also often form a vessel for other objects covering them (air-conditioning units etc.). A way to overcome this problem is by setting up a learning algorithm that keeps a record of previous openings, their shapes and positions relative to other building components. Where partial data is available for openings, the modelling algorithm has recourse to the stored data on other openings in similar positions. The learning algorithm uses features: area, aspect ratio, width, height, size relative to surrounding surface, distance from the sides, percentages of pixels within labelled as "E", "F" and "O" and inverted "U" shapes for doors.

OCCLUSION RECONSTRUCTION

The openings being determined in the previous step, the occluded regions are labelled and now filled in with an implanting algorithm. The algorithm stems from methods used to recover missing parts in damaged photographs and the 3D gap filling technique was proposed by Salamanca et al. (2008).

12.4.5 Implementation and results

12.4.5.1 Implementation and test results in case study A

Two residential houses and a small bank building were used to validate the proposed approach. In all three of these case studies, the point clouds were captured as comprehensively as possible and the scans used as raw data. The first subject was a Zero Net Energy Test House (ZNETH), as shown in Table 12.3, and the raw data contained 1,061,637 points. Data downsizing was run, and the leaf size vessel was set at 0.05 m. This is five times the resolution of the raw data (0.01 m). (This means the leaf size is 50 mm at a 10 mm point density.) The data sizing algorithm decreased the data set to 541,003 points, which is roughly 50% of the original raw data. The downsized data was segmented into plane clusters. A boundary and edge point detection algorithm was used to extract the inner and outer boundary points. A set of outer and inner boundary surfaces is the output from the previous algorithm.

A rule-based building envelope component algorithm is employed to categorise each boundary surface into a specific building element. Components extracted were 2 Door components, 39 Window components, 4 Roof components, 1 Underground Wall component, 1 Raised Floor component and 10 Exterior Wall components. Precision, recall and accuracy (Olson and Delen, 2008) were measured as a benchmark for element classification. "All recognised components except one window were correctly categorised." Of these recognised elements, the area for the modelled components was compared to the true dimensions and the absolute difference calculated for each recognised element. It is noticeable that the level of accuracy is strongly affected by the completeness of the raw data. That is, where the scanned points were missing for elements like the raised floor or reflective surfaces came into play, the accuracy dropped.

Table 12.3 Precision = TP/(TP+FP), Recall = TP/(TP+FN), Accuracy = (TP+TN)/(TP+TN+FP+FN)

Evaluation for case ZNETH and comparison between the recognised and the manually measured components

	TP	FP	FN	TN	Precision	Recall	Accuracy	Measured dimension (m^2)	Recognised dimension (m^2)	Error (m^2)	Error (%)
Exterior wall	10	0	0	46	100%	100%	100%	355.25	363.95	8.71	2.45
Window	39	0	1	17	100%	97.50%	98.25%	18.48	15.25	3.19	17.26
Door	2	0	0	54	100%	100%	100%	3.90	4.27	0.37	9.49
Foundation wall	1	0	0	55	100%	100%	100%				
Raised floor	1	0	0	55	100%	100%	100%	19.74	16.41	3.33	16.87
Roof	4	0	0	52	100%	100%	100%	156.74	143.48	13.26	8.46

Table 12.4 Precision = TP/(TP+FP), Recall = TP/(TP+FN), Accuracy = (TP+TN)/(TP+TN+FP+FN)

Evaluation for one-storey residential house and comparison between the recognised and the manually measured components

	TP	FP	FN	TN	Precision	Recall	Accuracy	Measured dimension (m^2)	Recognised dimension (m^2)	Error (m^2)	Error (%)
Exterior wall	4	1	0	20	80%	100%	96%	127.30	128.75	1.45	1.14
Window	14	0	0	11	100%	100%	100%	10.81	11.83	1.02	9.44
Door	2	1	0	22	67%	100%	96%	2.97	3.89	0.92	3.10
Roof	4	0	0	21	100%	100%	100%	137.50	148.60	11.10	8.07
Raised floor	1	0	0	24	100%	100%	100%	10.41	10.19	0.21	2.02

Table 12.5 Precision = TP/(TP+FP), Recall = TP/(TP+FN), Accuracy = (TP+TN)/(TP+TN+FP+FN)

Evaluation of one-storey bank building and comparison between the recognised and the manually measured components

	TP	FP	FN	TN	Precision	Recall	Accuracy	Measured dimension (m^2)	Recognised dimension (m^2)	Error (m^2)	Error (%)
Exterior wall	14	0	0	42	100%	100%	100%	347.70	324.74	22.97	6.61
Window	27	0	3	26	100%	90	95%	76.01	77.53	1.52	2.00
Door	3	0	0	53	100%	100%	100%	4.63	5.92	1.29	27.86
Roof	12	0	0	44	100%	100%	100%	1036.90	1054.00	17.11	1.65

The second test case was a one-storey residential house. The same procedure was followed evaluating the building element recognition and modelled area as opposed to true area. Tables 12.4 and 12.5 show the results.

12.4.5.2 Implementation and results in case study B

STUDY SITES AND DATA ACQUISITION

A pilot implementation was conducted for the proposed methodology at Engineering Research Park, Yonsei University, Seoul, South Korea. Two specific volumes were selected: The first was a corridor, being a relatively simple structure consisting of a long aisle and elements such as windows, doors and stairs. The other volume is an atrium and in contrast is complex with a large volume, with columns, walls and an over-bridge connecting two sides of the third floor. The roof is of steel frame and glass construction. Two laser scanners were used: Leica ScanStation 2 and FARO Focus3D. The ScanStation 2 time-of-flight scanner was first used, after which the Focus3D phase-shift scanner was introduced to collect the areas shaded or left out by the first scan. The Leica Cyclone 8.0 software package was used to register the various scans and remove noise.

GEOMETRIC DRAWING

The corridor-segmentation process yielded 60 plane components detected from 51.5 million points. The atrium produced 602 plane components from 111.5 million points. The RANSAC algorithm is used to segment the planes, and it is traced. The geometric drawing process reduced the data size from the original point cloud's 1.4 GB to 54.6 MB for the corridor and 3.4 GB to 134.3 MB for the atrium. This represents 3.8% and 4.3% of the original data size respectively. The reduction in data size enables the efficient manipulation of the 3D model in the BIM software.

MAIN CONTRIBUTIONS IN CASE STUDY B

The typically large size of a point cloud data set is reduced without loss of relevant geometrical information. As-built BIM for large and complex structures can be created under normal conditions, and extracted boundaries create a trace reference for the modeller to efficiently place parametric building components and so create the BIM.

12.4.5.3 Implementation and results in case study C

Two data sets are tested against the proposed method. The buildings tested are of different shapes and sizes.

VLAARDINGEN CASE

The Dutch town of Vlaardingen's central area is scanned with a Leica scanner from six different set-ups. The scans are registered with the ICP (iterative closest point) method and local control used to georeference in 2D plan orientation. An average point density of 500 points per square meter was achieved.

From the six semantic features identified in the study, the Vlaardingen case has identified walls, doors, windows and roofs fully automatically due to its large surfaces and sufficient resolution of points. Protrusions are recovered in a semi-automatic process. The method struggles to distinguish between noise segments and the building protrusions. The correct segments are therefore selected manually from a list of potential segments generated by the knowledge-based feature extraction process. The resulting model is checked against the original point cloud, and some errors are identified. The building dormer reconstruction is much smaller than reality. This happened because of the size of the panels in each dormer.

The frames are too narrow for the surface-growing algorithm to be effective throughout the dormer structure. The largest segments only are considered a dormer segment. Round corners in a protrusion were not recognised, as the method only processes planar surfaces. One of the windows had a curtain hanging in it, covering half the surface of the windowpane. This resulted in the area covered by the curtain not being recognised as a window opening, but as a wall segment. The wall has a 100 mm offset between outline and real position. This is caused by the upper boundary being calculated from the highest points on a left–right scan. If the building is bordered by another building with a slightly higher boundary, or the eaves on a higher wall, then the direction of the scan will cause offsets like this. Lastly, the roof outline edge is offset from its real position. This is also a consequence of the upper boundary scan.

ESSLINGEN CASE

The German town of Esslingen was scanned with a mobile scanning unit. The mobile unit is made up of Differential Global Positioning System (DGPS) and inertial measurement unit (IMU) sensors. The roof carries a laser scanning platform (Barber et al., 2008). The resultant point cloud is georeferenced by the scanning unit. The point density is about 100 points per square meter. Fully automated reconstruction took place in about five minutes. In this instance the window and door detection algorithms were disabled. This was done because of the low point density being insufficient to detect the features reliably. The testing phase was again conducted by overlaying the model with the point cloud and checking for discrepancies. The maximum outline offset with the raw point cloud was 50 mm on the facade outlines. This is not accurate enough for automatic texturing and further data manipulation. There is a measure of uncertainty about the roofs and dormers, which are only partly visible from the street. This means that knowledge-based modelling was done. A reasonable balance needs to be found between knowledge-based and data-based modelling.

DISCUSSION

The method described in this study has had success in modelling planar surfaces of buildings with relatively simple structural make-up. The authors feel that the method will fail when complex structures are introduced. Curved walls and protrusions or non-vertical walls will present a major obstacle. It is noted that better and more comprehensive building knowledge would aid to reconstruct more complex buildings. There is scope for improvement in the outline generation algorithm to eliminate the offset errors. The output model currently is still untextured wireframe models. The combination of registered digital imagery could be used in future to produce textured building models. This remains a challenging area of research, though, as the registration of 2D images within a 3D model space is not mature.

12.4.5.4 Implementation and results in case study D

The test facility is a schoolhouse consisting of two storeys and 40 rooms. The survey was done by a professional scanning service provider; 225 scan locations were processed with over 3 billion 3D measurements taken.

PATCH CLASSIFICATION

The scans were brought to a manageable size by subsampling to a factor of 13. A professionally created model (manual model) of the test facility was brought in to compare the results of the automatically derived model. The ground truth planar patch labels of the manual model were identified, and each point in the input data labelled according to the closest surface in the manual model. Patches were labelled in turn by the majority of points of a certain label contained in them. The accuracy of this process was manually validated. The data was split into two. The first floor's 23 rooms were used as a training and validation of the algorithm, and the 13 rooms on the second floor for testing. Rooms are being classified on an individual basis, and the regularisation pattern is chosen from a leave-one-out cross-validation. The training set produced 780 planar patches from 23 rooms, and the 13 testing rooms produced 581. The local features used are patch orientation (angle between normal and z-axis), area of bounding rectangle, height (max z-value), point density and aspect ratio. The choice of local features is led by their independence of each other and needs to discriminate in order for patches of different classes to be separate in the feature space. This study considered pairwise relations as follows: orthogonality, parallelity, adjacency and coplanarity. Contextual features are compared in different combinations of contextual features. The highest validation yields are achieved by coplanarity and orthogonality. Note that the combination with all the relations is not the best option. Three algorithms are compared: the focus of the study's stacking algorithm, conditional random field (CRF) and the LogR method, as shown in Table 12.6. The algorithms use the

Table 12.6 Results for three different algorithms

Precision	Clutter	Wall	Ceiling	Floor	R	Clutter	Wall	Ceiling	Floor	Average	
Stacking	0.87	0.84	1.00	0.87		0.95	0.66	0.82	0.81	Stacking	0.85
CRF	0.86	0.82	0.93	0.93		0.95	0.60	0.82	0.81	CRF	0.83
LogR	0.81	0.88	0.92	0.82		0.97	0.41	0.71	0.88	LogR	0.78

same validation, training and testing sets and above-described features. (Note that LogR doesn't use contextual features.)

Using the F1 score: $F_1 = 2 \dfrac{precision \cdot recall}{precision + recall}$ Average F1 score

The areas with the highest failure rate are between wall and clutter points. The reason for this is the similarity in local features due to occlusion. The stacking algorithm produces better results in recall on walls. This bodes favourably for the hypothesis that context can guide classification.

DETAILED SURFACE MODELLING

Two sets of experiments were undertaken. The first used surfaces produced by a stand-alone wall detection algorithm by A. Adan and D. Huber (2011), and the second the context-based algorithm. The experiment's results reflect a similar outcome, but the surfaces evaluated differ in choice. The context-based algorithm was configured using an independent evaluation. Some free parameters were established by setting the Hough transform threshold for the number of edge point votes in line detection to 90. The edge support threshold was set to 50%. Thresholds α and β were set at one and two times the voxel size.

Furthermore, the SVM (support vector machine) classifier that detects openings was trained on a data set of 370 examples that were split into openings and non-openings. The data was further divided into 95% training set and 5% validation set. The majority of the analysis was focused on understanding the opening detection performance and modelling steps. Two voxel sizes, 10 cm and 5 cm, were evaluated. With the 5 cm size, the memory usage limit was neared and the risk for missing wall sections became evident. The planarity of walls needs to be considered and tolerance left for imperfections. The resolution was set at 2 cm/pixel. Two aspects of performance were examined: reliability of opening detection and accuracy of modelling. The results of the algorithm are tested against the manually modelled ground truth model. The F-measure is used to calculate the best threshold for voxel resolution.

12.5 Discussion and conclusion

The use of planar normals as basic building blocks for segmentation is a common thread in all four studies. Patch detection and the further classification and growing of segments are the results of this method. Boundary creation tracing then forms the limits of segments that can be compared with semantic recognised attributes to further refine the geometry. This method is well known and uses the data in a three-dimensional model space. To detect a plane this is a necessary step, but data is also used in 2.5D and 2D in applications using raster graphics to indicate depth. In essence, the four case studies use region growing, brute force plane sweeps, Hough transforms, expectation maximisation and RANSAC algorithms to segment the point data. The size of the segments is varied by changing the constants in the equations and is mostly limited to the density of the point data and the minimum size of plane expected to be detected in the scene.

By obtaining the correct size of segment, the data can now be further examined. This is done by meshing the points up and using the neighbouring ratio of vertices. By voxelising the points, 26 different types of building element interdependencies are an easy handle to identify features such as openings for instance. All of this happens in 3D space though. Semantic feature recognition is a very interesting development. By using rudimentary information like the direction of gravity, the relationships between clusters of points can be refined to the point that building elements are recognised. Study C does this most efficiently, but it seems that accuracy is still a problem when greater areas are segmented and the boundary tracing discipline suffers. By developing the ability to create an outline or wireframe structure of the geometry, the data size can be reduced significantly in the subsequent phases. The same process of linear regression can be used to isolate normals that conform to a preset threshold when dealing with 2D data.

Studies B and C have followed the general five-step methodology, apart from the fact that study B didn't label building elements automatically. Both the studies focused on a speciality field. Study C has given particular emphasis to boundary tracing. The method of Tracing Grid Cells and algorithmic manipulation of the cell's border lengths and normal was interesting. The size of the chosen cell also determines the level of detail to be obtained up to the point where the cell becomes smaller than the spacing between points. This is the lower threshold for extracting planes and tracing boundaries.

Study C concentrated on learning algorithms to establish standard sizes for openings. The knowledge is applied in filling in gaps in the data caused by occlusions. This is a significant problem in manipulation of point data, and the solution seems a good one. By creating a learning set from the point data to be segmented, the algorithm has found a way to treat a specific building with generic measures and a bespoke solution for each data set is achieved. The ROC (receiver operating characteristics) curve classifying method for success would have been an interesting way of training the semantic feature recognition function. The almost bootstrapping-like method of choosing classifiers in the raw data makes the method ideal for employing machine learning.

References

Adan, A. Huber, D. (2011). *3D reconstruction of interior wall surfaces under occlusion and clutter 3D Imaging*, Modelling, Processing, Visualization and Transmission (3DIMPVT), Hangzhou, China.

Adan, A. Xiong, X. Akinci, B. Huber, D. (2011). *Automatic creation of semantically rich 3D Building models from laser scanner data*, Proceedings of the 28th International Symposium on Automation and Robotics in Construction, pp. 343–348.

Barber, C.B. Dobkin, D.P. Huhdanpaa, H.T. (1996). The Quickhull algorithm for convex hulls. *ACM Transactions on Mathematical Software* 22(4), 469–483.

Barber, D., Mills, J., Smith-Vossey, S., (2008), "Geometric validation of a ground-based mobile laser scanning system", ISPRS Journal of Photogrammetry and Remote Sensing, 63 (1), pp 128–141.

Barnett, V. Lewis, T. (1994). Outliers in statistical data. *Biometrical Journal* 37, 256. doi:10.1002/bimj.4710370219

Bohm, J. Haala, N. (2005). *Efficient Integration of Aerial and Terrestrial Laser Data for Virtual City Modelling Using LASER-MAPS*. IAPRS Vol. 36 Part 3/W19 ISPRS Workshop Laser Scanning, pp. 192–197.

Eggert, D. Lorusso, A. Fisher, R. (1997). Estimating 3D rigid body transformations: a comparison of four major algorithms. *Machine Vision and Application* 9(5–6), 272–290.

Fischler, M. Bolles, R. (1981). Random sample consensus: a paradigm for model fitting with applications to image analysis and automated cartography. *Communications of the ACM* 24(6), 381–395.

Gelfand, N. Mitra, N. Guibas, L. Pottmann, H. (2005). *Robust Global Registration*. Symp. Geometry Processing, pp. 197–206.

Hajian, H. Becerik-Gerber, B. (2009). *A research outlook for real-time project information management by integrating advanced field data acquisition systems and building information modelling*, Proc. International Workshop on Computing in Civil Engineering.

Hartley, R. Zisserman, A. (2003). *Multiple View Geometry in Computer Vision, 2nd ed*. Cambridge University Press, New York.

Isikdag, U. Zlatanova, S. (2009). Towards defining a framework for automatic generation of buildings in CityGML using building information models. In Isikdag, U. and Zlatanova, S. (eds.): *3D Geo-Information Sciences*, Lecture Notes in Geoinformation and Cartography. Springer, Berlin, Heidelberg, pp. 79–96.

Jain, A. Murty, M. Flynn, P. (1999). Data clustering: a review. *ACM Computing Surveys* 31(3), 264–323.

Johnson, T. Kwok, I. Ng, R. (1998) *Fast computation of 2-dimensional depth contours. Proc. KDD*, pp. 224–228.

Jung, J. Hong, S. Jeong, S. Kim, S. Cho, H. Hong, S. Heo, J. (2014). Productive modelling for development of as-built BIM of existing indoor structures. *Automation in Construction* 42, 68–77.

Kim, C. Son, H. Kim, C. (2013). Fully automated registration of 3D data to a 3D CAD model for project progress monitoring. *Automation and Construction* 35, 587–594.

Knorr, E. Ng, R. (1997). *A unified notion of outliers: properties and computation*, Proc. of the International Conference on Knowledge Discovery and Data Mining, AAAI Press, pp. 219–222.

Laefer D. Ryan, A. (2007). *UREKA participation in GUILD Alison Ryan Summer 2007*. Tech. rep., Report for Science Foundation Ireland.

Lee, G. Sacks, R. Eastman, C. (2006). Specifying parametric building object behaviour (BOB) for a building information modelling system. *Automation in Construction* 15(6), 758–776.

Maas, H. Vosselman, G. (1999). Two algorithms for extracting building models from raw laser altimetry data. *ISPRS Journal of Photogrammetry and Remote Sensing* 54(2–3), 153–163.

MacDonald, M. (2014). Building Information Modelling. [Online] mottmac.com. Available at: <https://www.mottmac.com/article/2385/building-information-modelling-bim> [Accessed 11 Dec. 2014].

Mitra, N. Gelfand, N. Pottmann, H. Guibas, L. (2004). *Registration of point cloud data from a geometric optimization perspective*. SGP '04, pp. 22–31.

Monteiro, L. Moore T. Hill, C. (2005). What is the accuracy of DGPS? *The Journal of Navigation* 58, 207–225.

Moravec, H. (1996). *Robot Spatial Perception by Stereoscopic Vision and 3D Evidence Grids*, Perception, September.

Musialski, P. Wonka, P. Aliaga, D. Wimmer, M. Van Gool, L. Purgathofer, W. (2013). A survey of urban reconstruction. *Computer Graphics Forum* 32(6), 146–177.

Olson, D. Delen, D. (2008). *Advanced Data Mining Techniques*, 1st edition. Springer-Verlag, Berlin Heidelberg, p. 138. ISBN 3540769161.

Papadimitriou, S. Kitagawa, S. Gibbons, P. Faloutsos, C. (2003) *LOCI: Fast outlier detection using the local correlation integral*. Data Engineering, 2003. Proceedings. 19th International Conference on, March 2003, pp. 315–326.

Pătrăucean, V. Armeni, I. Nahangi, M. Yeung, J. Brilakis, I. Haas, C. (2015). State of research in automatic as-built modeling. *Advanced Engineering Informatics* 29, 162–171.

Pottmann, H. Huang, Q. Yang, Y. Hu, S. (2006). Geometry and convergence analysis of algorithms for registration of 3D shapes. *International Journal of Computer Vision* 67(3), 277–296.

Pu, S. (2008). Generating building outlines from terrestrial laser scanning. *International Archives of Photogrammetry, Remote Sensing and Spatial Information Sciences* 37 (Part B5), 451–456.

Pu, S. Vosselman, G. (2009). Knowledge based reconstruction of building models from terrestrial laser scanning data. *International Archives of Photogrammetry, Remote Sensing and Spatial Information Sciences* 64, 575–584.

Salamanca, S. Merchan, P. Perez, E. Adan, A. Cerrada, C. (2008). *Filling holes in 3D meshes using image restoration algorithms*. Proceedings of the Symposium on 3D Data Processing, Visualisation and Transmission (3DPVT), (Atlanta GA).

San Jose Alonso, J. Martinez Rubico, Fernandez Martin, J. Garcia Fernandez, J. (2011). *Comparing time-of-flight and phase-shift. The survey of the royal pantheon in the basilica of San Isidoro (Leon)*. Remote Sensing and Spatial Information Sciences, Volume XXXVIII-5/W16, 2011, ISPRS Trento 2011 Workshop, 2–4 March 2011, Trento Italy.

Schall, O. Belyaev, A. Seidel, H. (2005). *Robust filtering of noisy scattered point data*. Proceedings of the Second Eurographics/IEEE VGTC Conference on Point-Based Graphics, SPBG'05, Aire-la-Ville, Switzerland, Switzerland, Eurographics Association, pp. 71–77.

Schuhmacher, S. Bohm, J. (2005). Georeferencing of terrestrial laserscanner data for applications in architectural modelling. *International Archives of Photogrammetry and Remote Sensing* 36(5/W17), 1–7.

Thomson, C. Boehm, J. (2014). Indoor modelling benchmark for 3D geometry extraction, *International Archives of the Photogrammetry, Remote Sensing and Spatial Information Science* XL-5, 581–587. doi:10.5194/isprsarchives-XL-5–581.

Triggs, B. McLauchlan, P. Hartley, R. Fitzgibbon, A. (1999). *Bundle Adjustment – A Modern Synthesis. ICCV '99*. Proceedings of the International Workshop on Vision Algorithms. Springer-Verlag, pp. 298–372.

Ulaby, F. Moore, R. Fung, A. (1982). *Microwave Remote Sensing: Active and Passive. Volume 2 - Radar Remote Sensing and Surface Scattering and Emission Theory*. Addison-Wesley Publishing Co., Advanced Book Program, Reading, Massachusetts.

Wand, M. Berner, A. Bokeloh, M. Jenke, P. Fleck, A. Hoffmann, M. Maier, B. Staneker, D. Schilling, A. Seidel, P. (2008). Processing and interactive editing of huge point clouds from 3d scanners. *Computer Graphics* 3(2), 204–220.

Wang, C. Cho, Y. Kim, C. (2015). Automated BIM component extraction from point clouds of existing buildings for sustainability applications. *Automation in Construction* 56(1), 1–13.

Wilkinson, B. Mohamed, A. Dewitt, B Seedahmed, G. (2010). A novel approach to terrestrial LiDAR georeferencing. *Photogrammetric Engineering and Remote Sensing* 76(6), 1–8.

Xiong, X. Adan, A. Akinci, B. Huber, D. (2013). Automatic creation of semantically rich 3D building models from laser scanner data. *Automation Construction* 31, 325–337.

13 HBIM and environmental simulation

Possibilities and challenges

Husam Bakr Khalil

While BIM is usually associated with new buildings, HBIM is associated with historic buildings, many of which are still in use, particularly those that are less than a hundred years old, which constitutes a significant portion of the building stock in many European and Middle Eastern cities.

The majority of such buildings have low energy performance; green retrofitting them can considerably reduce energy consumption and increase their market value.

Promising benefits of HBIM include modelling of thermal properties of different building components, thus facilitating environmental simulation of different alternatives, making the success of green retrofitting more likely. Challenges include adapting BIM to model historic buildings, high modelling effort and cost, and the uncertainty associated with existing condition of historic buildings. A review of the state of ongoing research to overcome such challenges follows.

13.1 Introduction

The inadequacy of natural resources and the increase of CO_2 emissions instigated a global concern for reducing energy consumption and CO_2 emissions. Every country is required to reduce its energy consumption and emissions. The building sector is responsible for a considerable portion of the global energy consumption and CO_2 emissions. In Europe, it is estimated that buildings consume about 40% of the total energy consumption,[1] yet in the USA, the ratio is about 49% of the country's total energy consumption, and 75% of its entire electricity consumption.[2] Reducing energy consumption in the building sector is thus considered a priority.

Studies indicated that the majority of the existing building stock have low energy performance and are in need of environmental retrofitting. More countries are recognizing the significant energy savings that can be achieved if existing building stock can be properly retrofitted. WBCSD[3] estimated that utility bills can be decreased by 40% to 60% if deep retrofits of existing buildings are achieved; Alajmi[4] estimated that retrofitting of existing buildings for energy conservation that involves significant capital investment can save up to 49.3% of their annual energy consumption.

Many countries are taking measures in that direction. The European Commission is targeting an energy reduction of 20% by 2020, and 25% by 2030.[5] The International Energy Agency (2008)[6] is targeting a reduction of 77% in the global carbon footprint by 2050. The municipal authority of Changchun in northeast China has implemented a compulsory refurbishment program to refurbish nearly half a million homes with wall and roof insulation and installation of energy-efficient windows and doors since 2010.[7] In light of the above, retrofitting and renovation of existing buildings, including historic buildings, can no longer be considered as voluntary, and retrofitting is likely to become one of the most important areas of business in the near future.

Achieving energy retrofitting of the existing building stock is not an easy task, due to a number of factors, the most important of which is the large number of buildings involved, as it is estimated that energy auditing of commercial building stock of the USA would require at least 13 years to complete.[8] In China, it is estimated that 570 million square meters of buildings would be in need of energy retrofitting before the year 2020, costing about 1.5 trillion Chinese yuan (about 230 billion US dollars).[9] It became apparent that there is a pressing need for a faster retrofitting process if the targeted figures are to be met.

13.2 Environmental performance of historic buildings: An overview

Historic buildings constitute a considerable portion of the building stock in many countries. In Europe, historic buildings constitute about 26% of the entire European building stock. In the UK, historic houses constitute 39% of the total housing stock.[10] It is estimated that more than 40% of European residential buildings have been built before the 1960s.[11]

The problem with historic buildings is that the majority of them have poor environmental performance, consume much energy, and contribute a considerable amount of CO2 emissions. The reason is that the majority of these buildings are deteriorating, and that these buildings were built in a time when energy conservation regulations were limited. These buildings constitute a chance for considerable energy saving potential. In Germany, for example, buildings constructed before 1967 consume an average of between 225 and 250 kwh/m^2, whereas buildings constructed after 2010 consume only an average of 53 kwh/m^2, which means that on average, historic buildings consume four to five times more energy than newly constructed buildings.[12]

Widström[13] identified differences that distinguish historic buildings from modern buildings including complexity in geometry, lack of insulation, cultural values and demand for reversible, non-destructive and non-invasive measures, heterogeneity, and ventilation.

Although historic buildings are currently exempted from the obligation of energy efficiency in many countries, their energy consumption will have to be reduced if they are to compete with new buildings and to remain in use, as Filippi[14] claimed that refurbishment of privately owned historic buildings can produce a significant increase in their market value. The European Construction Technology Platform is targeting that 30% of the building stock be retrofitted by 2030, and that the whole building stock be retrofitted by 2050,[15] implying that the entire stock of historic buildings in use will have to be energy efficient before 2050.

Although retrofitting options for historic buildings may seem limited compared to non-historic existing buildings due to restrictions that apply to intervention into historic buildings, many possible actions can be taken to improve their energy efficiency. Troi[16] claimed that an average reduction of 25% of the historic buildings' energy demand can be achieved by using the current available techniques and solutions. Troi and Bastian[17] claimed that a reduction by a factor of 4 to 10 of energy demand is feasible in historic buildings while still respecting their heritage value. BPIE[18] stated that minor to moderate retrofitting measures might be feasible in the case of heritage buildings. In a study involving 16 retrofitted historic buildings, Balson et al.[19] found that refurbished heritage buildings have achieved higher BREEAM ratings than non-heritage buildings, and in some cases, they even outperformed new buildings.

To encourage sustainable historic building retrofitting, Italy drafted a LEED version for historic buildings. Balson et al.[20] provided an overview of ways to achieve BREEAM ratings for historic buildings, based on analysis of 16 cases of historic buildings that had acquired BREEAM, and the outcomes of the study were used to make the BREEAM Refurbishment and Fit-Out scheme 2014 more appropriate for historic buildings.

The above indicated that considerable reduction in energy consumption could be achieved by retrofitting of historic buildings, and that the domain of historic building retrofitting is a candidate for future expansion.

13.3 Obstacles that face historic buildings retrofitting

To date, retrofitting of historic buildings for energy efficiency is sporadic due to a number of factors:

13.3.1 Historic buildings' value

Due to their historic and cultural value, and as a reaction to instances in which historic buildings retrofitting has been performed in ways incompatible with their conservation,[21] current regulations constrain retrofitting actions that would affect their architectural and historic value. The EU directive of 2010 exempted protected historic buildings and buildings of special architectural or historical merit from compliance with certain minimum energy performance requirements, if such compliance would unacceptably alter their character or appearance.[22] In practice, a distinction should be made between listed and non-listed buildings, as retrofitting measures for listed buildings and buildings of special merits should be reversible and non-destructive, which limits the scope of available choices of retrofitting strategies. Retrofitting alternatives for historic buildings are generally limited compared to existing buildings. Alternatives may include envelope air tightening, window isolation, repair of thermal bridges, reduction of mould growth, utilization of efficient heating and cooling systems, and proper utilization of daylighting.

13.3.2 Complexity of historic buildings

Historic buildings are relatively more geometrically complex than other existing buildings, as they are full of details, ornaments, projections, and recesses. They sometimes include elements such as domes, vaults, niches, irregular surfaces, non-parallel walls, and walls of varying thicknesses, which make both the modelling and environmental simulation process quite difficult, using the available technology. The decision has to be taken as to the degree of model complexity or simplicity that is suitable for the retrofitting purposes, as such a decision would affect the predictability of future energy performance.

13.3.3 Uncertainty of building condition

There is a great deal of uncertainty associated with historic buildings with respect to the conditions of building materials, fixtures, and building systems, due to deterioration and damage that might have altered material properties in one location differently than the same material in another location within the same building. To reduce the degree of uncertainty and to increase simulation accuracy, it is necessary to utilize non-destructive inspection techniques including material and texture-based recognition, radars, radiography, magnetic particle inspection, sonars, or electromagnetic waves.[23] Such inspections are usually performed by experts, which considerably increases the expense of retrofitting and makes it inaccessible to most small-sized historic buildings.

13.3.4 The inefficient retrofitting process

The current retrofitting process is expensive, time-consuming, and exhausting. Ma et al.[24] identified five major phases in a typical sustainable building retrofit program: pre-retrofit survey, energy audit and performance assessment, identification of retrofit options, site implementation and commissioning, and validation and verification of energy savings. The retrofitting process needs simplification if it is to become practical.

13.3.5 Shortage of data

Data collection is a prerequisite for any historic building retrofitting. Data required include accurate as-built drawings, information related to building materials and their properties, installations, weather data, energy consumption, and so forth. Information is sometimes difficult to gather and, if available, may require expensive data acquisition techniques – for example, obtaining accurate as-built drawings requires advanced, complex, usually expensive, and time-consuming surveying and modelling techniques. This is by far the most deterring obstacle that faces historic building retrofitting.

13.3.6 Inconsistent simulation applications

Studies indicated lack of consistency between different environmental simulation applications. A recent study by Schwartz and Raslan[25] investigated the extent of thermal load estimation differences between three widely used environmental simulation tools, namely Tas, EnergyPlus, and IES. They found considerable estimation differences of up to 35% between different applications, which they attributed to algorithm differences, differences in the required input data, or human error. This raises doubts concerning the ability of available simulation applications to credibly and consistently measure energy consumption.

For historic building retrofitting to become common practice, the above-mentioned obstacles need to be minimized and new technologies need to be introduced to make the retrofitting more rewarding. Advancement of energy simulation technologies and the utilization of BIM promise to achieve this goal. Each of these two fields is individually beneficial to historic building retrofitting, yet their successful integration would boost the practice of historic building retrofitting. A review of the state of these two fields and their merits, shortcomings, and constraints that limit their integration is presented in the following sections.

13.4 HBIM and energy simulation

HBIM and environmental simulation can play a pivotal role in facilitating historic buildings retrofitting. Combining BIM and energy simulation can result in efficient retrofitting process, as emerging technologies promise to make such a process more robust. Ahn et al.[26] stated that BIM-based energy simulation can significantly reduce the costs and time required for geometric modelling.

13.4.1 Benefit of integrating HBIM and environmental simulation for historic buildings retrofitting

The following is an account of how combining BIM and environmental simulation can benefit historic buildings retrofitting.

- HBIM models are intelligent parametric models that consist of building components such as walls, windows, doors, floors, roof, and so forth. These components have both geometric and descriptive data assigned to them, including thermal properties such as heat transfer coefficients (U), thermal resistance (R), thermal mass, absorption, roughness, solar heat gain coefficient, and light transmittance. This integration reduces the need for manual data entry in environmental simulation programs.
- Once an HBIM model is created, different types of environmental simulations can be performed, including energy consumption, airflow, natural lighting, assessing passive strategies to

reduce the need for mechanical systems, and assessing the effectiveness of renewable energy alternatives.
- Most BIM applications have built-in environmental simulation capabilities, which facilitate seamless and quick assessment of several alternatives without the need to export BIM models to external simulation applications, and this significantly reduces the effort and time needed to perform various environmental simulation tasks. In Autodesk Revit, for example, a single window type has as many as 120 alternative material combinations with different thermal properties and cost to choose from. Using internal simulation capabilities, different alternative combinations of window materials can be rapidly assessed and the optimum window materials would be easily decided.
- In many cases, reliance on internal BIM simulation capabilities might not be sufficient, and BIM models will have to be exported to stand-alone simulation applications. Interoperability is among the most important benefits of BIM, as it facilitates exporting BIM models to dedicated stand-alone simulation packages, using IFC standard interoperable data format or gbXML exchange file format. The required information can be extracted from BIM models to numerous simulation applications to perform detailed environmental analyses including energy simulation, daylighting analysis, solar gain, and photovoltaic (PV), which facilitate more accurate, detailed testing and comparison of different retrofit alternatives.
- HBIM provides a medium for collaboration between different professionals, which is necessary for the production of highly energy-efficient buildings, and for compliance with outcome-based design codes.

13.4.2 Challenges and limitations of HBIM

Despite the growing interest in integrating BIM and energy modelling, most energy modelling is currently done apart from BIM because there are still significant limitations in BIM energy modelling capabilities.[27] The following is a discussion of these challenges and the approaches to overcome them.

Achieving an accurate BIM model that adequately represents the reality is quite difficult due to several reasons, some of which were mentioned as retrofitting challenges, such as historic building complexity, lack of as-built drawings, and uncertainty of building conditions. In addition, there is a challenge related to the nature of existing BIM applications. Approaches to overcome these challenges are discussed in the next section.

Currently, several BIM applications are available, including but not limited to Autodesk Revit Architecture, Nemetschek Graphisoft ArchiCAD, Nemetschek Allplan Architecture, Nemetschek Vectorworks Architect, Gehry Technologies Digital Project Designer, Bentley Architecture, IDEA Architectural Design (IntelliCAD), CADSoft Envisioneer, Softtech Spirit, RhinoBIM (BETA), DataCAD, and Tekla BIMsight. Most of these applications, however, were built mainly for designing contemporary buildings, and are not well suited for historic buildings. Current BIM applications are not designed to deal with the asymmetric and geometrically irregular elements of many historic buildings. For example, walls in current BIM applications have fixed wall thickness; modelling of walls of variable thickness would be difficult. Moreover, current BIM families are mainly modern, and there is a lack of families of historic building elements.[28]

Fortunately, BIM developers have recognized the potential of the retrofitting sector and are developing BIM applications that would suit existing buildings retrofitting, which would improve the accuracy of BIM retrofit models, reduce time and cost of the process, and, hopefully, would integrate energy modelling and BIM seamlessly.[29]

Recent versions of BIM applications are more suited for retrofitting of existing buildings. For example, the latest version of ArchiCAD included a renovation, refurbishment, and retrofitting workflow, and Autodesk introduced a "retro BIM" workflow. However, these versions are mainly for designed for retrofitting of contemporary buildings. To date, no specific BIM applications have yet been developed that are dedicated to historic building retrofitting.

13.4.2.1 The HBIM modelling process

As mentioned earlier, the lack of accurate as-built drawing of historic buildings and their geometric complexity necessitate reliance on advanced surveying and modelling techniques that are expensive and time-consuming. This challenge is probably the most deterring obstacle that hinders historic building retrofitting.

Volk et al.[30] claimed that the scarcity of BIM application in existing buildings is due to the high effort needed for data capturing, conversion, and modelling, as well as the difficulty of information updating, and uncertain data handling. Availability of a reliable data capturing technique that could provide as-built BIM models at reasonable time and cost is a prerequisite for existing buildings to benefit from BIM.

Troi and Bastian[31] conducted the 3ENCULT project aiming to develop methods and tools to support a holistic approach for adapting existing retrofit solutions of historic buildings. Eight buildings were selected to demonstrate and verify the approach, representing different building functions, construction materials, historic eras, and climatic zones. Out of eight cases, an HBIM model was used for documentation purposes in only two cases. Digital models were used in three of the cases. Partial digital models of small components were used in some cases such as modelling of a lab for lighting simulation.

Creating an as-built BIM model of historic buildings from scratch is an extensive process, involving three main steps: acquisition, segmentation, and modelling. This process is sometimes referred to as reverse engineering, or scan-to-BIM.[32] The modelling process is discussed in the next section, and approaches to overcome the obstacles are reviewed.

A DATA ACQUISITION

Data acquisition is the stage where data about building geometry is collected by various surveying techniques. Volk et al.[33] identified the following data capturing techniques: manual techniques, image-based techniques, and range-based techniques.

1. Manual techniques include tape measures and calipers. Manual techniques may be adequate for simple buildings with few details, but may be time-consuming and require a considerable amount of effort and are not suitable for complex buildings.
2. Image-based or photogrammetric techniques depend mainly on images taken from different locations, resulting in point cloud and color information. Although photogrammetry is much simpler, faster, and less expensive than laser scanning, it has rarely been used as a main acquisition technique because of its relative lack of accuracy; it is usually used as a complementary technique for laser scanning. Future development may make photogrammetry more reliable and more accurate.
3. Range-based techniques include laser scanning, which is an acquisition technique in which a laser beam is projected onto the surface to be scanned, producing highly dense point clouds. Laser scanning is the most widely used scanning technique, mainly because of its high degree of accuracy. However, this accuracy is not without cost, as laser scanning is the most expensive

and time-consuming technique. To date, laser scanning point cloud files are quite large, as they consist of millions of points. Best results can be achieved by combining laser scanned point cloud with photos. There is a need for technologies to speed up the surveying process and make point cloud data simpler and more readily available, if laser scanning is to become practical for retrofitting purposes.

Selecting a suitable data acquisition technique depends on the degree of accuracy needed and the purpose of the HBIM model. For environmental simulation purposes, there may be situations where accurate and complete HBIM models may be needed, but for the majority of environmental simulation purposes, much simpler HBIM models may be adequate, or even preferred, as the majority of environmental simulation packages can only handle abstracted building models to be able to efficiently perform the required analysis. A white paper by Autodesk[34] claimed that a simple BIM model, produced in about two hours, could produce a high level of building performance predictability and attain reliable results that can assist in decision making with respect to approaches to follow for reducing annual energy consumption. Laser scanners deliver high levels of geometric detail, which may be too detailed and unnecessarily expensive for energy simulation purposes, particularly for small buildings.

B SEGMENTATION

Segmentation is the process of structuring the point cloud into sub parts and removing the unnecessary data from it. Segmentation may be conducted manually, semi-automatically, or automatically. Segmentation is not a straightforward process, as it requires complex algorithms and expertise to perform.[35] Current research focuses on simplifying this process and on making it fully automated and reliable.

C MODELLING

Modelling is the process of creating an as-built BIM model from the acquired data. The process involves three steps: the geometrical modelling of the component, the attribution of categories and material properties to the components, and, finally, establishing of relations between components.[36] Current BIM software does not have automatic element recognition capabilities, and constructing a BIM model has to be done manually. One of the problems that faces creating BIM models of historic buildings is lack of a library of parametric historic objects. Existence of such objects would facilitate automation of 3D object mapping. Arayici and Tah[37] emphasized the necessity of fast data capturing, automation of data processing, and pattern recognition to facilitate BIM modelling for existing structures to be suitable for environmental analysis and simulation.

Important research in this area is currently ongoing at Dublin Institute of Technology (DIT). They have created a BIM library of parametric objects of historic buildings, including Victorian and Georgian windows and classical Ionic, Doric, or Corinthian columns. They have also developed software that is capable of mapping the complex shapes of historic building details onto point clouds to speed up the BIM process.[38]

13.4.3 Recent developments

Currently, the scan-to-BIM is a barrier that hinders utilizing HBIM for historic building retrofitting, due to its complexity and amount of effort involved. UCL[39] evaluated the thermal performance of a Greenwich historic building. The evaluation process involved several steps including data collection, laser scanning, 3D BIM modelling, and environmental simulation. The process took 10 weeks.

For HBIM to become practical and widespread in historic building refurbishment, cheaper, faster, more reliable, and less complicated technologies have to be conceived.

There are, however, emerging third-party applications that include element recognition such as IMAGINiT, a scan-to-BIM add-on that works with Autodesk Revit and provides limited automation capabilities of architectural elements recognition such as walls and columns, and some mechanical, engineering, and plumbing (MEP) elements such as pipes, ducts, and conduits. EdgeWise BIM is a similar product by ClearEdge, with similar capabilities, but can extract more elements including structural elements, walls, windows, doors, pipes, conduits, and ducts automatically and transform them into Revit family objects. The software recognizes repeated items in point clouds and extracts them into the correct Revit family.

It is expected that within a few years the process would be fully automated, as current research focuses on simplifying the process. New software and scanning technologies are being developed that would make the production of accurate HBIM models much easier, cheaper, and faster. Lynn Bellenger, the current ASHRAE president, envisioned that by 2025, simulation tools would be robust enough to be able to assess various aspects of environmental performance, and that BIM models will seamlessly interface with energy models in a standardized format.[40]

Another step in the direction of simplifying the laser scanning process is that reference targets, which are usually distributed throughout a building to be automatically recognized by the software, are no longer needed. Newer applications, such as Autodesk's ReCap, are capable of aligning the survey data without the need for targets, which saves time on-site and allows on-site checking of the captured data on a laptop. ReCap is also capable of producing 3D models from point clouds, regular cameras, smartphones, or from a camera on a flying drone. The ReCap system was recently used to create a highly detailed 3D model of the historic Teatro Lirico in about four days.

Recently, Autodesk issued a beta version of a new software program called Memento, which can produce dense and accurate 3D meshes from photos or scans. It can export the resulting meshes to the ReCap software. It is, however, not yet compatible with BIM applications.

Another important development is that introduced by Bentley, which developed software that allows streaming of smaller portions of data directly into BIM platforms, instead of having to stream a point cloud of the entire building at once. This, if successful, would make the laser scanning technology handy, more powerful, and faster than ever before. It can run on mobile platforms, providing an incredible tool for working on retrofit projects.

13.5 Integrating environmental simulation and BIM

Environmental simulation is a rapidly evolving area of expertise that became necessary for any building that seeks compliance with current building directives, and is indispensable to historic building retrofitting.

There are currently many environmental simulation applications that can exchange BIM models as either gbXML files or IFC files, including Autodesk Green Building Studio, Autodesk Insight 360, Graphisoft EcoDesigner, IES Solutions Virtual Environment VE-Pro, Bentley Tas Simulator, Bentley Hevacomp, Design Builder, Sefaira, AECOsim, gEnergy, and Simergy.

13.5.1 Benefits of using environmental simulation

An AIA report mentioned a number of benefits of using environmental simulation at different stages of a building's life cycle, some of which are related to historic building retrofitting:[41]

1   Simulation provides numeric data about various aspects of buildings' environmental performance, which facilitate informed decisions about proper interventions.

2   Building codes are becoming more performance-based, as opposed to prescriptive-based, which necessitates the use of environmental simulation to quantify reduction of energy consumption.
3   Simulation facilitates dealing with complex phenomena, including estimating energy use of historic buildings, which involve several interrelating factors.
4   Simulation facilitates fast evaluation of various design alternatives and thus helps optimize building performance.
5   Simulation would help reducing initial, operational, and maintenance cost.
6   Simulation helps qualify buildings to receive energy efficiency financial incentives, by providing evidence of reduced energy consumption.

13.5.2 Limitations of environmental simulation

A number of limitations and challenges hinder utilization of environmental simulation in historic building retrofitting:

13.5.2.1 Simulation accuracy

Simulation accuracy means that results should resemble actual building performance and energy use data, under the same conditions.[42] Current research focuses on improving consistency of building energy outcomes. The lack of consistency between environmental simulation applications resulted in lack of accuracy, as Schwartz and Raslan[43] found estimation differences between different simulation applications of up to 35%. This might be due to a number of factors, including differences in the required input data, building complexity, prevalence of large numbers of interacting variables, difficulty of predicting occupants' behavior, weather change, the uncertainty of building condition, imperfect algorithms, lack of standardization, and human error. There is also a need for development and adjustment of simulation software to accurately represent the conditions of heritage buildings.

Accurate environmental simulations require a relatively accurate and complete BIM model, in terms of input data and building geometry. Simulation accuracy can be improved by complete and accurate modelling of different building components, including identification of building elements that cannot be subjected to invasive interventions, building layers, the materials of all surfaces, the thermal bridges and zones, the air infiltrations, condensation, and finally the voids and spaces that can be used for mechanical and electrical installations.[44]

Accuracy can also be improved by rigorous measurement and calibration of energy use before and after implementation.

a   **Measurement:** Measurement involves data collection on utility bills, use patterns, hours of operation, functioning of systems, and actual weather conditions of the existing building. Heikkinen et al.[45] estimated that between 2% and 5% savings on the total project cost could be achieved with the better capture of existing conditions.
b   **Calibration:** Calibration can also help improve simulation accuracy. Calibration is the process of comparing actual energy use with simulated estimates in order to adjust the simulation model and to determine causes of discrepancies between the two.[46] However, the above measures require a considerable amount of time, effort, and cost.
c   **Selecting a suitable simulation application:** Selecting suitable simulation software is a crucial decision that would affect simulation accuracy, yet it is not as easy as it may look, with more than 400 environmental simulation software applications available in the market.[47] Understanding the pros and cons of each would help in selecting the package that is more suitable for the

task. Some applications have difficulty in dealing with portions of buildings below grade; others do not have adequate light modelling capabilities; some others lack the capability of modelling natural ventilation.

13.5.2.2 Interoperability

Interoperability – the ability to exchange data between different applications – is still an issue. Although Industry Foundation Classes (IFC) and XML schemes have been introduced to resolve the interoperability problem, both of them still have limitations. Ahn et al.[48] stated that current IFC does not support conversion of irregular geometry such as circular, inclined, or curved plane of walls or roofs. Senave and Boeykens[49] investigated the relationship between BIM and energy simulation packages, and concluded that transferring BIM models to external simulation packages is complex, cumbersome, time-consuming, and prone to errors and misinterpretations. The process was not entirely automatic, as a number of steps had to be carried out manually. Current research focuses on improving export from BIM applications, standardizing the derivation of analysis geometry from IFC.[50] Future software should facilitate seamless transfer of building geometry and data from the design software to analysis software and vice versa. Ideally, it should be possible to create a single model that can be used on multiple software platforms in order to do various performance analyses. This is what HBIM is all about: creating a central model that integrates all sorts of geometrical and textual data from different domains, including properties of building elements and mechanical systems, making it possible to efficiently handle almost any level of information complexity and analysis.

13.5.2.3 Buildings' complexity

Historic buildings' complexity restricts modelling and environmental simulation. The available simulation technology is incapable of dealing with complex buildings. Since current environmental simulation applications can only deal with simplified geometric models, a decision has to be taken as to the degree of model complexity that is suitable for the retrofitting purposes, as there is no point to opt for a highly accurate yet expensive model that will eventually be abstracted to perform environmental simulation. This, however, is likely to change in the near future, as more capable HBIM application and environmental simulation software is currently being developed.

13.5.3 Selecting an environmental simulation application

The following considerations should be taken into account while deciding the software to use in a historic building retrofitting.

- **Retrofitting objectives:** Selection of simulation application is affected by building use, historic value, building condition, and retrofitting objectives. In residential historic buildings, reducing energy consumption may be the ultimate goal, necessitating a certain type of simulation to optimize energy consumption, whereas in office buildings, utilization of daylighting may be the ultimate goal. Some historic buildings that contain sensitive artwork require a specific microclimate to avoid sudden environmental changes that might damage the artwork in the long run, which necessitates the use of computational fluid dynamics (CFD) simulation.
- **Robustness:** The calculation engine should be robust enough to measure building performance and energy use for most traditional systems and components, as well as for low-energy systems and components. Some software may fall short of modelling some aspects of building

performance, such as passive design elements and life-cycle cost, yet the majority cannot adequately deal with complex geometry.[51]

- **Comprehensiveness:** It is easier to use an energy modelling tool that can model different environmental aspects rather than using several tools, each for a specific measurement.
- **Clear graphic output:** The application should represent performance results in an easily understandable graphic format.
- **Within BIM versus stand-alone:** Within BIM environmental simulation is much easier, yet it lacks the sophistication, depth, and level of details needed for many tasks. Autodesk Revit, for example, has internal energy analysis capabilities, linked to their cloud-based A360 drive, which is intended to provide insight into potential building energy use during early design stages. The results of such analysis are quite simple. However, exporting and importing to and from simulation packages lacks the convenience of within application analysis.

13.5.4 Recent developments

Environmental simulation software is becoming more compatible with BIM and with the way architects and retrofit specialists work. The exchange standards are still not perfect, but the ongoing development efforts promise to make the exchange of BIM models with external applications seamless in the near future.

As for within BIM simulation, a number of BIM energy modelling plug-ins were recently introduced, including Green Building Studio, Sefaira Architecture, and IESVE. Simulation capabilities of these plug-ins are improving.

Autodesk is developing an altogether new modelling workflow, which they call Rapid Energy Modelling (REM), to simplify the existing building retrofitting process. The process involves minimal basic data about the building, construction of a simplified BIM model based on photograph or satellite images of the building, and simplified energy simulation using Green Building Studio. A non-expert, in a matter of hours, can perform the process.[52] The US Department of Defense and Autodesk cooperated to estimate energy usage and identify potential energy savings for multiple buildings using REM; the process was 90% faster and 95% less expensive than an average audit. Excluding barracks, accuracy of predictions was 84% for electricity and 80% for natural gas on average.[53] If adopted at a wide scale, such an approach may significantly increase the number of existing buildings that undergo energy retrofitting, due to its low cost and speed. Research, however, is needed to test whether or not REM is suitable for historic buildings retrofitting.

13.6 Conclusion

A considerable reduction in energy consumption can be achieved by historic building retrofitting. However, retrofitting of historic buildings for energy efficiency is rare due to restrictions related to historic buildings' value, complexity, uncertainty of building condition, inefficient retrofitting process, shortage of data, and inconsistent simulation applications. Combining BIM and energy simulation can result in an efficient retrofitting process. HBIM models consist of intelligent components that have both geometric and descriptive data that facilitate performance of different types of environmental simulations, either within a BIM application or via stand-alone simulation applications. Simulations facilitate fast evaluation of retrofitting alternatives. The most challenging aspect that hinders HBIM utilization is the impractical as-built BIM process. However, new software and scanning technologies are being developed that would make the production of accurate HBIM models much easier, cheaper, and faster. Another concern is simulation accuracy, which necessitates improvement of simulation applications. Interoperability is still a problem that limits integration of HBIM and environmental simulation; however, research is ongoing to overcome that obstacle.

It is hoped that in the near future, utilization of HBIM in retrofitting of historic buildings will become a standard practice, and that HBIM models will seamlessly integrate with environmental simulation applications.

Notes

1 Buildings Performance Institute Europe (BPIE) (2011): *Europe's Buildings under the Microscope*, Buildings Performance Institute, Europe.
2 Fulton & Grady (2012): *United States Building Energy Efficiency Retrofits: Market Sizing and Financing Models*. The Rockefeller Foundation & Deutsche Bank.
3 World Business Council on Sustainable Development (2009): *Transforming the Market: Energy Efficiency in Buildings*. World Business Council for Sustainable Development. WBCSD.
4 A. Alajmi (2012): Energy audit of an educational building in a hot summer climate, *Energy and Buildings* 47 (2012) 122–130.
5 European Commission (2014): *Energy Efficiency and Its Contribution to Energy Security and the 2030 Framework for Climate and Energy Policy*. A EURELECTRIC position paper. EURELECTRIC.
6 International Energy Agency (2008): *Energy Technology Perspectives*. Scenarios & strategies to 2050. IEA.
7 Lo Kevin (2015): *The "Warm Houses" Program: Insulating Existing Buildings through Compulsory Retrofits*. Sustainable Energy Technologies and Assessments. 9, 2015, pp. 63–67.
8 Autodesk (2009): *Rapid Energy Modeling for Existing Buildings: Testing the Business and Environmental Potential through an Experiment at Autodesk*. Autodesk.
9 Tsinghua Building Energy Research Centre (THUBERC) (2010): *Annual Report on China Building Energy Efficiency*, THUBERC, Beijing, China.
10 Alexandra Troi (2011): Historic buildings and city centres – the potential impact of conservation compatible energy refurbishment on climate protection and living conditions. *Energy Management in Cultural Heritage*, 6.-8.4.2011, Dubrovnik, Croatia, UNDP Croatia.
11 Buildings Performance Institute Europe, BPIE (2011): *Europe's Buildings under the Microscope. A country-by-country review of the energy performance of buildings*. BPIE.
12 Ibid.
13 Torun Widström (2012): Enhanced energy efficiency and preservation of historic buildings – methods and tools for modelling. A Licentiate Thesis. School of Architecture and the Built Environment, KTH Royal Institute of Technology, Stockholm, Sweden.
14 Marco Filippi (2015): Remarks on the green retrofitting of historic buildings in Italy. *Energy and Buildings* 95 (2015) 15–22.
15 Alexandra Troi (2011): Historic buildings and city centres – the potential impact of conservation compatible energy refurbishment on climate protection and living conditions. Energy Management in Cultural Heritage, 6.-8.4.2011, Dubrovnik, Croatia, UNDP Croatia.
16 Ibid.
17 Alexandra Troi and Zeno Bastian (2015): *Energy Efficiency Solutions for Historic Buildings, a Handbook*. BIRKHÄUSER.
18 Buildings Performance Institute Europe, *Europe's Buildings under the Microscope*.
19 Kiruthiga Balson, Gavin Summerson and Andrew Thorne (2014): *Sustainable Refurbishment of Heritage Buildings – How BREEAM Helps to Deliver*. Briefing paper. BRE Global Ltd.
20 Ibid.
21 Enrico Genova, Giovanni Fatta and Tor Broström (2015): Categorization of the Historic Architecture in Palermo for the Purpose of Energy Assessment. CISBAT 2015, September 9–11, 2015, Lausanne, Switzerland.
22 Directive 2010/31/EU of the European Parliament and of the Council of 19 May 2010 on the energy performance of buildings. *Official Journal of the European Union*.
23 Rebekka Volk, Julian Stengel and Frank Schultmann (2014): Building information modeling (BIM) for existing buildings – literature review and future needs. *Automation in Construction* 38 (2014) 109–127.
24 Zhenjun Ma, Paul Cooper, Daniel Daly and Laia Ledo (2012): Existing building retrofits: Methodology and state-of-the-art. *Energy and Buildings* 55 (2012) 889–902.
25 Yair Schwartz and Rokia Raslan (2013): Variations in results of building energy simulation tools, and their impact on BREEAM and LEED ratings: a case study. *Energy and Buildings* 62 (2013) 350–359.

26 Ki-Uhn Ahn, Young-Jin Kim, Cheol-Soo Park, Inhan Kim and Keonho Lee (2014): BIM interface for full vs. semi-automated building energy simulation. *Energy and Buildings* 68 (2014) 671–678.
27 Ibid.
28 Stefano Cursi, Davide Simeone and Ilaria Toldo (2015): A Semantic Web Approach for Built Heritage Representation. G. Celani et al. (Eds.): The Next City: 16th International Conference *CAAD Futures 2015*, pp. 383–401, Biblioteca Central Cesar Lattes 2015.
29 The AIA Sustainability Discussion Group (2007): *50 to 50*. American Institute of Architects (AIA).
30 Volk, Stengel and Schultmann (2014): Building information modeling (BIM) for existing buildings—Literature review and future needs. *Automation in Construction* 38 (2014) 109–127.
31 Troi and Bastian, *Energy Efficiency Solutions for Historic Buildings, a Handbook*. BIRKHÄUSER.
32 N. Hichri, C. Stefani, L. De Luca and P. Veron (2013): Review of the "As-Built BIM" Approaches. International Archives of the Photogrammetry, Remote Sensing and Spatial Information Sciences, Volume XL-5/W1, 2013 3D-ARCH 2013–3D Virtual Reconstruction and Visualization of Complex Architectures, 25–26 February 2013, Trento, Italy.
33 Volk, Stengel and Schultmann (2014): Building information modeling (BIM) for existing buildings—Literature review and future needs. *Automation in Construction* 38 (2014) 109–127.
34 Autodesk white paper (2008): *Creating Models for Performance Analysis on Existing Buildings*. Autodesk.
35 Hichri, Stefani, De Luca and Veron (2013): Review of the "As-Built BIM" approaches. *ISPRS - International Archives of the Photogrammetry, Remote Sensing and Spatial Information Sciences* Volume XL-5/W1 (2013) 107–112.
36 Ibid.
37 Y. Arayici and J. Tah (2008): Towards building information modelling for existing structures. *Structural Survey* 26:3 (2008) 210–222.
38 Stephen Cousins (2014): *The BIM Journey . . . Destination Retrofit Sector*. In Construction Manager, March 2014. Website of the Chartered Institute of Building CIOB. http://www.bimplus.co.uk/technology/bim-journey-destination-retrofit/.
39 UCL Department of Civil, Environmental & Geomatic Engineering (2012): *Green BIM for UCL's Chadwick Building*. UCL Department of Civil, Environmental & Geomatic Engineering.
40 Rocky Mountain Institute (2011): *Building Energy Modelling Innovation Summit: Collaborate and Capitalize: Post-Report from the BEM Innovation Summit*. Rocky Mountain Institute.
41 The American Institute of Architects (2012): *An Architect's Guide to Integrating Energy Modeling in the Design Process*. AIA.
42 Ibid.
43 Schwartz and Raslan, Variations in results of building energy simulation tools, and their impact on BREEAM and LEED ratings.
44 Filippi, Remarks on the green retrofitting of historic buildings in Italy.
45 Pekka Heikkinen, Hermann Kaufmann, Stefan Winter and Knut Einar Larsen (2009): *TES Energy Façade Prefabricated Timber based Building System for Improving the Energy Efficiency of the Building Envelope*. Research project from 2008–2009, Woodwisdom Net.
46 Ibid.
47 Energy Efficiency and Renewable Energy (EERE): *Building Energy Software Tools Directory*, US Department of Energy. Available online: http://apps1.eere.energy.gov/buildings/tools_directory/
48 Ahn, Kim, Park, Kim and Lee, BIM interface for full vs. semi-automated building energy simulation.
49 M. Senave and S. Boeykens (2015): Link between BIM and energy simulation. Building Information Modelling (BIM) in Design, Construction and Operations. *WIT Transactions on the Built Environment*, 149 (2015). WIT Press.
50 The American Institute of Architects (2012): *An Architect's Guide to Integrating Energy Modeling in the Design Process*. AIA.
51 W. Ko (2014): Complex Geometry Facades in Building Energy Simulations and Standards. ASHRAE/IBPSA-USA Building Simulation Conference Atlanta, GA September 10–12, 2014.
52 Autodesk, *Rapid Energy Modeling for Existing Buildings*.
53 Jennifer Rupnow and David Scheer (2014): *Rapid Energy Modeling and Range Analysis, the Value of a Highly Educated Guess*. ACEEE.

14 Green BIM in heritage building

Integrating Building Energy Models (BEM) with Building Information Modeling (BIM) for sustainable retrofit of heritage buildings

Laila Khodeir

14.1 Introduction

What is Building Information Modeling (BIM)? How and why did BIM emerge? What problems does it tackle? How has BIM recently developed into other terms? What is Building Energy Modeling (BEM)? What is the emerging concept of Green BIM? How could these emerging concepts be implemented on heritage buildings conservation? What are the expected benefits and value added?

Although working on isolated islands has been a natural tendency in most construction projects around the world, the role BIM plays in this discipline is quite evident, as BIM is generally an integrated project design and delivery tool and management mechanism that allows collaboration between all stakeholders involved in the supply chain of the building. It was originally adopted to solve the issue of lack of collaboration among different disciplines and ensure they all work flawlessly. Recently, the term BIM has given rise to some other terms, like Urban Information Modeling (UIM), Existing Building Information Modeling (EBIM), Historical/Heritage Building Information Modeling (HBIM) and Green BIM. This trend has emerged as a trial to broaden the application of BIM to different types of construction projects (railway stations, bridges, etc.), in addition to landscape and urban projects. Consequently, the term HBIM emerged to express an integrated heritage building's delivery and management mechanism that incorporates all stakeholders with the objective of conserving the cultural sustainability of built heritage during their lifetime.

The power of BIM or HBIM in general is represented in their ability to integrate physical information representing building components (walls, floors, etc.) flows with layers of information including energy performance, daylighting, and so forth. Thus the integration between HBIM with BEM, entitled "the Green BIM", into one seamless process, although it encounters a number of challenges, is considered of great value. This could be made obvious especially when choosing to conserve heritage buildings through "sustainable retrofitting", which is a conservation approach that focuses on upgrading the systems and building services through sustainable solutions or, in other words, bringing the heritage building "back to life".

This chapter focuses on the potentials of adopting the Green Building Information Modeling (BIM) double paradigm through the process of sustainable retrofitting of heritage buildings. This approach aims at preserving these significant buildings from being obsolete under the rigorous building energy efficient regulations being currently introduced. Although the "Green Building Energy Models" (BEM) offer valuable insight into the use of energy in different types of buildings, the emerging term "Green BIM" refers to the integration of Building Energy Models (BEM) in correspondence with Building Information Modeling (BIM). This integration aims at achieving better support to likelihood of the growth of green projects. BEM mainly simulates the energy use of the building throughout an entire year of operation and performs a detailed analysis that outputs space cooling and heating loads, daylighting impacts, equipment energy use, resource consumption, energy costs and other performance-related parameters. On the other hand, BIM helps simulate the building elements and characteristics, such as materials, weight, thermal resistance and other physical properties that contribute to building performance.

Upon assessing heritage buildings from the environmental perspective, the main problem that faces these buildings is coping with energy policies including minimum energy requirements, especially in the case of retrofitting. Applying sustainable retrofit processes has the advantage of improving the energy efficiency, enhancing the overall building's condition and, as a consequence, increasing the building's value. However, this process should be supported with the application of the Green BIM tool in order to achieve sustainable refurbishment for heritage buildings.

In fact, the analogy between the case of Egypt, when applying BIM on new buildings or HBIM on heritage buildings, and other countries, especially in the Western world, is quite illogical. This is due to the lack of definite codes of practices, standards, guidelines or even sufficient and available technology. Nevertheless, similar standards should be generated in Egypt in order to be implemented within the Egyptian context.

Recently, a sustainable retrofitting process has been spreading worldwide with the utmost objective of achieving certification of all built environment, so as to be able to efficiently manage buildings, preserve the existing values in them and to extend their useful lifetime. Nevertheless, applying sustainable retrofit on built heritage has faced a lot of criticism, from the perception that it might negatively affect the historical and cultural value of the building. Thus it is recommended that the application of such an approach on heritage buildings, despite its advantages, should be based on a specific decision-making process that encompasses the value of the building, the legislations and laws that govern the way of treating the building and the phases of implementation of sustainable retrofitting with respect to the general project phases. The suggested process will highlight the role the HBIM plays, not only in the registration and documentation of the building, but also in achieving a better decision-making environment that calls for the collaboration of all stakeholders involved in the conservation of heritage buildings.

14.2 Setting the scene of heritage buildings within the Egyptian context

What are the values embedded in built heritage? Why does the built heritage in Egypt represent a unique case? What are the factors affecting such uniqueness? What are the laws and regulations that organize the methods of dealing with built heritage within the Egyptian context?

Built heritage generally represents an important constituent of cultural heritage. This is due to the role it plays in representing societal development along different historical eras, where cultural heritage represents inherited resources from the past that include all aspects of the environment resulting from the interaction between people and places through time (Dümcke, 2013).[1] The case of built heritage within the context of Egypt is quite unique in a number of ways. The following part discusses the factors affecting this uniqueness, including values embedded in them, the applied conservation strategies on these buildings, in addition to detecting legislative issues organizing their classification.

14.2.1 Values embedded in heritage buildings

Heritage buildings include different types of values that are embedded in them. These include evidential, historical, aesthetic and communal values. Evidential values are mainly concerned with evidences about central activities that took place within the building in significant time periods in the past. This type of value is evident in some types of buildings including "Madrasah" (school) and "Hammam" (public bathroom). Historical values, on the other hand, are more holistic, as they address wider remarkable historical matters, other than just evidences of building functions. Aesthetic values are evident in the visual characteristics of the heritage building; they include architectural style, construction methods and advances in architecture and construction. Finally, communal values are concerned with the tangible values included in the heritage building, such as economic, social and cultural values. (Cadw, 2011).[2]

However, in Egypt, the National Organization of Urban Harmony generated a different classification of the values embedded in heritage buildings, where historical, architectural, symbolic, urban, social and functional values are discussed in their guide. The urban value, according to this classification, is related to the importance of a whole urban context of the building (NOUH, 2010).[3] A more comprehensive classification of the value of heritage buildings was suggested by Khodeir et al. (2016).[4] According to this classification, cultural values form a major category that includes both evidential and historical values. The architectural and urban values are also included under this category. This was based on the definition of culture by Edward Taylor as: "that complex whole which includes knowledge, belief, art, morals, law, custom and any other capabilities and habits acquired by man as a member of society" (Moore, 2012).[5] Architectural and urban values, due to this classification, include the aesthetic values defined by Cadw (2011). It is a wider category that includes architectural style, planning approaches, construction methods and materials used (Riegl, 1996).[6]

14.2.2 Legislations and laws classifying heritage buildings in Egypt

To start setting the scene of heritage buildings in Egypt, a number of laws and legislations that arrange the classification of built heritage in Egypt should be highlighted. As a start, the law number 117 in the year 1983 and its modification in law number 12 in the year 1991 included the basis of dealing with conservation of historical buildings, with special reference to those buildings which lasted for a hundred years or more.[7] Buildings determined by this law are buildings that have a distinct architectural style, that is, architectural value connected to national history; historical value related to a historical character; symbolic value which represents a specific historical period; historical value which is considered as a touristic destination; and social and functional values. These buildings are protected according to these laws (MHUD, 2006).[8] These laws also gave the authority to the prime minister to apply them on any building that existed for less than a hundred years, as long as it is considered of national value.

Although the above-mentioned laws are considered obligatory, they do not mention any kind of penalties that are applied on building owners who do not yield to those laws, in case the buildings are privately owned. In addition, the conditions of classifying a building as "historical", according to this law, are vague and cannot be measured in all aspects. Moreover, the law number 144 in the year 2006 was applied; this law deals with the case of deteriorated historical buildings, where it prohibits demolition of such buildings without taking official permission. The identification process of such buildings according to this law is totally in the hands of the prime minister and the Ministry of Antiques.[9]

14.2.3 Evaluating the applied conservation strategies on heritage buildings in Egypt

What are the threats which face the conservation processes of heritage buildings? What challenges are there? What could be done to overcome threats and challenges?

Despite the great value of Egyptian built heritage, applying conservation strategies on them is faced with many problems and threats. Heritage buildings are subject to partial or full destruction, due to unawareness of the values of buildings or because of their deteriorated condition (Cairobserver, 2015).[10] The deterioration of these buildings is a direct result of improper maintenance, weak management and inability of implementing laws and policies. In addition, coping with energy policies, including minimum energy requirements, forms an extra challenge that is added in the case of reusing heritage buildings.

In fact, the application of conservation of heritage buildings in Egypt requires stronger legislative support. The Egyptian law which forms a keystone in this process, with its current form, gives the chance for building owners to contravene. In other words, it made it easier for them to avoid indicating their valuable buildings as heritage buildings. In addition to strictly monitoring the application of legislative implications, spreading awareness of the value of built heritage among common people is vital. The need to educate people and let them understand the value added from conservation of built heritage will provide a better community sense of responsibility. In addition, the National Organization for Urban Harmony in Egypt should effectively participate in spreading such awareness, guidelines and rules that facilitate the application of conservation and set rules for the collaboration of community members into this process (Khodeir et al., 2016).

14.3 Introducing the process of sustainable retrofitting of built heritage

What is the nature of the process of sustainable retrofit (SR)? Is it considered a type of conservation? What benefit should we expect from applying SR on heritage buildings? Should each heritage building turn into a museum? What impact does applying SR have on the values of a heritage building? What are the limitations of applying such process on heritage buildings? What are the computational tools that could enhance the process of sustainable retrofit?

14.3.1 The nature of sustainable retrofit

In general, the conservation process aims at protecting cultural heritage. Meanwhile, it allows for its accessibility to present and future generations (CEN, 2011).[11] From this view, sustainable retrofit process could be classified as a function of the conservation of cultural heritage in general, and of building heritage in particular. Retrofit process aims at improving energy and environmental performance of buildings through technical interventions to achieve benefits. It is mainly concerned with the building services. This is due to the nature of these services, which have remarkably shorter lifetimes than the building structure and fabric. Sustainable retrofit is defined as "any invention to adjust a building to suit new condition or request" (Sara, 2011).[12]

In fact, applying the sustainable retrofit process on heritage buildings varies from its application on modern buildings, where the former include different materials and exist in different structural forms compared with modern buildings and, consequently, they perform differently. They usually heat up and cool down more slowly. Moreover, they rely on sunshine, wind, heating and adequate internal ventilation. What makes the case of heritage buildings even more critical is that if changes to fabric performance, heating and ventilation are not correctly undertaken, this can change their balance and lead to problems of overheating, molds and bad indoor air quality.

14.3.2 Benefits of applying sustainable retrofit on heritage buildings

The process of sustainable retrofit has a number of benefits upon its application on heritage buildings. Generally, it enhances the overall environmental performance of such buildings, while self-guarding their social and economic sustainability. Enhancing environmental performance is achievable through extending service life and reducing energy consumption and $CO2$ emissions. On the other hand, social sustainability is achieved by increasing users' well-being, high-quality indoor air and comfortable space, more natural light and cleaner air and using healthy materials. Finally, economic sustainability is empowered by lowering operating costs as a result of efficient management of energy use, in addition to attracting economic return on investment (May, 2015).[13]

14.3.3 Examples of sustainable retrofitting of heritage buildings

In the coming part we will be able to better understand the nature of the sustainable retrofit process and to trace its benefits through two different examples, one in China and the other in Egypt.

The Chinese example of a retrofitted building is the Lui Seng Chun, Hong Kong, which represents a retrofit process on a type of Chinese building called "Tong-Lau", or the shop house. The building was originally built in the 1930s, and it accommodates a family business on the ground floor and a family house upstairs. The original business use of the building by the Lui's family was a Chinese bone-setting medicine shop (Lui Seng Chun, 2013).[14] In 2000, the Luis donated the building to the government. In 2012, the building was reopened with a new use that reflects its original historical use, a Chinese medicine and health care centre, by the Hong Kong Baptist University (Ho, 2015).[15] The values existing in this example included social values, which were represented in the typical living style of Chinese business families during the time to which the building belongs; economic values, where the building represents a typical economic activity of its time; and architectural value, where the building's architectural design is distinct from the standard design of other shop houses from this period, and it is one of the few remaining till now.

The new use of the building was intended to bring social and community benefits, provide education and raise awareness. Another objective was to apply some modifications, so the building will be accommodated to the current green building regulations and codes in China (AGC, 2010).[16] Among the new features added to the building are the glass panels added to the terraces in the main entrance. The enclosure of the terraces was a functional requirement to protect the building from environmental conditions and to allow people to use the terrace space away from the high temperature and noise. For safety requirements, a new staircase and fire protection measurements were added. A new elevator was installed to facilitate use by the disabled, and another new mechanical installation was added to the building. All installation considered energy saving, with the addition of solar panels for energy production (Ho, 2015). In general, retrofitting of this building showed a balance between considering the heritage value of the building, the functional requirements of the new use and the social benefits that can be driven from this new use.

The second example is the retrofit of Bassili Pasha Villa in the National Museum of Alexandria. It is an evident example of the wide potentials retrofit can provide for the conservation of built heritage. The objectives of the retrofit and reuse of this building into a national museum included the conservation of the main villa and enhancing the building services in order to accommodate the new use. Among the elements that were added during the retrofit process are the new glass showcases designed in diagonal orientation for the display of the artefacts, fire safety systems and new mechanical, electrical and acoustic installation (Elsorady, 2014).[17]

Values embedded in such a building are mainly social and architectural values. Social values are evident in the fact that the building represents a meeting place in which upper-class families lived and interacted. Architectural values of the building include the representation of multicultural interaction in Alexandria and how it affected architectural style; the building represents Neo-Renaissance style with classical decorative ornaments.

In general, retrofitting of these buildings focused on replacement of existing systems to accommodate the new use and the value of the building, without considering energy saving requirements or the accommodation of the green building codes in Egypt. Thus, this building did not completely follow the process of sustainable retrofit. Nevertheless, it represents a good example of well-adapted buildings of new projects that gained acceptance of both professionals and laypersons with different degrees (Elsorady, 2014). In addition, the efficiency of the decision of reuse of this building as a national museum is questionable and takes us to the question, "Should each heritage building be reused as a museum?" and also to the question of "How can we decide about the new use of a sustainably retrofitted heritage building?" To answer the latter question, we shall go through the different phases of applying sustainable retrofit.

14.3.4 Phases of sustainable retrofit and the role of building energy models

Upon deciding to apply sustainable retrofit on a heritage building, a number of questions should be posed: How is the building used? Can it be used more efficiently? What is the type of heritage protection needed? What scope do you have to make changes? What is the budget? What permissions should be taken?

The sustainable retrofit process starts with setting a retrofit scope then setting up the vision for the design. Afterwards, decisions on the strategy of the retrofit are settled. Finally, a number of mock-ups or alternatives are developed where the best overall comprehensive plan is selected (EHS, 2013).[18] So what role does building energy models play in all these steps? And how can their adoption allow for a better decision on the new use of the building? Although the application of the sustainable retrofit process includes a number of steps, the role of BEM is quite obvious in the preliminary planning and design steps, where they could lead to better and more efficient decision making.

At the beginning, BEM forms a keystone in the evaluation of the existing building condition, where the retrofit team analyzes the conditions and determines elements of the building which needs retrofitting (BCA, 2010).[19] In addition, condition audit is applied to determine the current condition and expected remaining life of the building's components. The areas have to be examined, including the structure, external walls and roof, thermal performance, water usage, daylighting, occupant satisfaction, materials, security and review of safety issues. Based on the evaluation of existing conditions, setting and preparing the objectives of the retrofit are planned. This is based on results from the analysis by the design team. The team has to set an approach after prioritizing the goal of retrofitting to determine items to be replaced, provide a base budget for the scope and identify the items that should be targeted for sustainability (Tobias and Vavaroutsos, 2009).[20]

The role of BEM comes clear again during the design process and evaluation of suitable techniques and strategies that can be applied to the building. These include energy efficiency retrofit (e.g. solar retrofit, lighting retrofit, passive design), indoor quality retrofit (e.g. internal shading, top-level skylights under floor supply) and water efficiency retrofit (e.g. low-flow water fittings and shower heads, low-flow plumbing equipment, water-efficient irrigation). After implementation of the retrofitting plan, the role of BEM comes up once again, where performance evaluation of implemented systems through BEM is performed.

14.4 Role of HBIM in conservation of heritage buildings

What is HBIM? How will it benefit conservation processes of heritage buildings in general? How can the perception of the value of the building be embedded in the HBIM process? How could it enhance sustainable retrofit in particular? What is Green BIM double paradigm? How can HBIM and Green BIM improve the decision-making process of sustainable retrofitting of heritage buildings?

HBIM (Historic Building Information Modeling), a plug-in for Building Information Modeling (BIM), is defined as a process for modeling historic structures from laser scan and photogrammetric data. HBIM is specially tailored to application on heritage buildings. The HBIM thus plays an important role in achieving conservation of heritage buildings, thus it automatically produces full engineering drawings for the conservation of historic structures and environments, including

3D documentation, orthographic projections, sections, details and schedules (Murphy, 2013[21]; Dore, 2012).[22]

It can also enhance the application of sustainable retrofitting of such buildings in particular, as HBIM provides a review of the building's exterior and interior, and it also eases the availability to survey renovations and changes that either took place through different time periods or changes that will take place upon applying the sustainable retrofit (Logothetis, 2015).[23]

14.4.4 Phases of applying HBIM on heritage buildings

The HBIM process involves a reverse engineering solution whereby parametric objects representing architectural elements are mapped onto laser scan or photogrammetric survey data. The HBIM process generally includes a number of phases, starting with collecting and processing of laser/image survey data, identifying historic details from architectural pattern books, building parametric historic components/objects and finally correlation and mapping of parametric objects onto scan data and the final production of engineering survey drawings and documentation (Murphy, 2013).

14.4.4.1 Creating an as-built BIM

Creating an as-built BIM is considered the first step of applying the HBIM process. It includes a number of steps:

a Data collection: in which dense point measurements of the facility are collected using laser scans taken from key locations throughout the facility. The product of the laser scan takes the form of a point cloud that represents the coordinates of the scanned building.
b Data processing: through this phase of the HBIM process, the sets of point measurements (point clouds) from the collected scans are filtered to remove artefacts and combine into a single point cloud or surface representation in a *common coordinate system*. This system is then textured from image data to create a virtual 3D model. Data processing includes surface meshing, texturing the point cloud, ortho-image and digital photo modeling processing separate to laser scan data (Tang, Huber, Akinci and Lytle, 2010).[24]
c Geometric modeling: in which the low-level point cloud or surface representation is transformed into a *semantically rich BIM*. The modeling of a BIM involves three tasks: modeling the geometrical form of each element, assigning object category and materials to each element and finally establishing relationship between building elements.

- Modeling the geometrical form: this process constructs simplified representations of the 3D building elements. In general, this process includes modeling of surfaces (planar surfaces, curved surfaces and extrusions), modeling volumes and modeling complex structures, like windows, doors and ornaments.
- Representation of knowledge: there are several techniques that are implemented for the modeling and recognition. In general, the as-built BIM model includes the representation of knowledge about the building objects shape, identities of the objects and their relationships. Regarding knowledge about the objects shape, shape representation could be explicit, implicit, parametric, non-parametric or global versus local. This type of knowledge representation is extremely vital in creating a BIM model, as it provides contextual information to assist in object recognition (Nuchter, 2008).[25]

- Relationship modeling: this includes establishing relationships between objects of the BIM model. The representation of knowledge about the relationship among different objects includes three different categories of spacial relationships: aggregation, topological and directional relationships (Tang et al., 2010).

14.5 A proposed decision-making framework for sustainable retrofitting of heritage buildings in Egypt

Why do we need an organized decision-making framework in Egypt? How are decisions in dealing with heritage buildings taken? Are there specific decision-making strategies? Does the community participation represent a key element in reusing built heritage? How can the approach of Green HBIM or integrated HBIM/BEM help in achieving decision making for retrofitting heritage buildings?

14.5.1 Decision making in an Egyptian context

To better understand the way decisions are made in Egypt, the case of retrofitting of "Al-Baron Palace", a famous heritage building in Cairo, will be discussed. Recently, the Ministry of Antiques published a public announcement in the official newspaper to call for a competition for renovating the building. The duration of the competition was about three weeks, while the required product was very detailed including complete drawings, plans, elevations, 3D shots and redesigning of the project layout. In fact, the way this competition was announced, the consultant office that shared in the announcement (a private one), the restricted timeline, and the large number of drawings needed shed light on the problems that face decision making and emphasized the lack of a definite design-making tool or framework. Another case is the Tahrir headquarters, where the governor of Cairo, Galal Mostafa, has recently announced the shutting down of the Mugamaa Complex in downtown Cairo, which forms, according to some, one of the ugliest office buildings in Egypt, by 2017.[26] That is the way the issue was discussed in the press, without mentioning any further details on their intentions for the future of the building, which is considered to have evidential values, although according to law it is not a historical building (its age is less than 100 years). Whether the building will be demolished, renovated or otherwise is unclear. This decision again reflects how decisions are taken out of definite plans and strategies.

14.5.2 A proposed decision-making framework

The proposed framework, shown in Figure 14.1, was originally based on a conceptual framework by Elaheh et al. (2013), where they based their framework upon integrating modern refurbishment technologies into the conventional procedure. Their aim was to explore how the potential of BIM could be integrated at an early stage with the refurbishment process to develop a systematic approach to make better decisions at the early stage of retrofitting (Elaheh, Stephen and Jalal, 2013)[27]. The first draft of the framework adapted the above-mentioned framework and replaced BIM procedures with HBIM procedures, which start with creating an as-built BIM model. The scope of the framework focused on the different gaps that face the implementation of HBIM and BEM in the case of sustainable retrofitting of heritage buildings in Egypt. This proposed framework was presented and discussed in the HBIM workshop that took place in March 2015 in Luxor, Egypt.

It primarily addressed three main gaps, as follows:

Gap (1): Loss of information: The first gap is a technical issue. It emerges between data collection and data processing phases, mainly during geometrical modeling, where the BIM design systems

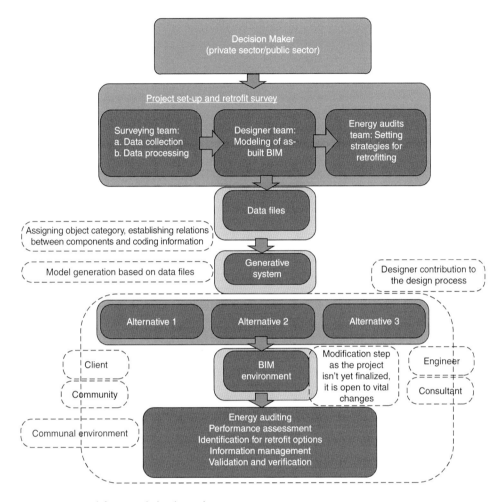

Figure 14.1 Conceptual framework, by the author.

cannot handle the massive data sets from laser scanners. As a result, modelers often shuttle intermediate results back and forth between different software packages during the modeling process, giving rise to the possibility of information loss due to limitations of data exchange standards or errors in the implementation of the standards within the software tools. This gap raises a number of questions like: How can we resolve limitations of data exchange standards within the software tools? How can a detailed as-built BIM be built for heritage buildings?

Gap (2): Interoperability Gap: Interoperability issues emerge when integrating BIM model with BEM. This is evident among software packages. This gap emerges between the BIM modeling team and the energy auditing team who are responsible for setting strategies for sustainable retrofitting. The question raised through this gap is: How can we work on creating a seamless workflow between BIM and BEM software?

Gap (3): Setting strategies for sustainable retrofitting: This gap is the main concern of the energy audit or sustainable retrofitting team. It mainly addresses the following questions: How can we work on setting strategies to sustainably retrofitting historic buildings?

Those three gaps generated from the framework were discussed with a number of professionals in the fields of IT, BIM, HBIM, remote sensing and conservation of heritage buildings. Further discussion of the framework led to its updating and modification, where the feedback that was given in the first round of discussion with professionals was dealt with. The issues that were highlighted included:

- **Issue 1:** The framework addressed many interrelated issues. Its scope is quite wide. The interoperability gap between different software programs for BIM and BEM is still a debatable issue and needs precise investigation from software developers.
- Action taken: In the updated version of the framework, the interoperability gap was taken away from the scope of proposed framework.
- **Issue 2:** The role of building employer or owner, either public or private, was not clear enough in the old framework. It was also not clear how the employer will decide about the new use of the buildings.
- Action taken: The updated framework highlights the role of the owner in the initiation phase, especially in setting the vision for the retrofit process based on the laws and regulations that organize the way of dealing with the building.
- **Issue 3:** Professionals in heritage buildings highlighted that the proposed framework did not take the type of value perception into consideration.
- Action taken: This issue was totally considered in all the phases of implementing the updated proposed framework.

14.5.3 Final updated decision-making framework

Thus the updated framework (Figure 14.2) made full use of the discussions and issues raised on its old version. This framework merges sustainable retrofit as a conservation approach, HBIM process, BEM as a computational tool and heritage values as guiding principles. The utmost aim of this updated framework is to make better decisions and integrate different stakeholders including owners, designers, HBIM modelers, the sustainable retrofit team and the energy simulation team. The framework is divided into five main phases, which represent the typical phases of any project: initiation, planning, evaluation of executed alternatives, implementation of best alternative, and assessment. In each of these phases, the phases of implementation of both HBIM and sustainable retrofitting were integrated.

The framework starts its initial phase by setting a vision for the conservation project. This phase includes value determination, where the decision about the building value is determined according to the defined law. The output of this process is a preliminary decision, which represents the type of heritage value and required protection level. Accordingly, the owner and any relevant decision maker can decide whether to proceed with the sustainable retrofit process or not. Moreover, this decision will logically lead to the start of the first phase of the sustainable retrofit process: setting scope and vision or, in other words, problem formulation. The role of the HBIM process is also initiated in this phase to support in collecting data using laser/image survey data and identifying historic details from architectural pattern books. In addition to collecting technical data, value-related data are also gathered. Built modeling requires information about all features of the building and the degree of allowed interventions.

The planning phase includes the design process of sustainable retrofitting. This phase starts by setting goals and strategies that can identify the building elements which need retrofitting and applying the suitable technique on the building on the detailed 3D model and the automated documentation in the form of engineering drawings that help in applying the main aim of retrofit. Planning and design provide different alternatives for the conservation of the building. The input of heritage value will be one important side of evaluation of this alternative, which forms phase 3 (execution

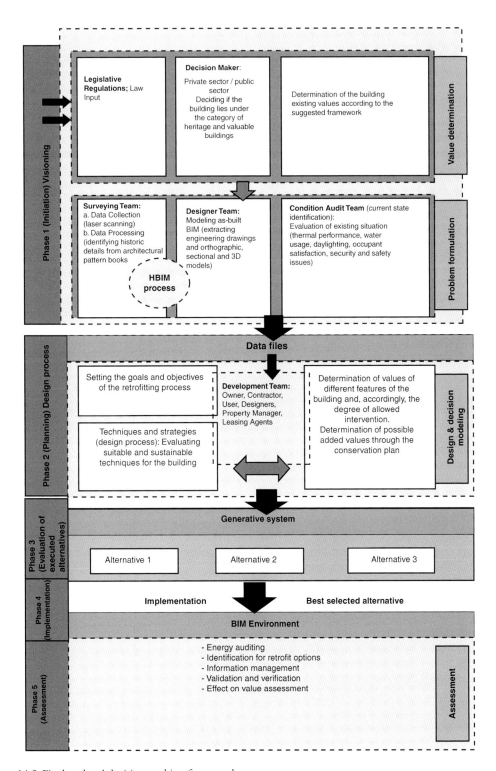

Figure 14.2 Final updated decision-making framework.

or evaluation). Thus, the output of the planning phase is the alternatives or mock-ups which will be easily executed through the HBIM, either by generating different models or using CloudCompare, which permits new repairs to be checked for fit against point clouds without added digital modeling.

Evaluation of executed alternatives includes selecting the alternative that has both minimum effects on heritage values and is most efficient in the sustainable retrofit. This is done by using BEM simulation tools embedded in HBIM as a testing tool that can help to compare different design options and make decisions about the solutions that will be applied further on the project. Achieving community cooperation and participation as well as good communication among the professionals responsible for this process are crucial in this phase. At the end of this phase, the best alternative should be selected, based on a number of factors:

- maximizing the benefits of the building;
- enhancing the building performance;
- highlighting the heritage values of the building;
- achieving sustainable design goals, including reducing energy use, costs and environmental impacts of the property;
- adding green improvements and optimizing financial performance.

The implementation phase represents the actual application of the best selected sustainable retrofitting design solution. It follows the evaluation of executed alternatives phase and is followed by the final assessment phase. The assessment phase aims at making sure of achieving high building performance. Different methods of assessment can be applied here, including sustainable performance assessment of the building, which includes process performance (e.g. integrated design, energy modeling, contracts, etc.), feature or system performance (e.g. energy/water, indoor environment quality, materials and resources, etc.), building performance (e.g. public benefits, sustainability compliance, flexibility and adaptability, etc.), market performance (e.g. operating costs, investor demand, space user demand, etc.) and financial performance (e.g. return on investment, risk and value).

14.5.4 Limitations of application of the proposed framework

It is worth mentioning that the suggested framework presented in this chapter is still a theoretical outline; it has not been applied yet. The goal was to integrate different sides of conservation in one framework that depends on a scientific background. This framework needs further evaluation and feedback from the operational perspective. Although energy efficiency retrofits have shown attractive returns on investment when applied in countries like the UK and China, this issue is highly debatable within the Egyptian context, as demolishing an old building and building a new multi-storey one can be more beneficial for the owner of the building (Khodeir et al., 2016). Therefore, among the limitations of the proposed framework is the return on investment of applying the HBIM process, in addition to the cost of applying sustainable retrofit processes on heritage buildings. Both previous factors could make severe changes in the decision-making process and are considered a primary determinant in selecting acceptable risk for evaluation standards and for selecting the minimum performance objective for retrofit public policy. This brings us to another fatal limitation, which is the employer information requirement, which represents the needs of the employer of the heritage building. These requirements represent a keystone when applying the proposed framework, and it is recommended that they should be collected and adjusted carefully to help in making better decisions and allowing for a lower number of variations within the project through its lifetime.

14.6 Conclusions

In conclusion, applying both HBIM and sustainable retrofit on heritage buildings in Egypt is still limited and faces a number of challenges. Among these challenges are the unavailability of equipment, limited availability of professionals, and funding and financial-related challenges. This calls for the need of inviting different international entities that are concerned with conserving worldwide heritage to share in the training of expertise and funding such projects.

Practices related to the decision-making process, regarding upgrading, adapting, renovating or sustainability retrofitting of heritage buildings in Egypt need to be more proactive and to adopt the concept raised by Stephen Covey in his book, *The 7 Habits of Highly Effective People*, "beginning with the end in mind".

Notes

1 Dümcke, C., Gnedovsky, M. (2013), "The Social and Economic Value of Cultural Heritage: Literature Review," *Eur. Expert Network. Cult.*, 34: 258–270.
2 Cadw (2011), Criteria for Assessing Historic Buildings for the Purposes of Grant Assistance. *Llywodraeth Cymru Welsh Government*. http://cadw.gov.wales/docs/cadw/publications/Criteria_assessing_grant_assistance_ENG.pdf.
3 National Organization of Urban Harmony (NOUH) (2010), *Principles and Standards of Urban Harmony for Heritage and Special Value Buildings and Areas: Reference Guide*, 1st ed. Cairo: National Organization of Urban Harmony – Egyptian Ministry of Culture.
4 Khodeir L., Ali D., Tarek S. (2016), Integrating HBIM (Heritage Building Information Modeling) Tools in the Application of Sustainable Retrofitting of Heritage Buildings in Egypt. *Proceedings of the conference Improving Sustainability Concept in Developing Countries*, Cairo, Egypt.
5 Moore, J.D. (2012), *Visions of Culture: An Introduction to Anthropological Theories and Theorists*, 4th ed. Lanham, MD: Rowman Altamira, USA.
6 Riegl, A. 1996. The Modern Cult of Monuments: Its Essence and Its Development. In: Stanley-Price, N. et al. (eds.), *Historical and Philosophical Issues in the Conservation of Cultural Heritage* (pp. 71, 77). Los Angeles: The Getty Conservation Institute.
7 Delta Law, Egyptian heritage legislation, 2016, http://deltalaw.blogspot.com.eg/2012/09/117–1983–12–1991.html, last accessed 12/Jan. 2016.
8 Ministry of Housing, Utilities and Urban Development (2006), "Implementation Regulations of the Law no. 144 Year 2006 about the Regulation of the Demolition of Structurally Safe Buildings and Architectural History Protection," *The Official Egyptian Journal*, 04-Nov.: 6–8.
9 National Organization for Urban Harmony (2016), Official Website, Egypt. http://www.urbanharmony.org/rule%20144%20for%202006.pdf, last accessed 12/Jan. 2016.
10 Cairobserver (2014), 11 Recent Cultural Disasters in Egypt. *Cairobserver*. http://cairobserver.com/post/75403717693/11-recent-cultural-disasters-in-egypt#.VfB43dJViko, last accessed 09/Sep. 2015.
11 CEN, EN 15898 (2011), *Conservation of Cultural Property – Main General Terms and Definitions*. Brussels: European Committee for Standardization.
12 Sara, W. (2011), Sustainable Retrofit Potential in Lower Quality Office Stock in the Central Business District. *Management and Innovation for a Sustainable Built Environment Conference*.
13 May, N., Griffiths, N. (2015), *Planning Responsible Retrofit of Traditional Buildings*. London: Sustainable Traditional Building Alliance.
14 Chun, Lui Seng (2013), Conserve and Revitalise Hong Kong Heritage Buildings. *Revitalising Historic Buildings through Partnership Scheme, Resource Kit*. http://www.heritage.gov.hk/en/doc/resource%20kit_lui_seng_chun.pdf.
15 Ho, W. (2015), The Impact of Climate Change in Hong Kong on the Strategic Planning for Built Heritage – A Case Review of Lui Seng Chun. *1st International Academic Conference on Climate Change and Sustainable Heritage 2015*.
16 AGC Design Ltd. Revitalisation Scheme (2010), Conversion of Lui Seng Chun into Hong Kong Baptist University Chinese Medicine and Healthcare Centre. http://www.lcsd.gov.hk/CE/Museum/Monument/form/20100427%20LUI%20SENG%20CHUN-HIA.pdf.

17 Elsorady, Dalia Abdelaziz (2014), "Assessment of the Compatibility of New Uses for Heritage Buildings: The Example of Alexandria National Museum, Alexandria, Egypt," *Journal of Cultural Heritage*, 15 (5): 511–21.
18 EHS Book Regulation (2013), *Green Building Regulation for Developments*. Under Dubai World Jurisdiction. June 2013.
19 BCA (2010), Building and Construction Authority. *Existing Building Retrofit*, Singapore, 2010.
20 Tobias, Leanne, Vavaroutsos, George (2009), *Retrofitting Office Buildings to Be Green and Energy-Efficient: Optimizing Building Performance, Tenant Satisfaction and Financial Return*. Washington, DC: Urban Land Institute.
21 Murphy, M., McGovern, E., Pavia, S. (2013), "Historic Building Information Modelling – Adding Intelligence to Laser and Image Based Surveys of European Classical Architecture," *ISPRS Journal of Photogrammetry and Remote Sensing*, 76: 89–102.
22 Dore, C., Murphy, M. (2012), Integration of Historic Building Information Modeling and 3D GIS for Recording and Managing Cultural Heritage Sites. *18th International Conference on Virtual Systems and Multimedia: "Virtual Systems in the Information Society"*, 2–5 September, 2012, Milan, Italy, pp. 369–376.
23 Logothetis, S., Delinasiou, A., Stylianidis, E. (2015), Building Information Modelling for Cultural Heritage: A Review, ISPRS Annals of the Photogrammetry, Remote Sensing and Spatial Information Sciences, Volume II-5/W3, 2015. *25th International CIPA Symposium 2015*, 31 August – 04 September 2015, Taipei, Taiwan.
24 Tang, Pingbo, Huber, Daniel, et al. (2010), "Automatic Reconstruction of As-Built Building Information Models from Laser Scanned Point Clouds: A Review of Related Techniques," *Automation in Construction*, 19: 829–843.
25 Nuchter, A., Hetzberg, J. (2008), "Towards Semantic Mobile Robots," *Journal of Robotics and Autonomous Systems (RAS)*, 56 (11): 915–926.
26 Egypt to Shut Down (2016), *"Most Hated Building": The Mugamaa*. www.egyptianstreets.com, January 7, 2016.
27 Elaheh, G., Stephen, S., Jalal, S., et al. (2013), Exploiting BIM in Energy Efficient Refurbishment: A Paradigm of Future Opportunities. *PLEA 2013–29th Conference, Sustainable Architecture for a Renewable Future*, Munich, Germany, pp. 1–2.

References

AGC Design Ltd. Revitalisation Scheme (2010), *Conversion of Lui Seng Chun into Hong Kong Baptist University Chinese Medicine and Healthcare Centre*. http://www.lcsd.gov.hk/CE/Museum/Monument/form/20100427%20LUI%20SENG%20CHUN-HIA.pdf.

BCA (2010), Building and Construction Authority. *Existing Building Retrofit*, Singapore, 2010.

Cadw (2011), Criteria for Assessing Historic Buildings for the Purposes of Grant Assistance. *Llywodraeth Cymru Welsh Government*. http://cadw.gov.wales/docs/cadw/publications/Criteria_assessing_grant_assistance_ENG.pdf.

Cairobserver (2014), 11 Recent Cultural Disasters in Egypt. *Cairobserver*. http://cairobserver.com/post/75403717693/11-recent-cultural-disasters-in-egypt#.VfB43dJViko, last accessed 09/Sep. 2015.

CEN, EN 15898 (2011), *Conservation of Cultural Property – Main General Terms and Definitions*. Brussels: European Committee for Standardization.

Chun, Lui Seng (2013), Conserve and Revitalise Hong Kong Heritage Buildings. *Revitalising Historic Buildings through Partnership Scheme, Resource Kit*. http://www.heritage.gov.hk/en/doc/resource%20kit_lui_seng_chun.pdf.

Delta Law, Egyptian heritage legislation, 2016: http://deltalaw.blogspot.com.eg/2012/09/117–1983–12–1991.html, last accessed 12/Jan. 2016.

Dore, C., Murphy, M. (2012), Integration of Historic Building Information Modeling and 3D GIS for Recording and Managing Cultural Heritage Sites. *18th International Conference on Virtual Systems and Multimedia: "Virtual Systems in the Information Society"*, 2–5 September, 2012, Milan, Italy, pp. 369–376.

Dümcke, C., Gnedovsky, M. (2013), "The Social and Economic Value of Cultural Heritage: Literature Review," *Eur. Expert Network. Cult.*, 34: 258–270.

Egypt to Shut Down (2016), *"Most Hated Building": The Mugamaa*. www.egyptianstreets.com, January 7, 2016.

EHS Book Regulation (2013), *Green Building Regulation for Developments*. Under Dubai World Jurisdiction. June 2013.

Elaheh, G., Stephen, S., Jalal, S. (2013), Exploiting BIM in Energy Efficient Refurbishment: A Paradigm of Future Opportunities. *PLEA 2013–29th Conference, Sustainable Architecture for a Renewable Future*, Munich, Germany, pp. 1–2.

Elsorady, Dalia Abdelaziz (2014), "Assessment of the Compatibility of New Uses for Heritage Buildings: The Example of Alexandria National Museum, Alexandria, Egypt," *Journal of Cultural Heritage*, 15 (5): 511–21.

Gholami, E., Kiviniemi, A., Kocaturk, T., Sharples, S. (2015), Exploiting BIM in Energy Efficient Domestic Retrofit: Evaluation of Benefits and Basrriers. *2nd International Conference on Civil and Building Engineering Informatics ICCBEI*, Tokyo, pp. 1–7.

Ho, W. (2015), The Impact of Climate Change in Hong Kong on the Strategic Planning for Built Heritage – A Case Review of Lui Seng Chun. *1st International Academic Conference on Climate Change and Sustainable Heritage 2015*.

Khodeir L., Ali D., Tarek S. (2016), Integrating HBIM (Heritage Building Information Modeling) Tools in the Application of Sustainable Retrofitting of Heritage Buildings in Egypt. *Proceedings of the conference Improving Sustainability Concept in Developing Countries*, Cairo, Egypt.

Logothetis, S., Delinasiou, A., Stylianidis, E. (2015), Building Information Modelling for Cultural Heritage: A Review, ISPRS Annals of the Photogrammetry, Remote Sensing and Spatial Information Sciences, Volume II-5/W3, 2015. *25th International CIPA Symposium 2015*, 31 August – 04 September 2015, Taipei, Taiwan.

May, N., Griffiths, N. (2015), *Planning Responsible Retrofit of Traditional Buildings*. London: Sustainable Traditional Building Alliance, Responsible Retrofit Series – STBA, England.

Ministry of Housing, Utilities and Urban Development (2006), "Implementation Regulations of the Law no. 144 Year 2006 about the Regulation of the Demolition of Structurally Safe Buildings and Architectural History Protection," *The Official Egyptian Journal*, 04-Nov.

Moore, J.D. (2012), *Visions of Culture: An Introduction to Anthropological Theories and Theorists*, 4th ed. Lanham, MD: Rowman Altamira.

Murphy, M., McGovern, E., Pavia, S. (2013), "Historic Building Information Modelling – Adding Intelligence to Laser and Image Based Surveys of European Classical Architecture," *ISPRS Journal of Photogrammetry and Remote Sensing*, 76: 89–102.

National Organization for urban harmony, Official Website, Egypt (2016): http://www.urbanharmony.org/rule%20 144%20for%202006.pdf, last accessed 12/Jan. 2016

National Organization of Urban Harmony (NOUH) (2010), *Principles and Standards of Urban Harmony for Heritage and Special Value Buildings and Areas: Reference Guide*, 1st ed. Cairo: National Organization of Urban Harmony – Egyptian Ministry of Culture.

Nuchter, A., Hetzberg, J. (2008), "Towards Semantic Mobile Robots," *Journal of Robotics and Autonomous Systems (RAS)*, 56 (11): 915–926.

Riegl, A. 1996. The Modern Cult of Monuments: Its Essence and Its Development. In: Stanley-Price, N. et al. (eds.), *Historical and Philosophical Issues in the Conservation of Cultural Heritage* (pp. 69–83). Los Angeles: The Getty Conservation Institute.

Sara, W. (2011), Sustainable Retrofit Potential in Lower Quality Office Stock in the Central Business District. *Management and Innovation for a Sustainable Built Environment Conference*.

Tang, Pingbo, Huber, Daniel, Akinci, Burcu, Lytle, Alan M. (2010), "Automatic Reconstruction of As-Built Building Information Models from Laser Scanned Point Clouds: A Review of Related Techniques," *Automation in Construction*, 19: 829–843.

Tobias, Leanne, Vavaroutsos, George (2009), *Retrofitting Office Buildings to Be Green and Energy-Efficient: Optimizing Building Performance, Tenant Satisfaction and Financial Return*. Washington, DC: Urban Land Institute.

15 HBIM, a case study perspective for building performance

Yusuf Arayici

15.1 Introduction

In recent years a variety of new and smart technologies have emerged. Our lives are now enveloped in software, connected by technologies such as smartphones and other gadgets. This in turn has become the stimulus for mankind to adopt new ideas and abandon old ones. The more these technologies become embedded, the more our lifestyles and the way we work change. For example, buildings were constructed in different ways depending on the era and the available technologies. Today, building performance, including energy efficiency and the tools by which to diagnose pathologies such as the causes of overheating, has become a priority for building design and construction. This chapter focuses on the drivers for and challenges identified in the recent exemplar HBIM-based refurbishment of Manchester Town Hall.

15.1.1 Heritage building

Heritage buildings may be distinguished by their style, date of construction, historic occupants or events, or by the building techniques applied. These give the building value and significance from a combination of natural, artistic, scientific, social, spiritual, ecological or historical perspectives. The restoration and conservation of such buildings is very important, and one of the most important factors to consider during this process is the energy efficiency of the building itself, as such buildings can present both advantages and disadvantages when compared to newly built structures in energy performance. Fabbri (2012) states that the relationship between energy performance and traditional buildings is important for the development of other disciplines such as urban planning and restoration and conservation. It is also believed by some architectural designers that some historical buildings conserve more energy than many newly built structures, which could be due to construction methods and processes.

The goal of performance-based design is to make a building sustainable. An increasingly significant proportion of new construction is indeed becoming carbon neutral and environmentally friendly. Sustainability, resilience and performance are current buzzwords in the construction industry, and heritage buildings pose particular challenges for sustainability and resilience, reflecting unique challenges and problems from a retrofit perspective.

15.1.2 Building energy performance

Building construction technologies developed during the twentieth century generally do not take into account energy behaviour (Tronchin & Fabbri, 2008). In the United Kingdom such construction technologies post-1950 have served to reduce the thermal inertia of buildings via wrapped structures and reduced wall thicknesses. Energy-consuming heating plants have been adapted to deliver the necessary performance that could have been resolved with insulation materials and a fabric first approach (Tronchin & Fabbri, 2008).

These factors, along with changes in the politics relating to climate following the 1973 energy crisis, led to the creation of EN832 and the EPBD (Energy Performance of Buildings Directive). These state that energy evaluation would rely on energy need and energy used during winter and summer for heating, cooling and ventilation; domestic hot water production (DHW); and electrical energy. Verification of the energy consumption of buildings also depends on variable factors, which include the geometry and materials, the local climate and seasonal variation, the habit of users and the DHW consumption. These variables are not comparable and standardisable.

Energy performance in buildings is the result of a complex set of interrelationships among the external environment, the shape and character of the building components, equipment loads, lighting systems, mechanical systems, building envelope, air distribution strategies, renewable energy options, operational protocols and the building occupants. Building optimization – achieving the greatest possible energy efficiency and environmental soundness with the least expenditure of resources – requires an understanding of these interrelationships. To achieve this, BIM-based software can be deployed as a most effective tool for the evaluation of heritage buildings, newly built structures and also for energy analysis as well as other processes during architectural design.

15.1.3 *Building information modelling (BIM)*

BIM is the process of development and use of a computer-generated parametric model to simulate the planning, design, construction and operation of a building facility (Azhar, 2009). The model is a data-rich, object-oriented, intelligent and parametric digital representation of the building facility, which can help to reduce time and cost associated with energy analysis throughout project conception and deliver tools for analysis and design alternatives in the early stages of the design process. These analyses are not possible with 2D tools.

BIM is currently the most common term for a new way of approaching the design, construction and maintenance of buildings (Bryde et al., 2013). It is also articulated as the set of interacting policies, processes and technologies generating a methodology to manage building design and derive project data in digital format throughout the building life cycle.

The integration of building performance analysis into the design and construction and operation of a building life cycle is crucial to reach for the demand for high-energy buildings. Building Information Modelling in conjunction with Building Energy Modelling (BEM) seeks to make this integration seamless throughout the process (Reeves et al., 2015).

The owners of heritage buildings are increasingly concerned about how they can save energy. Generally this is achieved via retrofitting. The question is how retrofit should be implemented and how performance can be enhanced in order to ensure greater sustainability in long-term use by owners or users. BIM is becoming more and more widely used at present to meet these ends, because construction and retrofit projects are both becoming increasingly complex. Thus, highly packaged digital information (BIM) is required to build better.

15.1.4 *BIM for heritage*

The concept of BIM is far more than a simple 3D digital model; it is a compilation of the electronic data related to a particular building. The history of using abstract drawings to represent the world has progressed through the years from simple flat 'CAD' drawings to 3D smart objects. These smart objects tend to be very complex and full of information aimed at increasing support for professionals in the construction area.

Heritage buildings built hundreds years ago have singular forms and components that make them valuable and significant. Taking, for example, a Corinthian column, it was handmade with extreme care and has survived for centuries. Today, if such an element is to be restored, we depend on technology to streamline the tasks and processes, since conservation expertise is in increasingly scarce supply. Such heritage components are often unique, due to deformation and degradation over the centuries, and hence can only be superficially represented in parametric libraries. Laser scanning and BIM are technologies that can underpin the restoration of such elements. BIM delivers an important tool promising new working methods for heritage professionals with enhanced accuracy. Laser scanning and BIM serve to greatly improve the accuracy and digital prototyping of heritage structures. The aim is to blend the digital and physical world for not only restoration but also new construction, geometric measurement and other analyses, and to be able to fix problems at a distance, even from the other side of the world, with almost 100% accuracy (Garagnani & Manferdini, 2013). In addition, such remotely operated tools enable collaboration, shared understanding, better public engagement and social innovation.

15.2 The Manchester City Hall exemplar

The recent exemplar implementation of BIM in the Manchester Town Hall Complex is a particularly good example that reveals the potentials and possibilities of using BIM for the extensive refurbishment of heritage buildings. The Manchester Town Hall Complex transformation programme (2008–2014) is one of the earliest high-profile uses of BIM digital modelling software on a set of heritage buildings in the UK. Owned by Manchester City Council, the complex contains the Manchester Town Hall Extension (1938) and the Manchester Central Library (completed 1934). Both buildings were originally won in competition by the architect E. Vincent Harris (1876–1971) from amongst the leading UK architectural practices of the day. The two buildings are noteworthy as examples of modern, twentieth-century heritage that are considered significant enough to merit legal protection by the UK government by being listed grade two star (II★). The immediate area surrounding them also contains other important listed items such as the Manchester war memorial designed by Sir Edwin Lutyens, the iconic British K2 red telephone box (listed grade II), public sculptures, undesignated heritage items such as early decorative cast-iron electrical distribution boxes and lamp standards. The whole complex is also protected by its location in the St Peter's Square Conservation Area.

Early consideration of a long overdue programme of maintenance and minor refurbishment in 2008 took on more ambitious scale with the city council's 'Transformation…' programme begun in 2009. This was going to controversially gut both listed buildings in pursuit of the UK government's 'Working without Walls' initiative, a process of 'deep refurbishment'. The Town Hall Extension was to have most of its internal walls removed to create large modern open-plan offices and enhanced public services at ground floor. The Central Library was also to be almost completely gutted in order to install new services and circulation, centred on the installation of a large new staircase and double lift, before floors were rebuilt and the original reading room interiors carefully reinserted. Both buildings were to be physically linked to create a seamless complex.

As the government's official advisory body on all matters related to the historic environment, English Heritage had to be involved in the approval process before any work could be done. Given the scale and complexity of the project, overall planning consent was agreed initially but delivered via a large number of planning conditions which needed to be 'signed off' by English Heritage on a regular basis. This close involvement was exceptional and involved regular weekly Heritage meetings to discuss problems, visit the site, and discharge the planning conditions over the three to four years the project was on-site. In this process the use of BIM digital modelling software was seen as essential to overcome the inevitable issues which arise on such large-scale projects.

Manchester City Council planned to use BIM extensively to better manage its facilities. 3D BIM-derived digital models were used in different meetings with many stakeholders to show design and refurbishment solutions. This helped them to understand the project better and helped the construction team via BIM to, for example, derive the dimensions and cost of new columns or slabs. Figure 15.1 shows the BIM digital model for the Town Hall and the Central Library.

The case study focused on the inter-organisational aspects of BIM, such as communication, collaboration, and data sharing between the project stakeholders during the design and construction planning. The use of BIM showed that changes in traditional document-based design and construction and FM are required for the stakeholders to operate and collaborate (Codinhoto et al., 2011).

From an owner perspective, to use HBIM successfully, it is necessary: i) to define organisational performance baselines such as problems and inefficiencies of services that the organisation uses and delivers; ii) to streamline processes before any financial commitment is made in technology and promote process changes and reset the baseline to identify which problems can be tackled by implementing BIM and what benefits can be achieved and the costs associated with the implementation.

From the design point of view, BIM digital modelling software helped tremendously in terms of visualisation to support decision making, and the automated model update was really easy and straightforward, leading to significant time and cost efficiency. Indeed, the contractors and the subcontractors identified that time was well managed, and the system supported necessary technical details, easy data exchange, project visualisation and better logistics. Major issues such as clashes in slabs, pipes and ducts in the existing walls were resolved through coordination and clash detection. For example, when the City Library model was integrated into the Town Hall model, clashes became apparent. For example, the Library model and the Town Hall were inadvertently assigned to different levels, causing 'clashes' between some slabs and walls. Figure 15.2 shows a clash analysis in the model.

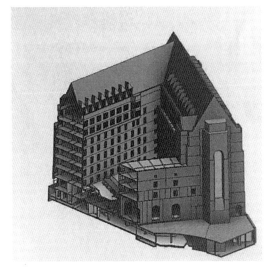 

Figure 15.1 Manchester Town Hall BIM model (Codinhoto et al., 2011).

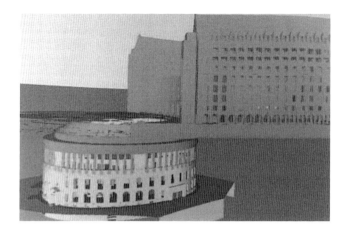

Figure 15.2 Clash detection on the Central Library and Town Hall (Codinhoto et al., 2011).

One of the key aspects observed from the case study was the collaborative environment. The core team consisted of around 100 people, including the client and key suppliers. Crucial factors for success are the ability of participants to use BIM as a standard process and the good team spirit.

However, additional questions are advised for further research from this case study HBIM project. These are (Codinhoto et al., 2011):

- Do FM and OM benefit from BIM?
- Does BIM enable integrated organisational facilities management?
- Does BIM enable better decision making in refurbishment projects?
- Does BIM enable better and integrated organisational decision making?

15.3 Conclusion

This example identifies some of the challenges of deploying BIM in a process of 'deep refurbishment' of heritage buildings. It demonstrated considerable cost and time savings, and led to a much smoother process. It remains to be seen how effective and useful the resulting BIM database will be in achieving better management of these facilities over the next decades, and the extent to which long-term energy efficiency can be ensured by these means, despite the likely continuing changes in technologies and usage. It shows that Heritage BIM can be delivered using conventional new construction digital BIM tools, although other chapters address the challenges in less recently constructed or less identifiably uniformly component-based structures.

References

Azhar, S., Brown, J., & Farooqui, R. (2009), BIM-based sustainability analysis: An evaluation of building performance analysis software. In Sulbaran, T. & Sterling C. (Eds.), *Proceedings of the Associated Schools of Construction (ASC) 45th Annual International Conference*, 1–4 April 2009, University of Florida.

Bryde, D., Broquetas, M., Volm, J.M. (2013), "The project benefits of building information modelling (BIM)". *International Journal of Project Management* (31) pages 971–980.

Codinhoto, R., Kiviniemi, A., Kemmer, S., Gravina da Rocha, C. (2011), BIM Implementation – Manchester Town Hall complex, Research report.

Fabbri, K., Zuppiroli, M., Ambrogio, K. (2012), "Heritage buildings and energy performance: mapping with GIS tools". *Energy and Buildings* (48) pages 137–145.

Garagnani, S., and Manferdini, A.M. (2013), "Parametric Accuracy: Building Information Modelling Process Applied to the Cultural Heritage Preservation", International Archives of the Photogrammetry, Remote Sensing and Spatial Information Sciences, Volume XL-5/W1, 2013 3D-ARCH 2013–3D Virtual Reconstruction and Visualization of Complex Architectures, 25–26 February 2013, Trento, Italy.

Reeves, T.J., Onlina, S., Issa, R.R.A. (2015), "Guidelines for using building information modelling (BIM) for energy analysis of buildings". *Buildings* (5) pages 1361–1388; doi:10.3390/buildings5041361.

Tronchin, L., Fabbri, K. (2008), "Energy performance building evaluation in Mediterranean countries: comparison between software simulations and operating rating simulation". *Energy and Buildings* (40) pages 1176–1187.

16 From LiDAR data towards HBIM for energy analysis

Lucía Díaz-Vilariño, Pawel Boguslawski, Miguel Azenha, Lamine Mahdjoubi, Paulo B. Lourenço and Pedro Arias

Knowledge about heritage buildings and structures is interesting for a wide variety of stakeholders, ranging from architects to operators or the public administration. Such knowledge includes a large variety of physical and functional characteristics of the building. Laser scanning allows efficient and accurate 3D digitalization of heritage sites and subsequent data processing towards the creation of geometrically and semantically rich models.

Parameterization and semantic enrichment of heritage building representations towards the creation of Heritage Building Information Models (HBIM) generally involves the use of point clouds and images as templates for manual mapping procedures in commercial software. Indeed, the information to include in BIM models depends on the requirements of the application it is intended to serve. In view of that, the applications presented in this chapter pertain to automated techniques that were implemented to parameterize point clouds towards models suitable for energy analysis purposes.

The challenge in automating the reconstruction of heritage buildings is to deal with their geometrical complexity and irregularity, meaning that the methodologies selected should be robust and efficient under these conditions. The resulting 3D semantically rich model enhances the knowledge of the heritage building, complementing other representations of the facility such as point clouds or handmade HBIM. The procedure is implemented and validated in a real case study: the Ducal Palace in Guimarães (Portugal).

16.1 Introduction

For a long time, drawings, schedules or reports generated in traditional ways have been the standard communication between building stakeholders, while 3D modelling was primarily used as a rendering tool, not as a project representation. The popularization of Building Information Modelling (BIM) in the early 2000s (Autodesk, 2002) established the beginning of a new way of thinking in the architecture, engineering, construction and facility management (AECFM) industry. BIM encloses the digital representation of both physical and functional characteristics of a facility, representing a shared knowledge resource as a tool for decision making throughout the entire building life cycle (Eastman *et al.*, 2008).

BIM was initially centred in the support to design and construction processes, and consequently, its major uses and research were concentrated in preplanning, design, construction and project delivery (Volk, Stengel and Schultmann, 2014). However, the large number of existing buildings and the need for knowledge about the built environment has motivated the AECFM industry to put efforts in the latest building life cycle stages. BIM for existing buildings (EBIM) has emerged in recent years as a new research trend. Quality control and monitoring under renovation processes (Bosché *et al.*,

2015), energy analysis (Díaz-Vilariño et al., 2013), indoor planning and navigation for building crisis management and emergency responses (Isikdag, Zlatanova and Underwood, 2013) are just some of the applications that make BIM reconstruction for existing buildings necessary.

The preservation of architectural knowledge is another application of BIM modelling. The DURAARK project (DURAARK, 2015) specifically deals with the semantical enrichment of 3D architectural models for the digital long-term preservation of built buildings. This purpose is especially relevant in the case of heritage buildings due to their singularity and value. The modelling of Heritage BIM (HBIM) is a specific domain within EBIM, which requires an accurate and complete documentation of the building, comprising both tangible and intangible knowledge (Simeone et al., 2014).

Non-destructive techniques such as laser scanning and photogrammetry have been widely used in the 3D digitalization of heritage because they allow efficient and accurate records of complex buildings. Even though point clouds are rather accurate, they are composed of massive raw information that requires post-processing in order to be converted to parametric components. For instance, a point cloud with several thousands of 3D points describing a façade is suitable for visualization, while it is not often suited for building analysis or planning.

The level of detail (LoD) and/or different levels of development of a BIM representation are still a controversial topic. Volk, Stengel and Schultmann (2014) consider that BIM can be seen from a narrow and a broader perspective. From a narrow perspective, BIM can be represented as attending to the LoD required for the applications it is intended to serve. From a broader perspective, functional, organizational and legal issues should be included in a BIM representation. The same controversy applies for HBIM. Independently of this controversy, building information, both geometry and semantics, should be ultimately processed to ensure its usability for an application. For instance, a curved wall can be accurately surveyed and parameterized, but it should be modelled as set of continuous planar polygons for energy analysis purposes.

This chapter specifically deals with a technique for highly automated reconstruction of façade components of heritage buildings from point clouds, with the final goal for application in energy analysis software. The result is a semantically rich 3D model represented under the paradigm of BIM, including both geometric and semantic information of the building as-is. Point cloud scans are processed by applying some rules in order to obtain parameterized models according to the Green Building XML (gbXML) schema (gbXML, 2015). The development of streamlined processes for obtaining semantically rich 3D models enables the conversion of point cloud scans into meaningful and interpretable data. The resulting HBIMs are, in any case, complementary to point clouds, 3D CAD models or non-automated high-level HBIMs, among other architectural representations, ultimately enhancing the knowledge about the building.

The challenge in automating the reconstruction of heritage building models is to deal with their geometrical complexity and irregularity. Although most of the reconstruction techniques have been applied to modern buildings, their use can be extended to heritage buildings, but they should be robust enough to deal with the complex and irregular shape.

This book chapter is organized in five sections, starting with the current introduction. Section 16.2 reviews the use of laser scanning for BIM reconstruction. Section 16.3 describes a set of simple approaches towards the reconstruction of building façades for energy analysis purposes while Section 16.4 deals with the results obtained from applying the methodology to a particular case study: the Paço dos Duques in Guimarães (Portugal). Finally, Section 16.5 concludes this work.

16.2 From laser scanning to BIM reconstruction

The BIM creation process differs if it is aimed towards an existing building or to a new building. Volk, Stengel and Schultmann (2014) schematize the BIM model creation process for new buildings and existing buildings depending on the availability of preexisting BIM and according to the life cycle (LC) stages of a building following ISO 22263:2008. Figure 16.1 shows a scheme wherein BIM models for new buildings might not initially be available and they need to be created from the first step of the building LC. EBIM are typically created to support maintenance and deconstruction stages, and HBIM is included in this group as a particular case of EBIM.

Heritage Building Information Modelling involves a reverse engineering solution, commonly named as "scan-to-BIM", whereby architectural elements are converted into parametric objects from data acquired with surveying techniques such as laser scanning and photogrammetry (Murphy, McGovern and Pavia, 2009). The development of the laser scanning technology in recent years, together with the increased accessibility to the use of this technology, as well as its cost reduction, have made the creation of as-built BIMs tractable, assisted by such technology (Tang *et al.*, 2010). The use opportunities of laser scanners for BIM reconstruction have been evidenced by several agencies and organizations, such as the General Services Administration through its program "3D-4D Building Information Modeling" (GSA, 2009) or even Building Smart Finland through COBIM (BuildingSmartFinland, 2012). Both programs include specific recommentations about laser scanner surveying for BIM reconstruction, where different aspects of the acquisition process are discussed such as the overlapping between scans, point cloud accuracy and spatial resolution, among others. Other remote sensing techniques such as thermography can be used to extract semantic information such as the thermal properties of the building materials (Lagüela *et al.*, 2013).

Most of the processes carried out to generate HBIM consist of mapping architectural elements onto laser scan or photogrammetric survey data using BIM software such as Autodesk Revit and Navisworks, Bentley Architecture, Graphisoft ArchiCAD or Tekla (Fai *et al.*, 2011; Brumana *et al.*, 2013; Oreni *et al.*, 2014). Point clouds and images are used as templates, and consequently, HBIMs are manually modelled by placing, positioning and extruding architectural elements. There is the need to streamline EBIM/HBIM reconstruction by using semi-automated and automated techniques (Tang *et al.*, 2010). Intense efforts have been made in recent years to automate the EBIM reconstruction. Specialized software in the area of reverse engineering incorporates developed tools to allow the rapid generation of BIM components from point clouds. Nonetheless, these solutions are far away from automated or semi-automated (Volk, Stengel and Schultmann, 2014).

Although literature on point cloud processing is extensive (Patraucean *et al.*, 2015), the issue is still an ongoing research topic because of the high variability of the building environment. In comparison with modern buildings, heritage buildings are characterized by their singularity, comprising irregular and complex components. Indeed, very little work has been done in terms of automatic HBIM reconstruction. In this context, the DURAARK project aims to generate semantically rich

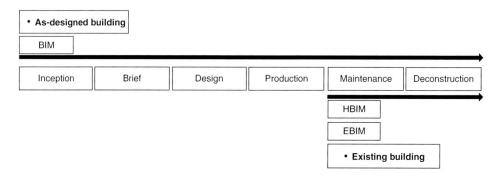

Figure 16.1 The creation process of BIM, EBIM and HBIM related to the life cycle stages following ISO 22263:2008 (adapted from Volk, Stengel and Schultmann, 2014).

as-built 3D models for the long-term preservation of the architectural knowledge (DURAARK project, 2015). They have developed several prototypes towards the automation of BIM reconstruction from point cloud scans. Similar to the approach of Adan and Huber (2011), the first prototype consisted of segmenting the point cloud into rooms and detecting doors by searching from visibility lines of sight between rooms (Ochmann et al., 2014). A second prototype improved the method, whereby walls were reconstructed including doors and windows (Ochmann et al., 2016). In all cases, point clouds were submitted to rules in order to correctly parameterize the building components, causing the loss of architectural details. In this regard, the project also proposes the joint indexing of the various types of architectural data ranging from point clouds to 3D CAD models and BIMs, since all of them are complementary to the knowledge about the building. Other works deal with these issues by texturizing the models with real-colour images for visualization (Xiao and Furukawa, 2012) or with thermal images for energetic purposes (Lagüela et al., 2013).

16.3 Methodology

In this section, a set of simple techniques for processing point clouds towards the automation of HBIM reconstruction for energy analysis purposes is presented. Figure 16.2 shows the general workflow of the methodology. It starts by segmenting the point cloud into planar regions in order to identify and to model the building envelope (Section 16.3.1.). Then, the planar regions corresponding to walls are rasterized with high resolution in order to detect openings, such as windows and

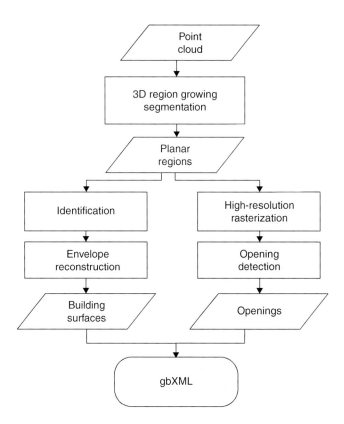

Figure 16.2 General workflow of the proposed methodology.

doors (Section 16.3.2.). Finally, the data extracted are represented according to the gbXML schema (Section 16.3.3.), which can be complemented with attributes obtained from other sources.

16.3.1 Façade segmentation

The proposed methodology is classified as top-down because it starts by selecting and parameterizing the general building elements such as walls. After such parameterization, a further refining is made towards the identification and modelling of more detailed elements such as, in this case, façade openings.

The complete point cloud is segmented into planar regions by implementing a seeded region-growing algorithm similar to the one used by Rabbani, Heuvel and Vosselman (2006). The point cloud is submitted to a curvature study for which the normal vector of each point is calculated by a principal component analysis (PCA) using the covariance method presented by Joliffe (2002). The eigenvector associated to the smallest eigenvalue can be assumed as the normal vector of the point. The study is performed by analyzing each point supported by a neighbourhood of its 50 closest points, as in Díaz-Vilariño *et al.* (2013). The algorithm chooses the point with the lowest curvature value as a region seed candidate. In each iteration, the algorithm analyzes the point cloud and includes all points satisfying two geometric conditions: planar fitting referring to the distance (d_p^r) between the point (p) and the planar region (r); and surface smoothness related to the angular difference (α) between the direction of its normal vector (np) and the region normal vector(nr) (Figure 16.3).

Given the irregular and complex elements of heritage buildings, segmentation thresholds (d_p^r and α) should be broad enough to ensure that façade points are classified as inliers in the segmentation process. For example, if the historic wall is built using non-plastered masonry, the mortar joints have different depths, being an intrusion from the wall plane. Moreover, historic walls are usually out-of-plumb and exhibit important curvature. While it is interesting to keep small intrusions conforming to the wall region, this planar fitting parameter can be useful for not including large protrusions such as those corresponding to openings.

Once the point cloud is segmented into regions that correspond to façade surfaces, they should be identified and intersected with roofs and floor in order to obtain the boundary points that enclose the points of façade planes. Figure 16.4 shows an example of this process by which segmented points describing a façade surface are enclosed by the boundary points obtained from the plane-to-plane intersection.

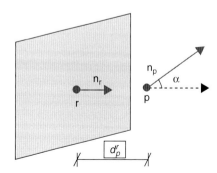

Figure 16.3 The geometric conditions to consider a point (p) included in a planar region.

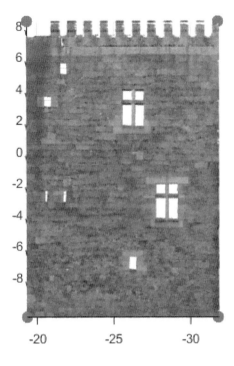

Figure 16.4 Image of a segmented façade region and the extracted boundary points enclosing them.

16.3.2 Reconstruction of openings

One of the most relevant challenges of modelling heritage elements is to deal with irregularity. Openings are commonly placed in façades following an irregular distribution as a result of the changes and renovations of the heritage building over time. Nonetheless, openings are generally characterized by being placed at a certain depth from the wall plane (intrusions) and, for this reason, they can be considered as holes in the wall, which in turn means areas without points after a planar segmentation process.

In a similar fashion to the work reported by Díaz-Vilariño *et al.* (2015b), the method to reconstruct windows and doors is based on the Generalized Hough Transform (GHT) and it consists of finding parametric shapes in edge images. Although a large variety of shapes can be parameterized (Khoshelham, 2007), rectangles are the most common shape for doors and windows.

The crucial step in this process is the correct generation of the edge images used as input for the GHT. Better edge images result in faster and more accurate results. In the approach of Díaz-Vilariño *et al.* (2015b), edge images were obtained from the reconstruction of real-colour wall images, since closed doors might be coplanar with the walls to which they belonged. In the work reported in this chapter, openings such as doors and windows are considered as holes in the point cloud, so no colour information is taken into account, and the edge images are obtained from raster wall images.

Each façade region is converted into a raster image by projecting the points on the wall plane defined by the façade boundary points. With this purpose, a rectangular matrix is created and each pixel is assigned with value "one" if the projected point cloud fills that specific pixel, or value "zero" otherwise. In this way, pixels corresponding to the wall are represented by "one" (in white) and points correspondent to an opening are represented by "zero" (in black), as shown in Figure 16.5a. If the raster image is created by assigning the value of point depth with regard to a reference point, the singularities of the façade can be represented, according to the image in Figure 16.5b. This second approach that represents singularities is outside the scope of this chapter. An alternative way to create a raster would consist in considering the intensity value of each point of the cloud returned by the laser scanner. In this regard, Riveiro *et al.* (2015) used the reflective intensity of laser light to isolate masonry blocks in the Guimarães Castle (Portugal).

For a successful rasterization, wall regions are previously rotated around the z-axis in a way that they are parallel to xz or yz axis of the point cloud coordinate system, as appropriate (Díaz-Vilariño *et al.*, 2015a). Finally, the binarized image is submitted to edge detection using the Canny method (Canny, 1986), based on finding image gradients.

The parameters involved in the detection process are represented in Figure 16.6. For each image, the origin of the xy coordinate system is established in the lower left pixel. The detection of rectangular candidates is carried out in two steps. Firstly, an R-table is created where the shape of a rectangle is stored and represented by its edge orientation (θi) and a vector defined (r, β) to an arbitrary centre point. Next, the orientation of each edge pixel (θi) is calculated and used to find the corresponding vector in the R-table. Using the vector and pixel coordinates, a centre point is reconstructed and used to cast a vote in a 4D accumulator array (a, b, Sx, Sy), where a and b are the

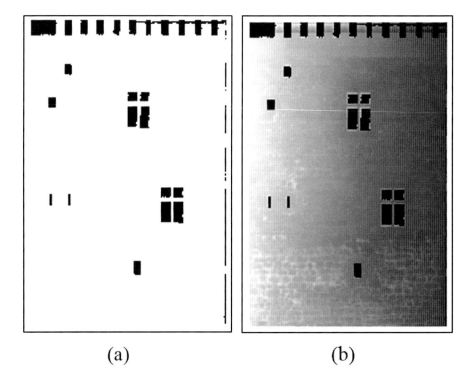

Figure 16.5 Image a represents rasterization considering the absence or presence of points. Image b shows rasterization taking into account the depth of the points with regard to the most external region point (according to the normal). In this case, mortar areas are distinguished from masonry units, especially in the lower part.

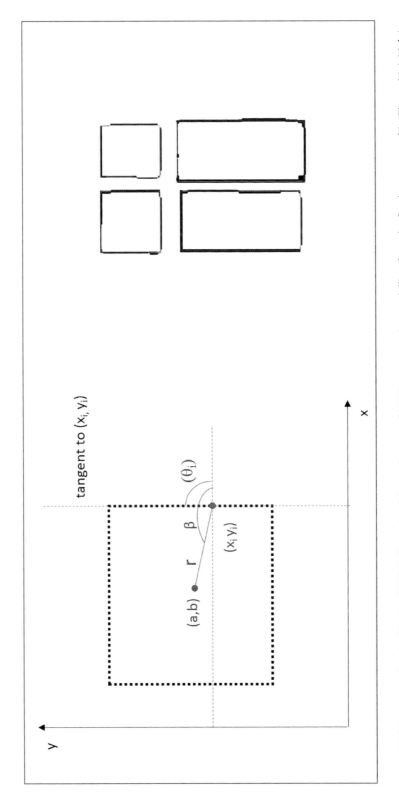

Figure 16.6 Parameters involved in the GHT (left); result of the application of GHT to an opening modelling from the façade represented in Figure 16.4 (right).

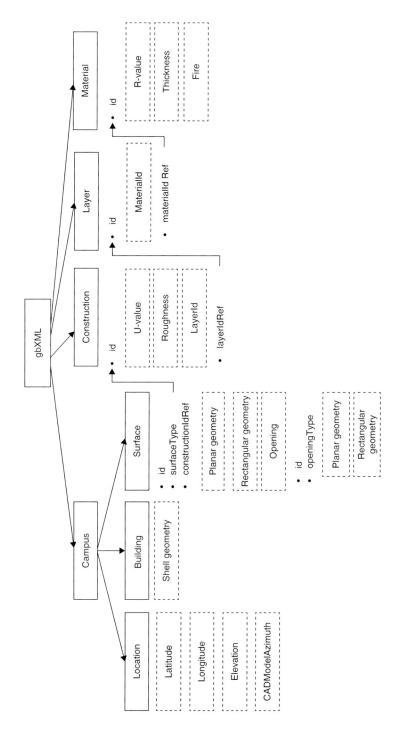

Figure 16.7 gbXML structure showing the relationship between the elements represented with contours and attributes represented with black dots.

coordinates of the centre point, and *Sx* and *Sy* are the scale parameters of the shape, corresponding to width and length of the rectangle.

Rectangles are selected by searching for maximum peaks (votes) in the accumulator array. Furthermore, the detection is enforced to a maximum and minimum opening width and height and overlapping rectangles are not allowed.

16.3.3 gbXML generation

The final step of the proposed methodology is the conversion of the extracted geometric data into a BIM-based schema for energy analysis: the Green Building XML (gbXML). It represents buildings with their geometry, semantics, topology and appearance. Because gbXML is based on the XML language (eXtensible Markup Language), it is structured as a tree of objects/elements with attributes. In Figure 16.7, the parent/child relationships among elements are represented for illustrative purposes. Geometry is contained in the "Campus" element, and the "Opening" element is a child of the "Surface" element. As shown in Figure 16.7, semantic contents such as construction, layer or material are related to the corresponding geometry by the element attributes (e.g. "id", etc.) (gbXML schema).

The schema was developed to support energy analyses, whereby surfaces are represented as simple geometries. Walls, ceilings and floors are normally modelled as planar surfaces, in most cases represented by four boundary points in their three-dimensional representation, given by its three Cartesian coordinates (x, y and z). A curved wall should be modelled as a set of continuous planar polygons. Openings need to be represented as rectangles coplanar with the wall to which they belong. These are just some of the requirements that have to be taken into account for modelling an existing building, whether being heritage or not, if the purpose is to adopt it for energy analysis, wherein parameterization provides a simplification of the building geometry. The model can be complemented with semantic attributes, such as those corresponding with materials of which they are made and their thermal properties such as R-value, thickness, conductivity and so forth. As in Lagüela *et al.* (2014), other remote measurement instruments such as thermographic cameras can be used to assist estimations of the thermal properties of building elements in a non-destructive way. Thickness could be obtained from the analysis of the point cloud if indoor rooms are acquired and registered in the same coordinate system as the façade.

Because BIM standards are mostly conceived for as-designed rather than as-built buildings, important characteristics such as thermal bridges cannot be represented in gbXML, for example. Even more, the attributes of a component are common for the whole component: the thickness value included in gbXML is unique for the entire wall, and variations of a wall thickness cannot be addressed. However, the schema could be easily extended. In this context, the critical issue would be the interoperability with applications for which the models are generated.

16.4 Case study

16.4.1 General overview

The Bragança Ducal Palace (Figure 16.8) is located in Guimarães, in northern Portugal. Its construction began during the fifteenth century, under order of the first Duque de Bragança, D. Afonso.

The palace was only permanently inhabited during the fifteenth century, and was then progressively abandoned until it became a ruin. Afterwards, in the nineteenth century, the palace was used

Figure 16.8 Image of the Palace of the Dukes of Bragança (from Google Maps).

Table 16.1 Technical characteristics of the FARO Focus3D X 330 laser scanning device according to the manufacturer datasheet (FARO Focus3D X 330 Tech Sheet, 2013)

Technical Characteristics	
Measurement range	From 0.6 m to 330 m
Ranging error (25 m, one sigma)	±2 mm
Step size (Vertical/Horizontal)	0.009°/0.009°
Field of view (Vertical/Horizontal)	300°/360°
Beam divergence	0.011°
Measurement rate (points per second)	122.000–976.000
Laser wavelength	1550 nm

for military purposes. The renovation of the palace was carried out between 1937 and 1959 based on a project by the architect Rogério de Azevedo. Currently, the Ducal Palace is classified as a national monument, and it involves a museum at the first floor, as well as a specific quarters in the second floor assigned to the President of the Republic.

Besides its historical importance, the palace is one of the most visited museums in Portugal. Due to its characteristics and geometric complexity, it was selected as a case study to assess the feasibility of the proposed methodology.

16.4.2 Instruments and data

The survey of the palace was carried out with a terrestrial laser scanner FARO Focus3D X 330 with the technical characteristics summarized in Table 16.1.

Data was acquired from different scan positions in order to cover the entire area of interest. Ten scans were carried out in order to complete the interior and exterior façade survey and they were placed to ensure an overlapping between consecutive stations, enabling the registration into the same coordinate system by manually selecting common reference points. Although colour is not necessary for implementing the methodology, it was acquired to ease the recognition of singularities in the point cloud, such as corners or windows, which are used as reference points between adjacent scans. Figure 16.9

shows the surveying strategy carried out for acquiring the complete façade of the palace, taking into account its dimensions and its surroundings. Roofs and buried parts of the palace are not acquired.

Points corresponding to exterior elements that do not belong to the palace itself, such as trees, grass, outer walls and so forth were deleted from the point cloud in order not to consider them for subsequent steps. Furthermore, the point cloud is submitted to an octree filter (Meagher, 1982) to ensure uniform density. Octree parameters are 0.15 m for x, y, and z directions, and the resulting point cloud contains 441,075 points. This process was carried out in RiSCAN PRO software. Figure 16.10 shows the point cloud obtained from registering all scans and cleaning exterior elements not belonging to the building. The origin of the coordinate system is established in the origin of scan position 1 (sp-1 in Figure 16.10).

16.4.3 Processing results

As explained in Section 16.3.1, segmenting the point cloud is the first step of the methodology. The segmented regions are the basis for the rasterization, so this step should be carefully addressed. Due to the large size of the point cloud, a two-step segmentation was implemented in this case study. Firstly, a rough segmentation was carried out to create general regions and to obtain the limits of the façades. In this case, the angular and distance point-to-plane thresholds were 45° and 0.5 m respectively, resulting in 553 regions. Regions were selected according to a minimum region size (5,000 points) resulting in 21 regions, which were classified in façades and floor according to their normal (Figure 16.11).

Secondly, a fine segmentation was implemented to refine wall regions. The objective was to deselect all the points that actually do not belong to the wall surface, such as window glass, reflections or window frames. For the purpose of obtaining a higher point density for creating the raster images, the information of the previous segmentation (i.e. definition of the location of each wall) was used

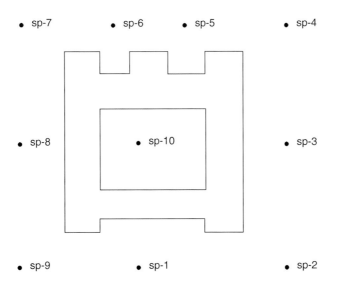

Figure 16.9 Schema of the surveying strategy for this case study.

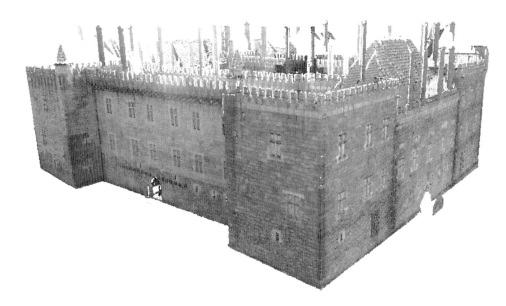

Figure 16.10 Point cloud of the complete façade of the palace, displayed in true-colour.

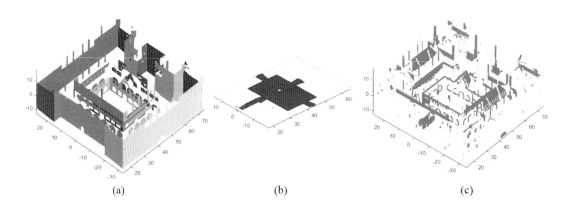

Figure 16.11 Planar regions are classified as walls (a), floor (b) and remaining regions (c) (Units: m).

for selecting the new input data from subsets of original point cloud with density of 1 point/0.01 m for X, Y and Z directions of the point cloud coordinate system. The angular threshold and distance point-to-plane were 45° and 0.10 m, respectively, in this case. With these conditions, if walls are slightly curved due to their large size, construction sequence or historic alterations, points far away from the centroid might fail in the point-to-plane condition. To avoid this, the fine segmentation was processed by using a mask of 5 m length along an horizontal plane, and covering the full height of the wall. An overlapping of 0.5 m was considered between adjacent masks.

From the refined wall regions, raster images were created with 0.06 m resolution. In the rasterization process, a pixel is considered as occupied if at least four points from the point cloud are inside the pixel (being density of the point cloud 1 point/0.01 m). This condition was established for avoiding falsely occupied pixels caused by the existence of one isolated point. Afterwards, the binarized raster is processed by the Canny edge detector. Figure 16.12b shows the rasterization result of the point cloud corresponding to the main façade (Figure 16.12a) and Figure 16.12c corresponds to the edge image of Figure 16.12b.

In this case, no morphological operations were implemented to improve the quality of the raster since point cloud density was sufficient to adequately fill all wall pixels at the resolution of 0.06 m. Resolution should be limited while ensuring that enough points are available to obtain a good edge image quality (Figure 16.12c). This strategy reduces the computational time and improves the quality of the openings detected, as openings with noisy edges are more difficult to detect by the GHT.

In the final step, the edge image is used in a rectangle search based on the GHT. As the number of rectangles is not known, a large number of peaks are searched, and opening candidates are selected by their vote ratio. Furthermore, the analysis is enforced to a minimum and maximum opening width and height. Figure 16.13 represents the results of an opening detection enforced to a minimum/maximum width and height of 130 cm/200 cm and 160 cm/250 cm, respectively. As the number of openings is assumed as unknown, a high number of peaks is searched (50 peaks), and then opening candidates are selected if their voting percentage is above the 75th percentile.

In this wall no false positives were detected, but they may be obtained in the presence of elements similar in size and shape to the objects that are being searched (Díaz-Vilariño et al., 2015b). With

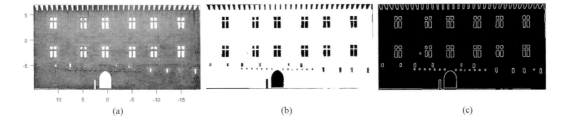

Figure 16.12 Main façade of the Ducal Palace: (a) point cloud; (b) raster image; (c) edge image (Units: m).

Figure 16.13 Results of applying the GHT to the main wall.

regard to false negatives, they can be obtained if previous steps produce a bad-quality edge image. For instance, a noisy edge can be caused by a bad segmentation due to the coplanarity of a window with the wall where it is contained, or due to the presence of additional structures such as window frames. The search can be extended to other parametric shapes and enforce the detection with different rules. For instance, if one wants to detect a door, the y coordinate of the centroid should be positioned to a certain distance from the floor. Finally, image coordinates of the openings are reprojected to obtain the 3D coordinates necessary to reconstruct the gbXML (Díaz-Vilariño et al., 2015b).

16.4.4 gbXML

Once the geometry of the parameterized components has been extracted, it is represented according to the gbXML schema (Figure 16.7). The global position of the building is defined in the element "Location", which is completed with "Latitude", "Longitude" (geographic coordinates in decimal degrees), "Elevation" (in meters) and "CADModelAzimuth" (the orientation from north). The "Building" element defines volumes enclosed by surfaces, typically indoor spaces. Each surface is represented by the "Surface" element, which is defined by several attributes that provide semantic information to the object, for example, attributes "id" or "surfaceType" attached to "ExteriorWall", "Roof", "Shade", "RaisedFloor", etc. Figure 16.14 (right) represents a fragment of the gbXML schema, wherein some elements such as "Building", "Surface" or "Opening" are shown.

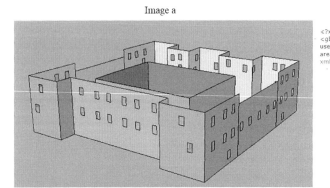
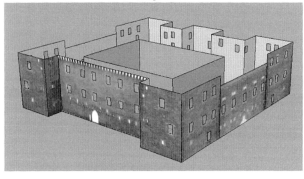

Figure 16.14 The gbXML schema is directly imported into SketchUp using the OpenStudio plug-in. Image a shows the model without texture and image b shows the model texturized with images from the point cloud (the roof has been deleted for visualization) (left). Fragment of the gbXML where windows from the main façade (Figure 16.12) are represented (right).

The semantically rich 3D model can be used for energy analysis. Image a from Figure 16.14 shows a capture of the model directly imported into SketchUp by using the plug-in OpenStudio from EnergyPlus, wherein the recognition of semantic attributes puts in value the resulting model with regard to other non-semantic models. On the other hand, image b shows the model texturized with images from the point cloud, representing different complementary levels of knowledge about the building.

16.5 Conclusion

This chapter deals with the use of laser scanning for documenting the real geometry of a heritage building and the use of automated techniques to parameterize the point cloud in order to define building façade elements according to energy analysis specifications.

Although laser scanners allow the accurate and efficient 3D digitalization of heritage, point clouds should be parameterized to create sets of components. The parameterization required for energy analysis software is highly specific since building elements should be represented as rectangular polygons. This form of representation is usually far from the reality, especially in the case of heritage buildings, which are often characterized by singularities associated to the architectonic style, the traditional way of construction and renovations over time. Methodologies and results presented in this chapter are not suitable for visualizing and documenting the heritage and cultural value of building elements. However, despite the fact that they do not represent a BIM in the broader sense, they represent a BIM (and particularly HBIM) in the narrow sense since the results fit with the purpose of modelling the heritage building for a specific application.

From the processing methodology, the following specific conclusions can be drawn:

- Segmentation parameters should take into account the irregularity of heritage building components to ensure a complete segmentation of wall façades.
- Rasterization resolution depends on the point cloud density for obtaining good-quality edge images.
- Higher resolution implies higher computational time and poses difficulties to the algorithms for detection of openings.
- After the first coarse segmentation, a second finer segmentation is necessary for deselecting windows and other openings. Noisy window edges make their detection more difficult, causing false negatives.
- Several parametric patterns can be used for detecting objects with similar shape in edge images. False positives can be detected in the case of the presence of other objects with similar size and shape.

The overall conclusion of this chapter is that simple automatic techniques for processing point clouds can be used in heritage buildings for creating a BIM model if the purpose is to parameterize and simplify the real geometry, according to the requirements of the application to which it is going to be submitted. The resulting 3D semantically rich model enhances the knowledge of the heritage building, complementing other representations of the facility such as point clouds or handmade HBIM.

Acknowledgements

The authors would like to thank the Ministerio de Economía y Competitividad (Gobierno de España) for the financial support given through their human resources grant (FPU AP2010–2969, ENE2013–48015-C3–1-R). This work is also partly supported by FEDER funds through the

IACOBUS program and by FCT (Portuguese Foundation for Science and Technology), within ISISE, project UID/ECI/04029/2013. The authors would also like to thank all the institutions and companies that have been involved in supporting and contributing to the development of this study, in particular, Isabel Fernandes and Flávio Vieira, director and staff member of Bragança Ducal Palace. Research of the second and fourth author is supported by a National Priority Research Program NPRP award (NPRP-06–1208–2–492) from the Qatar National Research Fund (a member of the Qatar Foundation). The statements made herein are solely the responsibility of the authors.

References

Adan, A., Huber, D. 2011. 3D reconstruction of interior wall surfaces under occlusion and clutter. *3D Imaging, Modeling, Processing, Visualization and Transmission (3DIMPVT)*, 16–19 May 2005, pp. 275–281.

Autodesk. 2002. *Autodesk BIM White Paper*. Available at: <http//www.laiserin.com/features/bim/autodesk_bim.pdf> [Accessed 14 June 2015].

Bosché, F., Ahmed, M., Turkan, Y., Haas, C.T., Haas, R. 2015. The value of integrating scan-to-bim and scan-vs-bim techniques for construction monitoring using laser scanning and bim: the case of cylin-drical MEP components. *Automation in Construction*, 49, pp. 201–213.

Brumana, R., Oreni, D., Raimondi, A., Georgopoulos, A., Bregianni, A. 2013. From survey to HBIM for documentation, dissemination and management of built heritage: the case study of St. Maria in Scaria d'Intelvi. *Proceedings of the 1st International Congress of Digital Heritage*. Marseille, France, 28 October–1 November 2013, pp. 497–504.

BuildingSmartFinland (2012), "Common BIM Requirements 2012", BuildingSmart Finland, http://www.en.buildingsmart.kotisivukone.com/3

Canny, J. 1986. A computational approach to edge detection. *IEEE Transactions in Pattern Analysis and Machine Intelligence*, 8, pp. 679–698.

Díaz-Vilariño, L., Conde, B., Lagüela, S., Lorenzo, H. 2015. Automatic detection and segmentation of columns in as-built buildings from point clouds. *Remote Sensing*, 7, pp. 15651–15667 (a).

Díaz-Vilariño, L., Khoshelham, K., Martínez-Sánchez, J., Arias, P. 2015. 3D modelling of building indoor spaces and closes doors from imagery and point clouds. *Sensors*, 15, pp. 3491–3512 (b).

Díaz-Vilariño, L., Lagüela, S., Armesto, J., Arias, P. 2013. Semantic as-built 3D models including shades for the evaluation of solar influence on buildings. *Solar Energy*, 92, pp. 269–279.

DURAARK (2015), "Durable Architectural Knowledge", EU funded DURAARK project, http://duraark.eu/

Eastman, C., Teicholz, P., Sacks, R., Liston, K. 2008. *BIM Handbook, a Guide to Building Information Modeling for Owners, Managers, Designers, Engineers, and Contractors*. Hoboken, NJ: John Wiley & Sons.

Fai, S., Graham, K., Duckworth, T., Wood, N., Attar, R. 2011. Building information modeling and heritage documentation. *Proceedings of the XXIIIrd International CIPA Symposium*. Prague, Czech Republic, 12–16 September 2011.

FARO Laser Scanner Focus3D X 330 Tech Sheet, 2013. Available at: <http://www.faro.com/> [Accessed 20 December 2015].

GbXML schema, 2015. Available at: < http://www.gbxml.org/ > [Accessed 20 December 2015].

GSA (2009), "BIM Guide for 3D Imaging", General Services Administration of USA, http://www.gsa.gov/portal/mediaId/226819/fileName/GSA_BIM_Guide_Series_03.action

Isikdag, U., Zlatanova, S., Underwood, J. 2013. A BIM-oriented model for supporting indoor navigation requirements. *Computers, Environment and Urban Systems*, 41, pp. 112–123.

Joliffe, I.T. 2002. Principal component analysis. *Series in Statistics*, 28, p. 487. Springer.

Khoshelham, K. 2007. Extending generalized Hough transform to detect 3D objects in laser range data. *Proceedings of the ISPRS Workshop on Laser Scanning and SilviLaser*. Espoo, Finland, 12–14 September 2007, 36, pp. 206–210.

Lagüela, S., Díaz-Vilariño, L., Armesto, J., Arias, P. 2014. Non-destructive approach for the generation and thermal characterization of an as-built BIM. *Construction and Building Materials*, 51, pp. 55–61.

Lagüela, S., Díaz-Vilariño, L., Martínez-Sánchez, J., Armesto, J. 2013. Automatic thermographic and RGB texture of as-built BIM for energy. *Automation in Construction*, 31, pp. 230–240

Meagher, D. 1982. Geometric modeling using octree encoding. *Computer Vision Graphics and Image Processing*, 19, pp. 129–147.

Murphy, M., McGovern, E., Pavia, S. 2009. Historic building information modelling (HBIM). *Structural Survey*, 27, pp. 311–327.

Ochmann, S., Vock, R., Wessel, R., Klein, R. 2016. Automatic reconstruction of parametric building models from indoor point clouds. *Computers & Graphics*, 54, pp. 94–103.

Ochmann, S., Vock, R., Wessel, R., Tamke, M., Klein, R. 2014. Automatic generation of structural building descriptions from 3d point cloud scans. *Proceedings of the International Conference on Computer Graphics Theory and Applications*, Lisbon, Portugal, 5–8 January 2014.

Oreni, D., Brumana, R., Georgopoulos, A., Cuca, B. 2014. HBIM library objects for conservation and management of built heritage. *International Journal of Heritage in the Digital Era*, 3, pp. 321–334.

Patraucean, V., Armeni, I., Nahangi, M., Yeung, J., Brilakis, I., Haas, C. 2015. State of research in automatic as-built modelling. *Advanced Engineering Informatics*, 29, pp. 162–171.

Rabbani, T., Heuvel, F.A., Vosselman, G. 2006. Segmentation of point clouds using smoothness constraint. *International Archives of Photogrammetry, Remote Sensing and Spatial Information Sciences*, 36, pp. 248–253.

Riveiro, B., Lourenço, P.B., Oliveira, D.V., González-Jorge, H., Arias, P. 2015. Document automatic morphologic analysis of quasi-periodic masonry walls from LiDAR. *Computer-Aided Civil and Infrastructure Engineering*. DOI:10.1111/mice.12145.

Simeone, D., Cursi, S., Toldo, I., Carrara, G. 2014. *BIM and Knowledge Management for Building Heritage*, pp. 681–690. Available at: <http://www.academia.edu/> [Accessed 20 December 2015].

Tang, P., Huber, D., Akinci, B., Lipman R., Lytle, A. 2010. Automatic reconstruction of as-built building information models from laser-scanned point clouds: a review of related techniques. *Automation in Construction*, 19, pp. 829–843.

Volk, R., Stengel, J., Schultmann, F. 2014. Building information modeling (BIM) for existing buildings – literature review and future needs. *Automation in Construction*, 38, pp. 99, 84–91.

Xiao J., Furukawa, Y. 2012. Reconstructing the world's museums. *Proceedings of the 12th European Conference on Computer Vision (ECCV2012)*, 7–13 October 2012, pp. 668–681.

17 Participatory sensing for community engagement with HBIM

John Counsell and Gehan Nagy

17.1 Introduction

This chapter focuses on the development of HBIM for active community engagement in valuing and conserving local architectural heritage. European and worldwide threats from climate change and peak resources are challenging us to sustainably but substantially repurpose and modify our occupied architectural heritage. Active community engagement is already identified as key for the conservation of world monuments and world heritage. The ICOMOS Burra Charter (Australia ICOMOS 2013) describes best practice standards for engaging communities in managing cultural heritage places. Although these are focused on Australia, its emphasis on community participation in establishing significance and value, and particularly on taking an active part in conservation and management, can be seen as a more universal imperative. Thus, the World Monument Fund also recently stated that "sustainable preservation only happens through local stewardship" (WMF 2015). Engaging communities with cultural heritage through HBIM and low-cost technologies is argued to be an essential stage towards such full and sustainable heritage community engagement.

17.2 A need to adapt occupied heritage to climate change

There is an increasing urgency to address the major global challenge of mitigating the effects of and adapting to unavoidable climate change. In Europe over one-third of total carbon-related emissions are from the stocks of millions of existing buildings. These stocks cannot sustainably be fully and swiftly replaced with new. Meijer et al. (2009) state, "in most countries the rate of new construction is around 1% of the total stock". So most of these stocks of carbon-emitting existing buildings will still be in use in 2100. Demolition and reconstruction is particularly undesirable where these buildings form part of our common heritage. Such common heritage buildings (dating from before 1919) form around 20% of the stock of 223 million (CRESME 2014) dwellings in Europe. Meijer et al. (op. cit.) give figures for pre-1919 heritage buildings that range from: Wales (34%); England (22%); France (20%); Austria (19%); Germany (14%); to the Netherlands (8%). Many of these are occupied. The challenge for climate change mitigation is to conserve their significance and value but minimise their carbon impact.

17.3 Challenges in sustaining such heritage buildings

While monuments and major works of built heritage are usually protected by legislation, in some countries that legislation slows change but does not preclude it. Much of the more commonplace architectural heritage that forms the backbone of heritage cities and landscapes is not protected. Where that heritage is evidently decaying, it is often undervalued, and is then more likely to become subject to clearance and reconstruction. The 'traditional' techniques and materials deployed prior to the First World War were often significantly different to those of today. For example, many masonry construction buildings had solid external walls, in which both the masonry and the lime mortar-based surfaces and bonding permitted moisture movement, reducing internal moisture and unhealthy mould growth. The uninformed application of modern impervious insulants, or conventional paints (also impervious) can seal moisture into the structure, leading towards reduced thermal resistance, accelerated frost damage and consequent decay. A similarly damaging process can occur by changes in the context of masonry buildings. The vibration from increasing volumes of heavy motor traffic is one challenge. In the Middle East, for example, the increasing discharge of water into streets and alleys around heritage buildings has led to the formation of what are termed 'perched' water tables, where the water ponds within the soil close to the surface. Without damp-proof courses such traditional buildings in these locations can decay from the ground up, due to the 'wicking' evaporative effect on the exposed surface of the wall aboveground combined with the destructive expansive force of hygroscopic salt crystal formation. Many additional vulnerabilities of a variety of forms of built cultural heritage to predicted climate change were identified in the EU-funded Noah's Ark research project (Sabbioni et al. 2009). This led to their publication of a vulnerability atlas, accompanied by a warning "that although climate change attracts wide interest at research and policy levels, little attention is paid to its impact on cultural heritage" (Sabbioni et al. 2012).

Such buildings become less commercially valuable as they decay, commanding lower resale prices or rents, and so enter a spiral of decline where they are increasingly less likely to justify substantial repair work. This decline similarly affects the value that local communities place upon the heritage they occupy. Protecting heritage buildings and landscapes and sustaining the cultural values of communities and conserving them for future generations is a real challenge for developers, architects and professionals, as awareness of cultural heritage and its values vary from one cultural context into another. Within the Middle East context, cultural heritage and heritage sites may be underestimated due to lack of awareness, which requires proactive mechanisms for engaging communities within the heritage sites. Settled communities may articulate the quality of the community bond, without identifying the effect contributed by their physical environment. Members of less settled communities may aspire to move into 'modern' buildings, unaware of the qualities or 'value' of the heritage around them.

Value is held to have always been "the reason underlying heritage conservation. It is self-evident that no society makes an effort to conserve what it does not value" (De la Torre & Throsby 2002). The New Zealand government describes a central issue as "an apparent undervaluing of historic heritage by the New Zealand public" (Thornley & Waa 2009). They go on to describe research to increase the perceived value of historic heritage among the New Zealand public, deploying social marketing interventions. They base this upon the 1992 ICOMOS New Zealand Charter definitions of the value of historic heritage, which states that heritage places: "have lasting values and can be appreciated in their own right; teach us about the past and the culture of those who came before us; provide the context for community identity, whereby people relate to the land and to those who have gone before; provide variety and contrast in the modern world, and a measure against which we can compare the achievements of today; provide visible evidence of the continuity between past, present and future." They report a study (Bell 2003) in which it was found that "participation in decision-making can motivate people to become involved". "The study concluded that citizen-led volunteering experiences were more likely to have committed volunteers, promote new voluntary action, and improve the skills of volunteers in contributing to conservation advocacy." These are similar findings to those about broader public engagement in planning, in which it has been found that individuals are far more likely to contribute if they believe their voice will be heard and their contribution will have an effect. Perkin (2010) describes the Bendigo project in Australia, where a bottom-up community approach is contrasted with the 'usual' authoritarian one. The key in this engagement was found to be the appointment of a 'project manager' held to have "an important promotional or marketing role that extends beyond their advocacy position".

Together with the attrition of previously valuable and valued heritage building stock there has been a long-term steady attrition of specialist conservation skills and craft knowledge among building professionals (Council of Europe 1998). This has not changed significantly; there remains a recognised EU need to "improve the educational base for heritage training" (CECIH – Transylvania Trust 2014). Given such a substantial skills and expertise gap, it is unlikely that this situation can be transformed fast enough to address the challenge of adapting heritage to climate change, while enhancing quality of life. We argue that new digital technologies offer great opportunities for engaging communities and enhancing their awareness of cultural heritage values, in addition to empowering them to gain and deploy requisite heritage conservation and adaptation skills and instrumentation, leading to more assured protection and rehabilitation, and a massively scaled-up approach.

17.4 'Deep renovation' focuses pose a challenge to built heritage

Even what is termed 'deep renovation' can threaten the heritage significance of such buildings. Adaptation and alteration for energy efficiency (and the associated carbon emissions reductions) is now often executed via such new construction techniques leading, for example, to inappropriate construction behind historic facades; the deployment of unsuitable new insulants or components; unsuitable construction processes; or even paints that inadvertently seal previously breathable walls. These buildings are also often occupied by the economically disadvantaged and fuel poor. It is clearly of utmost importance that the potentially disastrous impact of climate change on the poorest in society is mitigated. Consequently, there are emerging governmental top-down initiatives to alter and upgrade such existing buildings, such as the UK Green Deal. "In the funding period 2014–2020, the European Structural Fund is likely to more than double the funding available to co-finance national investments in energy efficiency – including buildings" (Renovate-Europe 2013). They then argue that "deep renovation of the EU building stock will kick-start the European economy by helping to create up to 2 million local, direct jobs by 2020. Buildings-related activity accounts for ~10% of EU GDP and ~8% of direct employment. Building and construction currently employs over 14 million workers – set to rise in the next few years in order to service the needed renovation" (ibid.). This construction GDP focus, without commensurate education and training, puts both heritage and tourism GDP at risk. Due to the prevailing lack of skills, there is grave risk that solutions that may suit recent buildings are widely applied in ignorance to such older stock that then fail to perform as expected, contribute to accelerated decay and so increase the difficulties of an already disadvantaged population.

17.5 Active community engagement is scarce

Active community engagement is not sufficiently recognised as important for the sustainable regeneration of the more 'mundane' occupied architectural heritage that forms the backbone of heritage cities and landscapes. Yet these buildings contribute substantially to the sense of place and identity of urban and rural areas, and thus enhance the well-being of local communities. They are often enormously important to maintaining healthy local economies and sustaining tourism GDP. While there have been a number of attempted technical solutions to modelling and valuing architectural heritage, most have not been driven by active local engagement, and probably for this reason have thus not achieved a long-term sustainable impact. Most of the prevailing approaches to garnering community support for heritage can be seen as 'passive', focused on 'secondary interpretation'. Primary interpretation is the purposeful analysis of heritage structures, usually by experts with a need for informed understanding prior to undertaking some conservation activity. Secondary interpretation filters that primary analysis into a readily graspable narrative that attempts to invoke in the audience a sense of the historic or social importance of the site. One could cynically term this prevailing approach as 'edutainment', intended to persuade the public that their ticket purchase has value or that their tax money is being spent wisely. While the active engagement of the immediate occupants is suggested as a key requirement, the Burra Charter (op. cit.) makes clear the importance of the active engagement of a wider 'community' of stakeholders. Thus, the full community that needs to be engaged is not physically co-located. Even in communities that are co-located, the pressures of modern life often preclude the regular face-to-face meetings that build consensus. So there is still a major research question to be addressed: How may emerging technologies and new social media (integrated into HBIM) assist stakeholders and local communities to actively engage with the challenges of enhancing the quality of life of the occupants and responding to climate change, while still valuing, caring for and conserving built cultural heritage?

17.6 Active community engagement is important

The OECD (2001) described the emergence of new forms of representation and public participation 'in all our countries'. In their document on "Citizens as Partners: Information, Consultation and Public Participation in Policy-Making", public involvement in government decision-making is defined in several stages, ranging from commonplace information to rare, but highly desirable, active participation:

- Information, by which government informs citizens in a one-way relationship;
- Consultation, that forms a two-way relationship in which citizens are asked to contribute their views within a framework defined by government;
- Active participation, that creates a relationship based on partnership between citizens and government, in which citizens join in setting the agenda, shaping policy options, and shaping the policy dialogue.

(OECD 2001)

Kingston et al. (2005), in analysing the above, further posed the question: "Should users be able to upload their own spatial data rather than just comments and ideas?" That theme is at the heart of this chapter.

17.7 Challenges of the 'digital divide'

The deployment of a fully digital system for community engagement via HBIM carries risks. A particular risk is described as the 'digital divide'. Particularly in older heritage urban centres, the occupants of these local communities are often diverse, with both short- and long-term occupants, old and young. Where the heritage is decaying and the occupants lack amenities and are at risk of fuel poverty, they are also often on the wrong side of the digital divide, not due to lack of information literacy or search skills (Mossberger et al. 2008), but due to lack of affordable access, risking increasingly unequal online participation in the future (Brandtzæg et al. 2010). There is also a prevailing skills gap, which has been identified as creating another form of digital divide. Brandtzæg et al. (op. cit.) go on to identify that in 2005 across a representative sample of European countries, "60% of the population was found to be either Non-Users or Sporadic Users" (whereas only 12% could be described as advanced users). The UK and Norway were described as having better Internet penetration and so higher use. Two years later Mossberger et al. (op. cit.) quoted research in the USA where 74% were occasional Internet users. The recent Eurostat (2015) figures give frequency of Internet use in 2014 (% of individuals aged 16 to 74) as 75%, and broadband access by households as fractionally over 75% for the same period, in the EU 28. While there is likelihood of further penetration of Internet use since 2014, one must also consider that there may remain a proportion of such populations without telephones of any sort, let alone Internet.

17.8 Need for mechanisms to capture information at source into HBIM

Dr Nagy was the manager of the Rotary-funded Remal Foundation project of renovating and upgrading slums in the Warsha area of Manshiet Nasser. This dense urban settlement lies at the heart of the historic centre of Cairo. It has been described as a slum, occupied by almost a million inhabitants, many of them among the poorest in Cairo. In many cases they are thus limited in access to the Internet or the wider range of digital technologies. This project engaged with the local community to co-design a public spaces enhancement scheme, since substantially implemented. It used digital technologies to generate images to show what the transformed spaces would look like. The face-to-face community consultations and feedback were recorded in digital images and transcripts (not georeferenced). The German government (Arab German Yearbook 2010) similarly funded participatory processes in the area, including an infrastructure upgrading project focused on water supply, sanitation and roads, to reduce health risks and further enhance quality of life. "The key element is that residents are involved in the planning processes and that local democracy is promoted; the residents are encouraged to put forward their own solutions in order that aid projects can be tailored to their needs. The sense of ownership of the improved facilities by the beneficiaries will thus guarantee the sustainability of the project" (ibid.). It is claimed that not only has this worked, local inhabitants do feel a sense of ownership, but that similar schemes are now being promoted elsewhere in Cairo and Alexandria.

Thus, while the EU states, "The development of the information society is regarded as critical to meet the demands of society and the EU economy," not everyone has access, so it is often argued that hybrid systems are still needed. Indeed, out of three e-participation case studies reported by the Balance Project (2007), the best result was that 25% of the total number of stakeholders who took part engaged by digital means. The report did conclude that these were more complete than those by other means, and also "were the most precise and accurate responses". While they then concluded that "e-participation is an additional tool and should not replace other, more personal participation techniques," they also concluded that "e-participation enables involvement of all societal groups at any time and any location."

However, it can instead be suggested that if HBIM is to fully function from the start, it should be a repository that contains the input and outcomes of the active participation of a fully representative group of citizens and stakeholders, whether digitally enabled or not. This implies that where there is a physical process for those without digital tools, such as those in Manshiet Nasser, it is most important to capture that process digitally, to georeference the information, and to make it available digitally for broader 'involvement of all societal groups at any time and any location'. Hence community interactive engagement and co-design could be enhanced by digital recording of the process, with associated georeferencing to the physical environment forming the focus of the enquiry, using, for example, methodologies such as 'Planning 4 Real' (MIT 2001) or 'Enquiry by Design' (Wates & Thompson 2013) (which originated in the UK, but are now used more globally). Possible 'volunteered' approaches to this recording process are outlined below.

17.9 Forecast citizen engagement in BIM, additional requirements for HBIM

The UK Department for Business, Innovation & Skills (BIS 2015), developing specification for BIM level 3 (post 2016), forecasts "dramatically better use of current and future built assets" empowering "citizens and stakeholders with the capability and confidence to make the most of the opportunities these technologies provide". It may be argued that citizens and stakeholders in HBIM are all the Burra Charter (op. cit.) 'associated groups and individuals', and that tools should be integrated into HBIM that promote that 'local stewardship'. Others argue for "public engagement in the construction and appropriation of archaeological knowledge . . . participative archaeology" (NEARCH 2014). For full engagement using HBIM in such activities, tools will be required that permit computer-supported collaborative work (CSCW). CSCW has long been an interdisciplinary research field. Some of the challenges being addressed in this research should be applicable to community engagement in HBIM, provided there are also tools that address the wider heritage-focused requirements such as arriving at community consensus on significance and value.

17.10 Wider engagement in HBIM by communities

In the HBIM workshop in Luxor in March 2015, we discussed, 'What would make HBIM use widespread?' UNESCO (2003) in its charter on the "Preservation of Digital Heritage" defined it as consisting of "unique resources of human knowledge and expression. It embraces cultural, educational, scientific and administrative resources, as well as technical, legal, medical and other kinds of information created digitally, or converted into digital form from existing analogue resources. Where resources are 'born digital', there is no other format but the digital object." We would argue that this is already a community activity, but that communities are not aware of their contribution to the creation of or stewardship of 'digital heritage'. There are currently also few tools to support the geo-tagging of 'digital heritage artefacts' to relevant physical locations in cultural heritage landscapes, buildings or structures. UNESCO also points out that such artefacts are "frequently ephemeral, and require purposeful production, maintenance and management to be retained", but that it is important to do so, since many "have lasting value and significance". (Chapter 3 discusses the need to georeference digital and intangible heritage via HBIM tools.) The New Zealand government also pointed out that the most immediate step towards wider valuation of heritage is to raise "awareness of historic heritage within the community". Figure 17.1 shows the digital technologies for cultural heritage awareness.

We suggest that 'raising cultural heritage awareness' can be affordably achieved by digital means. There is a need for a well-designed framework, developing from simple processes to a more complex and complete HBIM. One simple process would focus on tools to permit individuals and communities to assess the value and significance of their digital heritage, and to georeference it. Another would be to deploy the range of simple-to-use emerging technologies for use by communities at their local heritage sites. These approaches include gaming, augmented reality for navigation guidance and semantic information, mixed reality and avatars for storytelling and immersion. Such applications would help the visitors to interact with and experience the site and be better informed about the intangible culture related to the site and thus better value and assess its significance. Webcams can now be deployed to give a similar experience via the web, and begin to build towards a network of virtual museums. Such a fusion of augmented and virtual reality with real-time video, applied to heritage landscapes and structures, was explored in the EU-funded Framework 5 'Valhalla' project

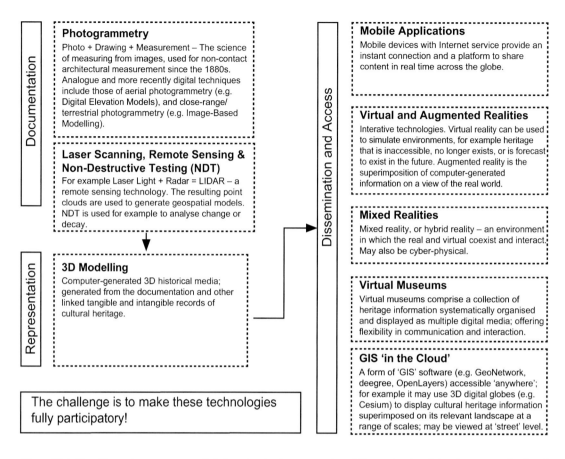

Figure 17.1 The illustration represents digital technologies supporting different levels of development of cultural heritage awareness via digital heritage.

(Counsell et al. 2003). Figure 17.2 illustrates a framework for community engagement using digital technologies.

17.11 Motivating HBIM use by such communities

In the Luxor workshop the question was also raised, 'What motivation would be required to get widespread community engagement?' This chapter suggests that participatory sensing is going to become a key part of the answer, via increasingly widespread affordability of what has up to now been expensive scientific tools, and active community engagement. It is already argued that there is a more general massive expansion in our power to digitise the world via "volunteered geographic information"(Geocrowd 2014). The Geostream (2015) FP7 research project described this as the "oncoming geospatial data tsunami", while the ongoing Canadian Geothink research project pertinently argued that "Geospatial Web 2.0 is Reshaping Government-Citizen Interactions" (Geothink 2013). We do not suggest that there is an effective active community engagement focused HBIM process or ICT platform yet. However, we do suggest that there are promising indicators, and that promise is imminent. These are discussed below.

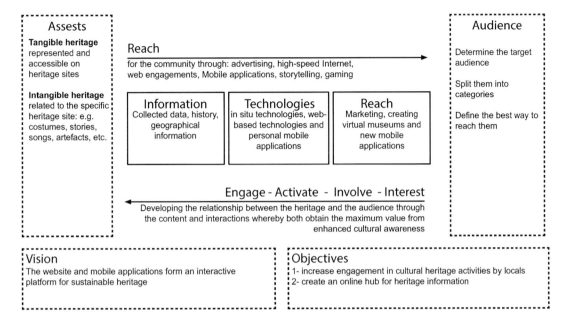

Figure 17.2 A framework for community engagement with digital heritage, based upon Visser & Richardson (2013).

17.12 Participatory sensing

"The sensing process can be broken down into the following steps, each of which is facilitated by corresponding technology: Coordination, Capture, Transfer, Storage, Access, Analysis, Feedback, and Visualization" (Goldman et al. 2009). In this model, particular special challenges for HBIM are to: A) enthuse participants (coordination step); B) identify and integrate appropriate low-cost sensor technologies into smartphones and similar low-cost devices in the capture step; C) to calibrate that data, and develop and make widely available expertise in the analysis and visualisation steps; and D) Storage standards. For example, "Failing to prepare for the future of their photographs could potentially leave future conservators without records from our time" (Beck 2013).

17.13 Coordination step, via integrated social media

It is argued that the "whole e-participation process is assisted by social media tools and visualization techniques that offer … essential information at a glance" (Sonntagbauer 2014). The VEP3D (2008) Interreg NEW-funded project (2003–2008) examined e-participation in planning, and included a case study of a heritage building reconstructed by the National Museum of Welsh Life. The case study explored the range of different heritage information available and how to deploy this in an interactive visual form on the web, in conjunction with tools for community engagement and commentary. For fully informed and empowered local stewardship it is suggested that new and emerging social media need to be integrated into a fully accessible interactive heritage repository or HBIM, and that "volunteered information" will form a key part of this, with particular challenges for data validity and reliability. For those not fully motivated to participate, others suggest effective mechanisms for active engagement, such as gamification (Budde 2014a). Social media may well be the most promising avenue with which to develop the necessary mechanisms for recording the digital outcomes of physical consultation and engagement processes such as Planning4real.

17.14 Capture step, via citizen sensing and volunteered geo-information

It is argued that "Participatory Sensing . . . has enabled a multitude of novel applications in recent years, ranging from collaborative noise pollution maps . . . cyclist experience reports . . . to automatically characterizing places" (Budde et al. 2014b). There have been a number of exploratory projects seeking to engage local communities or volunteers in recording (and analysing) heritage using low-cost technologies. For example, the English Heritage Rock Art project successfully engaged volunteers in photogrammetry using off-the-shelf cameras and low-cost software (ERA 2008). More recent research projects predict that smartphones will enable "lay people to become more involved in their local heritage, an important aspiration identified by English Heritage" (Kirchhoefer et al. 2011). Other projects have already found that engaging 14-year-olds in recording their environment using participatory sensing achieved "16.2 percent more students" that "viewed education and careers in science and technology as a potential option for themselves after high school" (Heggen 2012).

17.15 Transfer, storage, and access steps

Researchers and community activists are already exploring the potential of smartphones and their integrated sensors for participatory sensing. These almost invariably use cell phone and web technologies for transfer and access, while online storage is becoming increasingly accessible. These approaches have the capacity to be increasingly ubiquitous via the Internet of Things. While these applications rely on the currently integrated sensors in the more common smartphones, there is a range of sensors available that can link to smartphones that cover much or most of the recording process required as the first stages of HBIM, before there is more ubiquitous access to this Internet of Sensors (Things). Thus, McKinsey argues that "future smart devices will incorporate new mobile sensor types, such as biometric, pressure and environmental sensors" (McKinsey 2012). Google's Project Ara is developing a modular smartphone to accept new sensors as they are developed (Mail 2016). There are also challenges being addressed in a similar manner to measure the resilience of the environment to man-made and natural catastrophes. For example, "new sensor technologies have greatly reduced the cost of lower-resolution strong motion seismometers . . . Two examples include the Quake-Catcher Network and the Community Seismic Network that install low-cost sensors in homes, businesses, and schools" (USGS 2014). The cost of sensing airspeed is also predicted to plummet. A research project in the University of Berkeley stated that the "team expects the new sensors to be disruptively inexpensive ($20–$100 retail) and sophisticated, with accuracy to equal or exceed that of any alternative airflow sensing technology" (CBE 2014). Effective thermal imaging survey attachments are already available for smartphones (Stone 2014b).

Deeming environmental factors, such as pollution, there is a need to capture the detailed 3D data that defines the geometry and can then be used for 3D georeferencing, and to create visualisable 3D models (an analytic process that can add semantic data to the geometry). The challenges of the 3D model are discussed in Chapter 3, but the 3D model does offer opportunities for wider virtual interaction, and so may form a key ingredient of community and citizen engagement, particularly asynchronous or off-site. Google supply their Project Tango, currently an Android tablet, to developers with the intent of achieving real-time sensing of 3D environments via future smartphones (Tango 2016). This is one of a number of emerging low-cost 3D scanning technologies. The Structure

Sensor (Structure 2016) is a 3D scanner attachment for smartphones and tablets. Quanergy (Spar 3D 2016) have launched a prototype solid-state low-cost Lidar attachment.

17.16 Analysis and visualisation steps

There are emerging tools that support the view that large communities can actively engage in 3D modelling, analysis, and interactive engagement. Linden Labs (2013) claimed that in the previous 10 years more than 36 million users had created accounts for Second Life, and that over 1 million use it regularly. Sequeira and Morgado (2013) analyse the use of Second Life for virtual heritage modelling: "The Second Life viewer includes building and programming tools, designed with amateurs in mind . . . it is considerably easier to create content." "Second Life (and its open source equivalent, Open Simulator) has been a test bed for different approaches in presenting historical reconstructions in virtual worlds" (ibid.). Over 20 million users have paid for Minecraft, presumably justifying its more recent acquisition by Microsoft, and it has over 100 million registered users. While the models may be somewhat crude depictions of real buildings and contexts, it still teaches modelling skills and interaction in an engaging and accessible manner. It has also been used for heritage models (Rowan County 2014) and for active community engagement in planning future landscapes (Siebert 2015). A number of recent participatory sensing projects use 3D visualisation over the web, and one constructs predictive visualisation of building responses to earthquake shaking, which may be adaptable to flood and similar events. "The visualization of building response to earthquake shaking is accomplished through construction of geometric models of the buildings. We developed software packages and functions that produce 3D and 4D visualizations of building response during an earthquake using acceleration data recorded in buildings instrumented with multiple community-hosted accelerometers. The entire process for generating the visualizations is contained within two software packages, SketchUp and MATLAB which can be modified for any building and earthquake. The scripts available here can be customized for nearly any other structure" (CSN 2011).

Others have proposed the use of virtual or augmented reality. For example, the "eViz Project, a multi-centre initiative researching visualisation for carbon reduction, has been addressing the maturity and usability of real-time Virtual Reality (VR) techniques as a means of, in effect, making 'the invisible visible'" (Stone 2014a). While such models contain explicit semantic information, that is recognisable building elements, there is a need to be able to add implicit information. This was explored over four years in the VEPs3D (op. cit.) research project, leading to the development of a comment markup language (termed COML), georeferenced to a web-based VR interactive model via simple tools within a web browser. Other information will need to be developed and added using similar means, once analysed via civic or citizen science supported participatory action. Public Lab (2015) is one example of such an open source accessible initiative, not currently georeferenced in 3D. An ongoing initiative seeks to further "the development of social communities in digital public space and help to provide criteria for how the digital public space is modelled and visualised" (Creative Exchange 2015). The Creative Exchange also has a cluster focused on "Making the Digital Physical" which includes an approach to active engagement of communities, and a cluster focused on "Stories, Archives and Living Heritage" which includes an approach to consensus on value and significance.

17.17 Feedback step

A further key question has been whether the resulting data can be relied upon. Studies into participatory sensing, for example, for air or noise pollution monitoring, have identified techniques to improve calibration and reliability of the data. Piedrahita et al. (2014) state that "collocation and collaborative calibration will be a valuable tool in the next generation of air quality monitoring." Jones (2014), in a recent study, argues further that "high precision measurement in the wrong place has less value than a poor/indicative measurement in the correct place." Calibration starts with seeking to ensure data quality to begin with, rather than eliminating data problems. Reddy et al. (2010) identify the potential for scoring on (previous) reputation as a reliable indicator of ability to deliver high-quality data in participatory sensing, that is feedback from previous input leads to a reputation score. D'Hondt et al. (2013) claim that "the combination of calibration and statistical averaging over large datasets allows to counteract inaccuracies and imprecisions inherent to measurements taken by mobile phones." In this study they focused on noise pollution measurement, and claimed that the resulting noise maps had "error margins comparable to those of official simulation-based noise maps". Birkin et al. (2011) describe active citizen engagement in calibration via work to create "a model of a new commuting charge in one England's major conurbations (the Manchester Congestion Charge) and then crowd-sourced the response of local people to calibrate that model, with a view to further articulation and evaluation of policy options". We suggest that feedback and evaluation are a key aspect of continuous improvement, and that automated approaches to calibration may be used effectively for such feedback, informing a tutoring and learning system to support the next phase of users.

The authors have initiated several exploratory studies in this area, which in various ways seek to enable citizen and community engagement with low-cost tools in support of quality of life, regeneration and heritage. These are beginning to show the benefits of integration of the various approaches outlined above. The first was conducted over the summer of 2015, with funding from Cardiff City Council (and other agencies), who were seeking innovative solutions to enhancing and upgrading the energy efficiency of their heritage stock. The team, from Stride Treglown Architects, Cardiff Metropolitan University and Swansea University, first undertook laser scanning, then 3D modelling to ensure the best fit without waste, then fabricated and installed innovative 50 mm thick pre-dried internal wall insulation panels made of moisture-permeable hemp-lime. These were used to replace the previous poorly retrofitted internal plaster coatings, which as a result had become impermeable, causing the solid masonry wall of the case study building to become saturated, and hence very cold. Heat box analysis of the hemp-lime showed an improvement of overall thermal performance from a calculated 1.55 W/sq m deg C to 0.8237 W/sq m deg C. A low-cost Raspberry Pi was configured as a data logger and connected to Wi-Fi and to low-cost moisture and humidity sensors inside, outside, and embedded in the tiles. Data logged at five-minute intervals over a season showed substantial improvement in drying out the wall without affecting permeability. The cost of monitoring with these sensors was less than £400. The householder was able to reset and restart the logging in the event of powercuts or a change of router.

The Cadw Heritage Cottage (discussed in Chapter 2) monitoring cost, by contrast, more than the purchase price of the cottage, although evidently was more comprehensive. In another ongoing research project that started in February 2016, funded internally at Cardiff Metropolitan University, school buildings are being surveyed using two methods. The first is a snapshot in a one- or two-day visit, for calibration, using expensive equipment that demands skills and training, such as thermography, and laser scanning. The second is to provide staff and students with low-cost equipment that replicates and extends much of the expert equipment, and some support, to engage them in the citizen science learning activities identified by Heggen (op. cit.). The equipment includes the Google Tango and Structure Sensor (op. cit.), together with low-cost Arduino-based portable and personal sensors that can be monitored in real time using smartphones, and uploaded and collated on the web. This is based upon a pioneering approach developed by Australia Aircasting (2015) in the USA. The purpose is to identify the environmental factors that affect well-being, productivity, creativity and learning. A third ongoing project is a two-year collaboration that commenced in January 2016 between universities in the UK and Egypt, which has resulted from a follow-up bid for funding following the HBIM workshop that triggered this book. That project is also seeking to actively engage with occupants in heritage buildings and carry out a similar expert 'snapshot' study, primarily for

Participatory sensing 253

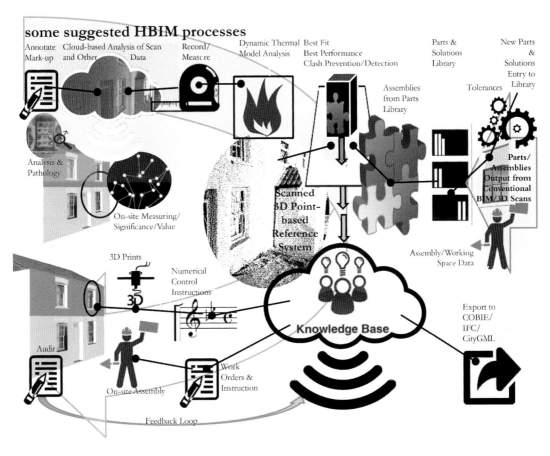

Figure 17.3 We proposed at the workshop a cloud-based workflow for analysing best fit and marrying good practice solutions to 3D point cloud scans; even 3D printing is becoming widespread.

calibration and to acquire data for modelling, and then to install low-cost sensors that upload data that can be displayed in the context of the 3D modelling via the web, driven by a prototype HBIM platform that is under development. Figure 17.3 shows a cloud-based workflow for the analysis of best fit and marrying good practice solutions.

17.18 Conclusion

This chapter has argued that there is a major challenge, and there is a significant skills and resource gap. The 20% approximately of pre-1919 heritage buildings in Europe emits over one-third of the EU 28's carbon. This constitutes approximately 45 million buildings. It can be argued that it is highly unlikely that a construction industry that already only achieves just over 2 million new buildings per annum in Europe has the additional capacity to swiftly transform heritage buildings in Europe to assist the overall EU target of 80% below 1990 levels by 2050. Yet tackling existing building stock was described as 'low hanging fruit' by the newly established UK Climate Change Committee in 2008, who went on to state, "Aggregate emissions reduction potential from existing buildings is likely to far exceed that for new buildings given that the former will continue to dominate the building stock in the period to 2020 and beyond" (CCC 2008). We therefore argue that without substantial, highly active community engagement, meeting the emissions targets while also enhancing and preserving our cultural heritage and enhancing quality of life for occupants appears to be unlikely; an effective HBIM platform needs to incorporate and facilitate all the stages and aspects of participatory sensing, citizen science and volunteered information.

References

Arab German Yearbook (2010). 'Improving the Living Conditions of the Poor in Manshiet Nasser'. Online at: http://egypt-urban.net/wp-content/uploads/2010/07/Arab-German-Yearbook-2010_Article-on-MN.pdf.

Australia Aircasting (2015). 'About AirCasting'. Online at: http://aircasting.org/about.

Australia ICOMOS (2013). 'The Australia ICOMOS Charter for the Conservation of Places of Cultural Significance – (The Burra Charter)'. *Australia ICOMOS*. Online at: http://australia.icomos.org/wp-content/uploads/The-Burra-Charter-2013-Adopted-31.10.2013.pdf.

Balance Project (2007). Interim Report no 22 on 'E-Participation as tool in Planning Processes'. Baltic Sea Region Interreg III B.

Beck, L. (2013). 'Digital documentation in the conservation of cultural heritage: finding the practical in best practice'. *ISPRS-International Archives of the Photogrammetry, Remote Sensing and Spatial Information Sciences*, 1(2), 85–90.

Bell, K. (2003). 'Assessing the Benefits for Conservation of Volunteer Involvement in Conservation Activities'. Department of Conservation, New Zealand Government, Wellington, New Zealand.

Birkin, M., Malleson, N., Hudson-Smith, A., Gray, S., and Milton, R. (2011). 'Calibration of a spatial simulation model with volunteered geographical information'. *International Journal of Geographical Information Science*, 25(8), 1221–1239.

BIS (2015). 'Digital Built Britain Level 3 Building Information Modelling – Strategic Plan'. *Department for Business, Innovation & Skills*. Online at: www.gov.uk/government/uploads/system/uploads/attachment_data/file/410096/bis-15–155-digital-built-britain-level-3-strategy.pdf.

Brandtzæg, P., Heim, J., and Karahasanović, A. (2010). 'Understanding the new digital divide – a typology of Internet users in Europe'. *International Journal of Human Computer Studies*, 69(3), 123–138.

Budde, M., Zhang, L., and Beigl, M. (2014a). 'Distributed, low-cost particulate matter sensing: scenarios, challenges, approaches'. In *1st International Conference on Atmospheric Dust, DUST 2014* (pp. 230–236). ProScience 1.

Budde, M., De Melo Borges, J., Tomov, S., Riedel, T., and Beigl, M. (2014b). 'Leveraging spatio-temporal clustering for participatory urban infrastructure monitoring'. In *Proceedings of the First International Conference on IoT in Urban Space* (pp. 32–37). ICST.

CBE (2014). 'Development of Low-Cost MEMS-Based Ultrasonic Airflow Sensors for Rooms and HVAC Systems'. *Centre for the Built Environment, University of Berkeley*. Online at: www.cbe.berkeley.edu/research/mems-airflow-sensor.htm.

CCC (2008). 'Building a Low-Carbon Economy – The UK's Contribution to Tackling Climate Change'. The Stationery Office, Whitehall Place, London.

CECIH – Transylvania Trust (2014). 'Celebrating European Cultural Intangible Heritage for Social Inclusion and Active Citizenship – CECIH'. Online at: www.transylvaniatrust.ro/index.php/en/programs/celebrating-european-cultural-intangible-heritage-for-social-inclusion-and-active-citizenship-cecih/.

Council of Europe (1998). 'Strategies for vocational training in architectural heritage skills'. In *Proceedings: Symposium Organised by the Council of Europe, International Heritage Fair, Le Carrousel Du Louvre, Paris*, 13 April 1996.

Counsell, J., Smith, S., and Aldrich, S. (2003). 'A role for 3D modelling in controlling webcams and interpreting real-time video'. In *Geometric Modeling and Graphics, 2003. Proceedings. 2003 International Conference* (pp. 2–7). IEEE.

Creative Exchange (2015). 'Building Social Communities – Dynamic Structures for Growth'. Online at: http://thecreativeexchange.org/clusters/building-social-communities-dynamic-structures-growth.

CRESME (2014). 'Euroconstruct Summary Report'. Online at: www.euroconstruct.org/publications/publications.php.

CSN (2011). 'Community Seismic Network'. *California Institute of Technology*. Online at: http://csn.caltech.edu/about/.

De la Torre, M., and Throsby, D. (2002). Assessing the Values of Cultural Heritage. Research report. Getty conservation institute.

D'Hondt, E., Stevens, M., and Jacobs, A. (2013). 'Participatory noise mapping works! An evaluation of participatory sensing as an alternative to standard techniques for environmental monitoring'. *Pervasive and Mobile Computing*, 9(5), 681–694.

ERA (2008). 'England's Rock, the Prehistoric Rock Art of England: Recording, Managing and Enjoying Our Carved Heritage'. ISBN 1–873402–28–7. Online at: http://archaeologydataservice.ac.uk/era/section/record_manage/rm_projects_nadrap_home.jsf.

Eurostat (2015). 'Information Society Statistics – Households and Individuals'. Online at: http://ec.europa.eu/eurostat/statistics-explained/index.php/Information_society_statistics_-_households_and_individuals.

Geocrowd (2014). 'Call for the 3rd ACM SIGSPATIAL GIS Conference'. Online at: www.geocrowd.eu/workshop_2014.

Geostream (2015). 'Resources'. Online at: http://geocontentstream.eu/geostream/resources.html.

Geothink (2013). 'Newsletter Issue 1'. Online at: http://geothink.ca/wp-content/uploads/2015/10/Geothink-Newsletter-Issue-1.pdf.

Goldman, J., Shilton, K., Burke, J., Estrin, D., Hansen, M., Ramanathan, N., and West, R. (2009). 'Participatory sensing: a citizen-powered approach to illuminating the patterns that shape our world'. *Foresight & Governance Project, White Paper*, 1–15. Online at https://web.ohmage.org/~nithya/publications/ParticipatorySensingScenarios%209–19–08.pdf.

Heggen, S. (2012). 'Integrating participatory sensing and informal science education'. In *Proceedings the 14th ACM International Conference on Ubiquitous Computing (Ubicomp 2012), (Doctoral Colloquium)*. Online at: www.sheggen.com/sites/default/files/publications/ubicomp2012dc.pdf.

Jones (2014). 'Low Cost Sensor Networks for Air Quality Monitoring in Welsh Air Quality Forum'. Online at: www.welshairquality.co.uk/documents/seminars/458141002_4_Rod_Jones_Welsh_Air_Quality_Meeting(RLJ)v3.pdf.

Kingston, R., Babicki, D., and Ravetz, J. (2005). 'Urban regeneration in the intelligent city'. In *Proceedings of the 9th International Conference on Computers in Urban Planning and Urban Management*, CASA, UCL, London.

Kirchhoefer, M.K., Chandler, J.H., and Wackrow, R. (2011). 'Cultural Heritage Recording Utilising Low-Cost Close-Range Photogrammetry'. CIPA. 23rd International Symposium, 12–16 September 2011, Prague, Czech Republic.

Linden Labs (2013). 'Infographic – 10 Years of Second Life'. Online at: www.lindenlab.com/releases/infographic-10-years-of-second-life.

Mail (2016). 'Step aside Google'. Online at: www.dailymail.co.uk/sciencetech/article-2856245/Step-aside-Google-Finnish-firm-plans-rival-Project-Ara-modular-swap-Puzzlephone.html.

McKinsey (2012). 'Making Smartphones Brilliant: Ten Trends'. *McKinsey & Co Inc*. Online at: www.mckinsey.com/~/media/mckinsey/dotcom/client_service/high%20tech/pdfs/making_smartphones_brilliant_march_2012.ashx.

Meijer, F., Itard, L., and Sunikka-Blank, M. (2009). 'Comparing European residential building stocks: performance, renovation and policy opportunities'. *Building Research & Information*, 37(5–6), 533–551.

MIT (2001). 'Urban Upgrading – Issues and Tools'. *Massachusetts Institute of Technology*. Online at: http://web.mit.edu/urbanupgrading/upgrading/issues-tools/tools/Planning-for-Real.html.

Mossberger, K., Tolbert, C.J., and McNeal, R.S. (2008). 'Digital Citizenship: The Internet, Society, and Participation'. MIT Press, Cambridge, MA.

NEARCH (2014). 'What is NEARCH?'. *EC Culture Programme* (2013–2018). Online at: www.nearch.eu/what-is-nearch.

OECD (2001). 'Citizens as Partners: Information, Consultation and Public Participation in Policy-Making'. OECD, Paris, France.

Perkin, C. (2010). 'Beyond the rhetoric: negotiating the politics and realising the potential of community driven heritage engagement'. *International Journal of Heritage Studies*, 16(1–2), 107–122.

Piedrahita, R., Xiang, Y., Masson, N., Ortega, J., Collier, A., Jiang, Y., Li, K., Dick, R. Lv, Q. Hannigan, M., and Shang, L. (2014). 'The next generation of low-cost personal air quality sensors for quantitative exposure monitoring'. *Atmospheric Measurement Techniques Journal*, 7, 3325–333. Online at: www.atmos-meas-tech.net/7/3325/2014/.

Public Lab (2015). 'About Public Lab'. Online at: https://publiclab.org/about.

Reddy, S., Estrin, D., and Srivastava, M. (2010). 'Recruitment framework for participatory sensing data collections'. In *Pervasive Computing* (pp. 138–155) in Proceedings of the 8th International Conference on Pervasive Computing. Springer, Berlin Heidelberg.

Renovate-Europe (2013). 'It Pays to Renovate'. Online at: http://renovate-europe.eu/wp-content/uploads/2015/09/RE_IT_PAYS_TO_RENOVATE_brochure_v05_spreads.pdf.

Rowan County (2014). 'My Heritage My Future'. Online at: https://myheritagemyfuture.wordpress.com/morehead-historical-buildings-minecraft-interactive/.

Sabbioni, C., Brimblecombe, P., and Cassar, M. (2012). "Atlas of Vulnerability to Climate Change". Anthem Press, Strasbourg.

Sabbioni, C., Cassar, M., Brimblecombe, P., and Lefevre, R.A. (2009). 'Vulnerability of cultural heritage to climate change'. *Pollution Atmospherique*, 202, 157–169.

Sequeira, L. and Morgado, L. (2013). 'Virtual archaeology in second life and open simulator'. *Journal for Virtual Worlds Research*, 6(1), 1–16.

Siebert, B. (2015). 'Govt Enlists Minecraft for Youth Consultation'. Online at: http://indaily.com.au/arts-and-culture/design/2015/04/21/govt-enlists-minecraft-for-national-parks-design/.

Sonntagbauer, P., Nazemi, K., Sonntagbauer, S., Prister, G., and Burkhardt, D. (2014). 'Handbook of Research on Advanced ICT Integration for Governance and Policy Modeling'. IGI Global, Hershey.

Spar 3D (2016). 'Quanergy's Solid-State LiDAR: Details from CES'. Online at: www.spar3d.com/news/hardware/vol14no2-quanergy-solid-state-lidar-details/.

Stone, R. (2014a). 'Exploiting gaming technologies to visualise dynamic thermal qualities of a domestic dwelling: pilot study of an interactive virtual apartment'. In *Proceeding of BEHAVE 2014 Conference*. Online at: http://behaveconference.com/wp-content/uploads/2014/08/F_Robert_Stone_University_of_Birmingham.pdf.

Stone, R. (2014b). 'Low-Cost Thermal and Panoramic Imaging Systems for Drone-Based Architectural Surveys'. Online at: www.birmingham.ac.uk/schools/eese/news/architectural-surveys.aspx.

Structure (2016). 'The Structure Sensor'. Online at: http://structure.io/.

Tango (2016). 'Project Tango'. Online at: www.google.com/atap/project-tango/.

Thornley, L. and Waa, A. (2009). 'Increasing public engagement with historic heritage: a social marketing approach'. In *Science for Conservation 294, Department of Conservation*. Online at: www.doc.govt.nz/Documents/science-and-technical/sfc294entire.pdf.

UNESCO (2013), "The UNESCO Guidelines for the selection of digital heritage for long term preservation", The UNESCO Task Force, http://www.ifla.org/files/assets/hq/topics/cultural-heritage/documents/persist-content-guidelines.pdf

USGS (2014). 'Next Steps Improving the Sensor Network'. *US. Geological Survey*. Online at: http://earthquake.usgs.gov/research/earlywarning/nextsteps.php.

VEP3D (2008). 'The Virtual Environmental Planning System Project'. Online at: www.veps3D.org.

Visser, J. and Richardson, J. (2013). 'Digital Engagement in Culture, Heritage and the Arts'. Online at: http://digitalengagementframework.com/digenfra3/wp-content/uploads/2016/02/Digital_engagement_in_culture_heritage_and_the_arts.pdf.

Wates, N. and Thompson, J. (2013). 'The Community Planning Event Manual: How to Use Collaborative Planning and Urban Design Events to Improve Your Environment'. Routledge, London.

WMF (2015). 'Engage Your Community with Watch Day!' Online at: www.wmf.org/2014-world-monuments-watch www.wmf.org/content/nomination-guidelines (accessed 26/02/2016).

18 Development of OntEIR framework to support heritage clients

Shadan Dwairi and Lamine Mahdjoubi

Employers of heritage projects always have a main requirement, which is for the project to be completed within the time scale, budget and with the highest quality. Evidence suggests that for clients of heritage buildings, various factors led to unfortunate increases and out-of-control costs, such as poor definition of requirements and lack of capacity to engage effectively with suppliers.

Employer Information Requirements (EIR) are seen as central to delivering the Heritage BIM (HBIM) agenda. The successful and clear identification of the specifications of the employer and heritage building requirements is an important, if not the most important, phase in delivering successful heritage projects, and is referred to as the Employer Information Requirements (EIR). Despite the various research efforts, the specification of employer's information requirements is still underdeveloped. Indeed, previous research confirmed that due to complexities in identifying and conveying accurately employers' actual needs and requirements to the heritage project's team and the immense magnitude of project information, project briefings may not truly reflect employer's requirements.

This chapter reports on an ongoing study designed to support heritage clients in making informed and sound decisions in defining their requirements. The aim of this research is to investigate the development of an ontology-based, BIM-enabled framework for EIR. This research examines innovative approaches and methods for conveying employer information requirements (EIR) using a framework to capture, analyse, and translate these requirements based on an ontology model. This framework will enable the heritage team to capture these requirements and convert them to constructional terms understood by all stakeholders. It is the contention of this chapter that this process will save time, effort and cost, and will provide an informed basis for delivering a successful project that satisfies both the employer and the supply chain.

18.1 Introduction

The importance of OntEIR emerges from the importance of EIRs and the fact that Building Information Modelling (BIM) was mandated by the UK government to be used in all public buildings by 2016.

One of the key pillars of BIM (PAS 1192–2:2013, 2013) produced by the BIM Task Group proposed setting out the EIR, as part of the Employer's Requirements document, which is incorporated into tender documents. Such documents provide information that is mandatory for suppliers to be able to produce the Heritage BIM Execution Plan (HBEP), in which the proposed approach, capability and capacity can be evaluated. This information includes requirements from the client in addition to key decision points and project stages.

EIR documents are designed to be included in the tender documents for the procurement of both the design team and the constructor (Employer's Information Requirements Guidance Notes, 2013). The importance of such documents is derived from being the first step in the information delivery cycle of the project, and is an important element of project BIM implementation, in addition to being the guideline for the BIM protocol implemented through the HBEP, as shown in Figure 18.1 below.

The development of EIR starts from the early stages of the project. Initially it might take the form of a simple process map which identifies the key decisions that will need to be made during the project to ensure that the solution developed satisfies the business needs and defines in very broad terms the information that will be needed to make such decisions. In the report published by the Government Construction Strategy Cabinet Office (2011), it was announced that the government's intention was to require collaborative 3D BIM (with all project and asset information, documentation and data being electronic) for its projects by 2016. It should be mentioned that in the BIM plan of work, defining EIR is the main phase because of its importance in setting out the information required by the employer aligned to key decision points or project stages, and thus enabling suppliers to produce an initial HBEP from which their proposed approach, capability and capacity can be evaluated.

18.2 State of the art

Refurbishment and restoration projects account for about 46% of the total construction output in the construction industry in the UK (Lee and Egbu, 2005). These kinds of projects hold higher risk

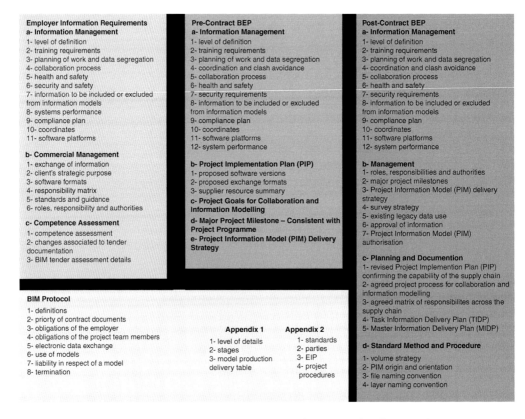

Figure 18.1 Impact of EIR on the construction process (adapted from Michael Earley, www.bim-manager.net).

and are more complex and require more coordination than a new build project; that is why clear identification of requirements and constraints may pose a difficult task. Studies have been carried out in the field of requirements, of which clear definition is important for the success of both refurbishment projects, including restoration and heritage projects, and for new projects.

The term "client" or "employer" will be used in this study to refer to the owner or representative of the owner of a building, who has the authority to comment or make a decision on the restoration process.

Research work in the field of requirements, postulated by Kamara *et al.* (2000), advocated construction briefing as "client requirements process" within the discipline of concurrent engineering for life cycle design and construction. Bruce and Cooper (2000) highlighted the importance of understanding both hard and soft processes when developing requirements for employers. The document that contains the written instructions/requirements of the employer is referred to as the "brief", which should include the following information:

- the background, purpose, scope, content and desired outcomes of the project
- the functions of the intended facility and the relationships between them
- cost and time targets, instructions on the procurement and organisation of the project
- site and environmental conditions, safety, interested third parties and other factors that are likely to influence the design and construction of a facility (Kamara and Anumba, 2001)

Kamara *et al.* (2002) have also described the requirements for any type of project to be:

- Client requirements: which describe the facility that satisfies his or her business need; incorporates user requirements, those of other interest groups and the life cycle requirements for operating and maintaining the facility
- Site requirements: which describe the characteristics of the site on which the facility is to be built
- Environmental requirements: which describe the immediate environment
- Regulatory requirements: which are building, planning, health and safety regulations and other legal requirements
- Design requirements: which are requirements for design, a transition of the client needs, site and environmental requirements
- Construction requirements: which are requirements for actual construction, which derive from design activity

Another model that has been implemented in this area is the Client Requirements Processing Model (CRPM), which adopts structured methods in translating the "voice of the client" into the "voice of the designer". The model has three main stages: define client requirements, analyse client requirements, and translate client requirements. These stages subdivide further into activities and utilise appropriate information-gathering tools, decision support tools and quality assessment tools (e.g. Quality Function Deployment) to develop solution-neutral specifications. CRPM is computerised within a software system called ClientPro and has been received as generally satisfactory in effectiveness. Test feedback reports that requirements generation, prioritisation, clarity and visibility were adequately supported within the formal process. Kamara and Anumba maintain that client requirements be:

- precisely defined, with as little ambiguity as possible, and reflective of all the perspectives and priorities represented by the client body

- stated in a format that is solution-neutral (i.e. not based on any design concept that could serve as a solution to the client's problem) and which makes it easy to trace and correlate design decisions to the original intentions of the client

ClientPro was evaluated by four industry practitioners and rated relatively low in areas such as the facilitation of communication among members of the processing team, the usefulness of the software to the overall construction process, and the ease to use the system (Kamara et al., 2002).

Another tool introduced for processing requirements is the Quality Function Deployments (QFD), which can be used for understanding and tracking requirements and improving communication among product development team members (Kamara et al., 1999). This method is based on representing the requirements through matrixes as well as documenting. However, the use of QFD has been very modest in construction (Dikmen et al., 2005). Limitations of the use of QFD in construction as pointed out by Lima et al. (2008) are that it is time-consuming to process this information, particularly if the proportions of the matrix become very large, and it is not easy to involve product development team members in the processing stages that are necessary to produce the matrix.

In another attempt to enhance the definition of the roles of employers in construction projects, the National Building Specifications (NBS) has adopted the Ministry of Justice's approach, which advocated a set of plain language questions that an employer is expected to answer at each stage of a construction project. Key decisions such as whether to proceed to the next work stage or not will be made based upon the answers to these questions. Each stage (from 0–7) has its own questions, according to what will be completed in the stage. Although the questions are written in plain language and are easy to interpret and answer, they are not able to fully capture the employer's information requirements. Clearly there are many other important aspects that should be covered in order for the client to be able to deliver a complete and comprehensive requirement document for the construction team. According to PAS 1192–2:2013 (2013), the EIR should include information regarding three main aspects – information management, commercial management, and competence assessment – in addition to the employer's requirements and the vision the employer has for the project.

18.3 Purpose

Despite the various research efforts, the specification of EIR is still underdeveloped. Indeed, previous studies conducted by Graham (1983), Hudson et al. (1991) and Barrette et al. (1999) confirmed that due to complexities in identifying and conveying accurately employer's actual needs and requirements to the project's team and the immense magnitude of project information, project briefings may not truly reflect employer's information requirements.

In order to improve the specification of requirements, Kiviniemi (2005) suggested that it is essential to develop IT tools to provide some degree of automation for requirements management. However, the use of IT in that task poses important challenges, such as the difficulty of capturing both implicit and explicit requirements, maintaining information up to date, and storing different requirements from distinct stakeholders throughout the product development process (Leinonen and Huovila, 2001). An ontology-based approach has the potential to improve both requirements elicitation and management. Indeed, ontology as defined by Gruber (1995) is a specification of a conceptualisation; that is that an ontology is a description of the concepts and relationships that can exist in the domain. This definition is consistent with the usage of ontology as set-of-concept-definitions, but more general. The relationship of the concepts existing in the domain is seen through the hierarchy of these concepts, where a domain is fragmented into classes and each class into subclasses until the instances are reached which are at the lowest rank of the hierarchy in the system.

The aim of this ongoing study is to develop an ontology-based, BIM-enabled framework for Employer Information Requirements (OntEIR), for facilitating the identification and representation of requirements for construction and heritage projects. It seeks to provide a holistic OntEIR tool designed to assist employers to capture their requirements, in a language that is understandable and easy to interpret, which leads to improved quality information requirements in construction projects

EIR is the cornerstone for a successful project. The importance of OntEIR is derived from the fact that defining adequate EIR is an important step in the forming of the HBEP, which will have the most influence on the project outcome. Another key reason for considering this system to be critical is, in its novelty in being addressed to main key players of the different disciplines involved in the BIM project, it seeks to provide answers and address questions and issues that will be of great importance for the formulating of the project programme for all disciplines.

18.4 Methodology

Well-defined EIRs will lead to the production of well-written and comprehensive essential documents in the success of heritage projects. Due to the large number of stakeholders and domain experts involved in the collaborative process of the project, in addition to the many types of heritage projects, ontology is considered the best option for this process due to what it offers in:

- semantic heterogeneous description of requirements
- the ability to retrace requirements
- the ability to instantiate the framework to any construction project

The framework described in this study will have the aim to facilitate the identification, clarification and representation of employer requirements in construction projects.

18.5 Why ontology?

In reference to Gruber's (1995) definition, ontology is a specification of conceptualisation. Domain ontology specifies a generic of a specific domain with all its relevant concepts, relationships and axioms given in a context.

Ontology defines a common vocabulary for researchers who need to share information in a domain. It includes machine-interpretable definitions of basic concepts in the domain and relations among them. Noy and McGuinness (2000) summarise the reasons we need ontology in the following points:

- to share common understanding of the structure of information among people or software agents
- to enable reuse of domain knowledge
- to make domain assumptions explicit
- to separate domain knowledge from the operational knowledge
- to analyse domain knowledge

In other words, ontology is a formal representation of an abstract; it is a simplified view of a domain that describes the objects, concepts and relationships between them.

In construction projects, and to ensure that projects are more employer-oriented, EIRs should be processed rigorously and coherently before the start of conceptual design. This should be done through a structured framework which provides for the definition, analysis and translation of EIRs into design specifications that are solutions-neutral (Kamara et al., 1999).

To be able to form a clear and understandable brief by all parts of the construction team, EIR should go through a process that begins with eliciting the EIRs in plain language and converting them into constructional terms to be understood by all members of stakeholders.

The ontology introduced in the research will specify the concepts and relationships in the EIR domain. Ontology will facilitate the sharing of knowledge and understanding between relevant parties by defining domain terminologies, vocabularies and relationships in the same way for all stakeholders. The ontology introduced provides the common terms and vocabulary relations for modelling the EIR, where the ontology will lay the semantic understanding foundation.

In addition, the structuring and decomposition of requirements in a hierarchal way can facilitate the understanding of tracing requirements (Kott and Peasant, 1995; Ulrich and Eppinger, 2008), which will be tracked easier with the use of ontology. To evaluate attributes of criteria as well as compare requirements of each alternative in the system, it is necessary to use an appropriate data modelling and analysis methodology. In this system, the ontology-based approach is adopted. This method is one of the fastest-growing areas in the last several decades, and it is used to support decision-making in challenges related to several disciplines (Chen and Nugent, 2009).

Ontology provides a shared and reusable piece of knowledge about a specific domain, and has been applied in many fields, such as Semantic Web, e-commerce and information retrieval. More and more researchers are beginning to pay attention to ontology research. Until now, many ontology editors have been developed to help domain experts to develop and manage ontology (Li et al., 2005).

18.6 OntEIR

The first stage of the study was to develop a conceptual framework. The proposed conceptual framework identifies the predominant research issues, which have significant implications on EIR, works in the EIR and construction field such as PAS 1192–2:2013, RIBA Plan of Work, and the EIR guidance notes were studied as reference to the OntEIR framework; this framework seeks to draw together existing research and to provide a foundation for future work in this field. Within each of these categories, a series of attributes are examined. The framework is evaluated in a real-life situation, and the findings and results of this evaluation are extracted and elicited and thoroughly documented; this will be used to improve and update the framework.

The initial framework developing process will go through a series of stages that attempt to define functions, classify functions and develop function relations based on the literature review previously conducted.

1 – Generating high-level goals

Using mind mapping, the problem domain is elicited and analysed. The data visualised in the mind map is actually a representation of elicitation and analysis of requirements done with relevant stakeholders and domain experts via interviews. High-level needs are formulated as a result.

The visualisation of the initial EIR framework is done through mind mapping that consists of all aspects of the EIR, which are broken down to reach high-level goals. This will help us in determining and generating requirements more clearly and precisely, and help the employer body in determining their more specific requirements and concerns regarding the project.

2 – Requirement specification

After root goals are reached in the goal hierarchy generated by the mind map, requirements are used where relevant. If the requirements do not cover a root goal, then new requirements will have to be generated.

3 – Requirements validation

Goals generated from the mind map will be validated with relevant stakeholders and domain experts in the construction industry.

EIR is divided into two types of requirements, static requirements and dynamic requirements, as shown in Figure 18.2.

- Static requirements (which include management information and technical information):

  This type of requirement looks at the whole process as one system and delivers requirements in that sense; they are unchangeable through the different phases of the process, and contain definitions for terms that will be used during the process and in the different data drops that are specified in detail in the dynamic requirements. These requirements organise the whole process and define the standards and the guidelines that will have to be met, in addition to the technological issues of the BIM process.

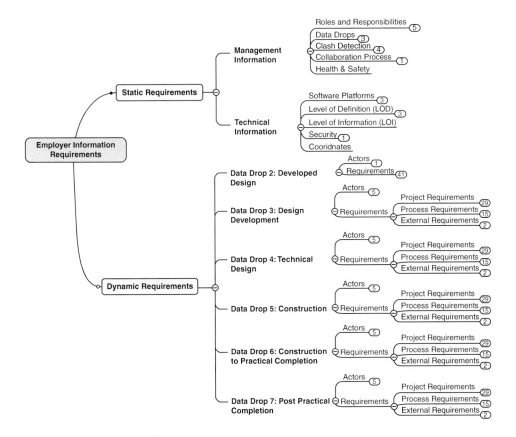

Figure 18.2 Heritage Employer Information Requirements.

Static requirements will cover the following issues:

- First of all, the employer should specify the different stages he or she would prefer the process to go through (data drops).
- The employer should define the different levels of development, and levels of information, which will be delivered at the different data drops.
- Software platforms that will be used during the HBIM process, including version numbers that will be used by the supply chain to deliver the project, and any particular constraints set by the employer on the size of model files.
- Standards, protocols and regulations to be used during the process.
- Coordination and clash detection.

Management requirements will detail high-level roles and responsibilities, standards, data security, the key decision points and the information to be available at each one. Technical requirements will cover issues like information format and file types, the minimum levels of definition at each stage and the software platforms to be used for exchanging information.

- Dynamic requirements (which include the commercial information):

    This set of requirements looks at the process as different data drops; each drop having its own set of requirements and information to be delivered, it includes details on the client's strategic purpose, the defined deliverables and a competence assessment for those looking to tender the heritage project and demonstrate their ability to deliver the requirements of the EIR.

These requirements are more specific to data drops, what is required by the project team to be delivering at each stage, who will be involved, and how will that be done, they will answer the major part of the plain language questions. These requirements include:

- Main outcome of the data drop
- Project requirements for each data drop
- Level of detail, and level of information (as defined in the management information) for each data drop
- Actors
- Standards and guidance needed
- Software formats to be used
- Security

The breaking down of the requirements to reach high-level goals is the first step in specifying requirements, and helps the employers determine their more specific concerns regarding the project, as shown in Figure 18.3 below, which represents the visualisation of the breaking down of Data Drop 2 in the dynamic requirements of the EIR.

After reaching the mind map that holds the necessary information – EIR domain knowledge, problem domain knowledge, solution domain knowledge and stakeholder information – OntEIR is developed to create an ontology-based system for EIRs, which will create quality requirements in terms of consistency, completeness, correctness, traceability and the ability to be instantiated to any project.

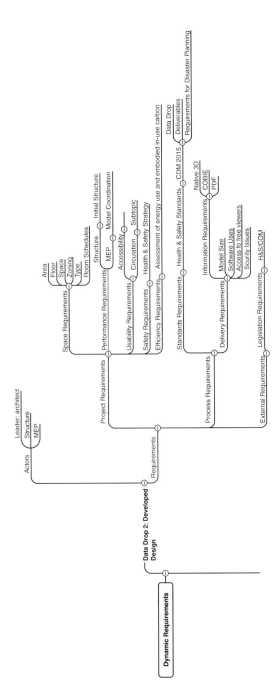

Figure 18.3 Dynamic requirements in EIR.

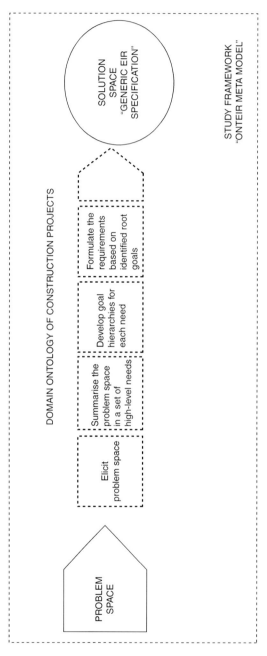

Figure 18.4 Domain ontology of construction projects.

The approach has the ability to classify and prioritise the requirements, determine the relations between them and deliver them in terms understandable for the construction team, which is the reason behind using ontology as the method for capturing and processing EIR.

The approach for reaching the generic ontology meta model for specifying EIR for construction projects includes (see Figure 18.4):

- Elicit problem space
- Summarise the problem space in a set of high-level needs
- Develop goal hierarchies for each need
- Formulate the requirements based on identified root goals

As part of the framework, a meta model for specifying EIR is reached. The meta model basically describes the components of the EIR processing model and the relation between the different classes in it; analysing, constructing, and developing the rules and constraints between them.

18.7 Concluding remarks

This ongoing and novel study of creating an ontology-based, BIM-enabled framework for defining EIRs will have potential in delivering a better defined, comprehensive and consistent EIR, on which the HBEP will build. Above all, a clearer EIR will lead to a more successful heritage project.

The success of OntEIR is due to the following characteristics:

- Stakeholder and expert contribution to the study assists in understanding the need in heritage projects, which leads to defining better requirements in OntEIR to bridge the gap between clients and execution in heritage projects.
- OntEIR is able to cover all aspects of a well and completely defined EIR as specified in PAS 1192–2, in a clear and understandable form for both the employer and the project team.
- Hierarchy and tractability offered by ontology will make it possible for OntEIR to be initiated for different types of projects.
- OntEIR is able to clearly answer the plain language questions and requirements that are to be met before moving from one stage to the next.
- OntEIR will assist employers in defining clear and adequate requirements, and at the same time will be easy for the design team to interpret and translate into building terms, and thus create better HBEP.
- OntEIR saves time in specifying requirements for projects and gives excellent results in terms of quality and consistency.

References

Barrett, P.S. and Stanley, C. (1999) *Better Construction Briefing*. Oxford: Blackwell Science.
Bruce M. and Cooper R. (2000) *Creative Product Design: A Practical Guide to Requirements Capture Management*. Chichester: Wiley & Sons, Ltd.
Chen, L. and Nugent, C. (2009) Ontology-based activity recognition in intelligent pervasive environments. *International Journal of Web Information Systems*. 5 (4), 410–430.
Dikmen, I., Birgonul, M.T. and Kiziltas, S. (2005) Strategic use of quality function deployment (QFD) in the construction industry. *Building and Environment*. 1, 245–255.
Earley, M. (2015) *BIM Level 2 Standards for Business*. Available from: http://www.bim-manager.net/.
Employer's Information Requirements Guidance Notes (2013). BIM Management for value, cost, and carbon improvement.
Graham, P. (1983) Reading the client's mind. *Building*. 30 September, 22–30.

Gruber, T.R. (1995) Toward principles for the design of ontologies used for knowledge sharing. *International Journal of Human Computer Studies*. 43 (5–6), 907–928.

Hudson, J., Gameson, R.N. and Murray, J.P. (1991) The use of computer systems to improve communication between clients and construction professionals during the briefing process. In: Barret, P. and Males, R., eds. *Practice Management, New Perspectives for the Construction Profession* (pp. 138–1143). London: Chapman & Hall.

Kamara, J. (1999) *Client Requirements Processing for Concurrent Lifecycle Design and Construction*. PhD Dissertation, University of Teesside.

Kamara J. and Anumba C. (2001). A critical appraisal of the briefing process in construction. *Journal of construction research*, World Scientific, Hong Kong. 1(2), 13–24.

Kamara, J.M., Anumba, C. and Evbuomwan, N. (2000) Establishing and processing client requirements – a key aspect of concurrent engineering in construction. *Engineering, Construction and Architectural Management*. 7 (1), 15–28.

Kamara, J.M., Anumba, C. and Evbuomwan, N. (2002) *Capturing Client Requirements in Construction Projects*. London: Thomas Telford Ltd.

Kamara J.M., Anumba C.J. and Carrillo P.M. (2001). Selection of a knowledge management strategy for organisations, Proceedings of the Second European Conference on Knowledge Management, D. Remenyi, ed., Bled, Slovenia, 8–9 November, 243–254.

Kiviniemi, A. (2005) *Requirements Management Interface to Building Product Models*. PhD Dissertation, Stanford University.

Kott, A. and Peasant, J.L. (1995) Representation and management of requirements: the RAPID-WS project. *Concurrent Engineering: Research and Applications*. 2 (2), 93–106.

Lee, C.C.T. and Egbu, C. (2005) Capturing client needs in refurbishment projects. In: Khosrowshahi, F. (Ed.), *21st Annual ARCOM Conference, 7–9 September 2005, SOAS, University of London. Association of Researchers in Construction Management*, Vol. 2, 865–74.

Leinonen and Huovila (2001), "Requirements Management tool as a Catalyst Communication", ECCE Conference, June, Espoo.

Li, M., Du, X.Y. and Wang, S. (2005) Learning ontology from relational database. In: *Proc. of the 4th International Conference on Machine Learning and Cybernetics*, Guangzhou, China.

Lima, P., Formoso, C. and Echeeveste, M. (2008) Client requirements processing in low-income house-building using visual displays and the house of quality. In: *Proceedings of 16th Annual Conference of the International Group for Lean Construction*, Manchester, UK.

Noy, F. and McGuinness, D. (2000) *Ontology Development 101: A Guide to Creating Your First Ontology*. Stanford, CA: Stanford University.

Ulrich, K.T. and Eppinger, S.D. (2008) *Product Design and Development*. 4th ed. New York: McGraw-Hill.

19 Conclusion

Yusuf Arayici, John Counsell and Lamine Mahdjoubi

In this book, we attempt to establish the field of Heritage Building Information Modelling from heritage building management, maintenance and improvement perspectives, including the meaning of conservation, understanding and managing significance, surveys and inspections, energy efficiency and sustainability, building maintenance and repair, managing buildings and project management, managing quality and appropriateness of work and consent applications.

BIM is ultimately about creating the correct and most appropriate information in a satisfactory format for it to be shared by all parties, brought together and managed. If the heritage sector is to reap some very obvious benefits of BIM, then it needs to have some independent vision that is realistic, that where appropriate looks for benefits outside the heritage sector and also looks for the means by which BIM can improve the way we deal with historic buildings.

Within the heritage context, the building or site already exists and therefore the exact condition of what lies beneath the surface cannot always be known – not without invasive investigation. HBIM, therefore, implies a degree of uncertainty. The sources of materials will not always be known, and neither will the building's original condition. The idea of the original is itself problematic: the historic building has a past; it has been occupied; it has very probably been the subject of change; and, importantly, the site will have accrued cultural value (stories, associations) which may become of greater significance than the built fabric itself. Places cherished for their event value (something once happened there) or their personality value (someone once lived there) are such examples – they have become heritage sites for reasons other than any intrinsic, measurable qualities of the site itself.

The HBIM research detailed a series of key emerging areas for further research and practice. The introductory chapter, together with Chapters 2 ("It's BIM – but not as we know it!") and 3 ("What are the goals of HBIM?"), outlines some of the challenges in defining HBIM, and the recent progress that has been made, while clearly outlining challenges yet to come. Further distinctions between BIM and HBIM are discussed in Chapter 4, "Heritage and time". Chapter 5, "From history to heritage", and Chapter 6, "Integrating value map with building information modelling approach for documenting historic buildings in Egypt", identify ways in which changes wrought over time may be identified, captured, stored and interpreted, and argue for the importance of the role of interpretation and preserving the potential meaning of a site as a key element of HBIM, as compared with its BIM equivalent. The gaps between 'parametric parts library-based' commercial BIM and HBIM are discussed. Value mapping with the digital tools and processes underpinned building information modelling to be considered in the context of heritage. Although Heritage BIM will inevitably describe any given site in exacting detail, and the resolution of digital scanning is certainly impressive, HBIM must embody an attitude of openness, interpretation and ambiguity that standard BIM does not. The model must be accessible to a range of audiences and contributors, and any mapping of a site must consider the nature of mapping itself – as well as the physical contours, both idealised and actual, there are memories and stories which can be mapped. This makes HBIM an open source and open-ended project, embodying inputs ranging from the quantifiable techniques of conservation to the vagaries of politics and identity. Heritage is, after all, a tool through which societies tell stories about themselves. Heritage Building Information Modelling, in order to reach maturity in its own right rather than position itself as a subset of BIM, must include a careful consideration of information. Chapter 6 proposes a particular process of parametric modelling as an analytical tool and a repository for the heterogeneous digital data about heritage that UNESCO claims is so important to manage and conserve. One research challenge in this area is that there are still different schools of thought about parametric modelling related to laser scanning.

Chapter 7, "Capturing heritage data with 3D laser scanner", and Chapter 8, "Evaluation of historic masonry", reveal that current BIM practice still appears predicated upon the one-way push of information from expert to participant, inadequately supporting wide interactive participation and inclusion. On the other hand, GIS practice has long enabled users to engage with data and perform their own analysis. Chapter 7 further examines detailed tools and techniques for reducing 'survey subjectivity' by means of a novel system for analysing three-dimensional data from non-destructive laser scanning and photogrammetry, for more focused and sustainable conservation treatments and repair. With the increasing use of mapping services online, web-users are becoming increasingly accustomed to performing their own analysis, but more in 2D than 3D. It may be that the step change forecast soon from the increasingly low-cost virtual reality and augmented reality technologies will at last result in widespread familiarity with 3D usage. Common standardised access to data remains a major barrier. Even at the inventory level, heritage data generally remains in silos, without common standards.

The use of laser scanning and photogrammetry can aid in recording a very high level of details in the field for heritage documentation. The HBIM process can be automated using accurate parametric objects that can be altered and mapped to heritage survey data. As a result of using HBIM, several documents such as plans, sections and elevations can be produced. The integration to GIS can provide further capabilities for linking the 3D heritage model to information systems. The integrated models help in performing efficient management and analysis that is required for maintaining important cultural heritage sites.

Current practice in surveying and the evaluation of historic masonry shows that measurement of these forms of heritage structures is made difficult by various factors including:

- The lack of regularity in this type of structure historically led to the use of ad hoc, approximate work measurement methods.
- The lack of access compounds measurement difficulties and increases survey costs. Access problems are common for these structures due in part to their complexity of evolution in building form and materials.
- The current methods of assessment rely significantly on the experience of surveyors. Surveyor variability and subjectivity in evaluation can lead to further cumulative error in assessments.
- Condition assessment requires the consideration of multiple data and information sources obtained on-site but also from third parties and from desk studies. Structuring and managing this information over time using traditional (paper-based) practices is challenging.

However, surveyors are now increasingly considering novel technologies in two areas: (1) novel reality capture technologies, in particular TLS and PG; and (2) ICTs for the structuring and management of this data, including HBIM.

Chapter 9, "HBIM applications in Egyptian heritage sites", elaborates on the HBIM dimensions through heritage case study examples and highlights that significant progress is required in this area if the above technologies are to be widely embraced as a new standard form of practice. It argues that a range of 3D technology is needed, not only to document but also to monitor the buffer zones of those historic districts which are in critical proximity to the valuable heritage areas. Valuable information could be automatically extracted from such data, including the detection of individual stones, better estimation of the amount of mortar in a linear-metre form, and the quantification of the amount of pinning required.

Similar to BIM, HBIM is seen as the way forward to resolve the information management related problems associated with heritage structures, including bill of quantities for the conservation process, time scheduling and cost evaluation. A friendly environmental approach of sustainability should be adopted when dealing with monuments as the sixth dimension (sustainability). As for the seventh and eighth dimensions, they are already considered in the form of preventive conservation.

On the other hand, Chapter 10, "Planning of sustainable bridges using building information modeling", widens the vision for HBIM for civil engineering heritage structures such as Bridge Information Modelling and management systems for existing and heritage bridges, where BIM and GIS integration would be rationalised for building inspections and queries on the heritage structures and build up arguments for knowledge storage mechanisms and analytical tools. The visualisation and defect analysis can lead to timely and accurate decision making. Research on Bridge Information Modelling and management defined an assessment methodology for the structural condition of bridges.

There is still debate about the extent to which parametric modelling, the basis of most current commercial BIM packages, can be deployed for HBIM. In the case of applying the scan-to-BIM method to Jeddah's historical buildings, Chapter 11, "Jeddah heritage building information modelling (JHBIM)", analyses the time-consuming process of applying commercial BIM software, dependent for ease of use on existing libraries of premade parametric library parts, on heritage buildings for which no such parts exist. It became evident that HBIM would provide many benefits, such as a better understanding and context of heritage buildings, knowledge of materials and also construction techniques. On the other hand, it seems there are a lot of challenges facing those applying BIM to the historical buildings in Jeddah, including organisation, technical obstacles and site issues. Another important point regarding applying BIM to the historic field is determining suitable levels of detail or the levels of development (LoD). In the case of Jeddah's historical buildings, one of the most significant points is to rebuild the past, which requires a very high level of detail. This level of detail can be reached via implementing advanced technologies such as laser scanning and photogrammetry. These technologies provide a very rich point cloud model, which can be used as a base for the HBIM model.

Chapter 12, "Algorithmic approaches to BIM modelling from reality", introduces state-of-the-art methods and approaches for automating the scan-to-BIM methods; the principles and features of algorithmic development to automatically create parametric BIM models from 3D point clouds is also explored. Region growing, brute force plane sweeps, Hough transforms, expectation maximisation and RANSAC algorithms to segment the point data are elaborated through their implementation in projects. The size of the segments is varied by changing the constants in the equations and is mostly limited to the density of the point data and the minimum size of plane expected to be detected in the scene.

Semantic feature recognition is a very interesting development. By using rudimentary information like the direction of gravity, the relationships between clusters of points can be refined to the point that building elements are recognised. However, accuracy is still a problem when greater areas are segmented and the boundary tracing suffers. By developing the ability to create an outline or wireframe structure of the geometry, the data size can be reduced significantly in the subsequent phases. The same process of linear regression can be used to isolate normals that conform to a preset threshold when dealing with 2D data.

Learning algorithms to establish standard sizes for openings are also introduced as another method for openings. The knowledge is applied in filling in gaps in the data caused by occlusions. By creating a learning set from the point data to be segmented, the algorithm has found a way to treat a specific building with generic measures and a bespoke solution for each data set is achieved.

Chapter 13, "HBIM and environmental simulation", and Chapter 14, "Green BIM in heritage building", discuss challenges of energy modelling in the background heritage building stock that is still occupied and so has an energy and carbon footprint, which is a particular challenge for a retrofit and refurbishment process that still values heritage and conserves it. They claim that a considerable reduction in energy consumption can be achieved by historic building retrofitting. However, retrofitting of historic buildings for energy efficiency is rare due to restrictions related to historic buildings' value, complexity, uncertainty of building condition, inefficient retrofitting process, shortage of data, and inconsistent simulation applications. HBIM models consist of intelligent components that have both geometric and descriptive data that facilitate performance of different types of environmental simulations, either within the BIM application or via stand-alone simulation applications. Simulations facilitate fast evaluation of retrofitting alternatives. The most challenging aspect that hinders HBIM utilisation is the impractical as-built BIM modelling process. However, new software and scanning technologies are being developed that would make the production of accurate HBIM models much easier, cheaper and faster. Another concern is simulation accuracy, which necessitates improvement of simulation applications. Interoperability is still a problem that limits integration of HBIM and environmental simulation. Furthermore, applying both HBIM and sustainable retrofit on heritage buildings in Egypt is still limited and faces a number of challenges. Among these challenges are the unavailability of equipment, limited availability of professionals, and funding and financial-related challenges.

The overall, simple automatic techniques for processing point clouds can be used in heritage buildings for creating a BIM model if the purpose is to parameterise and simplify the real geometry, according to the requirements of the application to which it is going to be submitted. The resulting 3D semantically rich model enhances the knowledge of the heritage building, complementing other representations of the facility such as point clouds or handmade HBIM. Chapter 15, "HBIM, a case study perspective for building performance", identifies some of the challenges of deploying BIM in a process of 'deep refurbishment' of heritage buildings. The case study also demonstrated considerable cost and time savings, and led to a much smoother process. It remains to be seen how effective and useful the resulting BIM database will be in achieving the better management of the facilities over the next decades, and the extent to which long-term energy efficiency can be ensured by these means, despite the likely continuing changes in technologies and usage.

There is still much research to do to optimise the addition of semantic information to point cloud data, and no clear agreement on the best way forward. Automated approaches have been tried, but yet have at best less than 90% success, or often involve gross simplification of the shape and form of the actual heritage structure, thus seen by some as denying the history of the artefact over time, including the dimensional distortions that emerge as aging and settlement take place. Nevertheless, Chapter 16, "From LiDAR data towards HBIM for energy analysis", argues for the importance of appropriate applications of automated approaches for use in energy analysis, where geometric simplifications are often used. This chapter deals with the use of laser scanning for documenting the real geometry of a heritage building and the use of automated techniques to parameterise the point cloud in order to define building façade elements according to energy analysis specifications. Although laser scanners allow the accurate and efficient 3D digitalisation of heritage, point clouds should be parameterised to create sets of components. The parameterisation required for energy analysis software is highly specific since building elements should be represented as rectangular polygons.

The wider dimensions, challenges and importance of passive and active stakeholder engagement were discussed in Chapters 17 and 18. Chapter 17, "Participatory sensing for community engagement with HBIM", shows that the analysis of such varying user needs would be a precondition for identifying different levels of appropriate access and security, together perhaps also setting a user brief for appropriate data mining and spatial analytical tools. This user brief is discussed with regard to employers in Chapter 18, "Development of OntEIR framework to support heritage clients". It was highlighted that there is a major challenge, and there is a significant skills and resource gap and that, without substantial highly active community engagement, meeting the emissions targets while also enhancing and preserving our cultural heritage and enhancing quality of life for occupants appears to be unlikely; an effective HBIM platform needs to incorporate and facilitate all the stages and aspects of participatory sensing, citizen science, and volunteered information. Chapter 17 further addressed a challenge in integrating data from different sources over time and at differing levels of accuracy and quality. This challenge is compounded when one considers how one might integrate, for example, data from the crowd-sourced open street map with government-sponsored mapping, and yet ensure that the whole was more than the sum of the parts. The potential and the challenges of this form of crowd-sourced data as an active participatory process by a wide range of stakeholders is discussed.

Index

Note: Information in figures and tables is indicated by *f* and *t*.

3ENCULT 195

access 82
accuracy 80–1
aerial photogrammetry 84
Al-Baron Palace 210
ANSYS 114, 117, 125, 126, 128
ArchiCAD 51, 136
Arkwright Building 72, 73*f*
as-built Building Information Modeling 51, 154–5, 209–10
Australia 243–4
authenticity 33–41, 36*f*, 37*f*, 39*f*, 40*f*
AutoCAD Plant 3D 157
Autodesk 123 Catch 8
Autodesk Memento 53, 56
Autodesk ReCap 138, 143
Autodesk Revit 104, 109, 112, 139, 145–6, 226
Autodesk Revit Architecture 194
automated segmentation 24
automated vectorising 156–7
AVEVA Laser Model Interface 157
Ayerst, Frances 35, 36*f*, 38

Bahla Fort 26–7
Baron Empain 106–12, 107*f*–111*f*
Bassili Pasha Villa 207
Bath Abbey 35, 36*f*, 37*f*, 38
BEM *see* Building Energy Modeling (BEM)
Bendigo project 243–4
Bentley Architecture 194, 226
BIM *see* Building Information Modeling (BIM)
boundary detection 175–7, 176*f*
boundary point extraction 175
boundary tracing 175–7, 176*f*
BREEAM 191
bridge(s): importance of 115; management system 115–16; structural integrity of 124–5
Bridge Information Modeling (BrIM): 3D Bridge Information Model in 117–18; database module in 118–22, 119*f*–124*f*; data extraction in 118–19; features 116–17, 117*f*, 118*f*; geographic information systems in 121–2, 123*f*, 124*f*; structural advanced analysis in 125; structural condition assessment in 124–6, 126*f*–128*f*; structural health in 116–17; visualization in 117, 119–21, 120*f*, 121*f*
British Standard BS 7913 6–7, 12–13
BS 7913 6–7, 12–13
Building Energy Modeling (BEM) 208; for existing buildings 224
Building Information Modeling (BIM) 1; as-built 51, 154–5, 209–10; environmental simulation in 197–200; Heritage Building Information Management *vs.* 35–7, 43*t*; maturity levels 7–8
bundle adjustment 163–4
Burra Charter 18

C# 118, 125
CAD *see* computer aided design (CAD)
CADSoft Envisioneer 194
Cairo 106–12, 107*f*–111*f*, 210
calibration, in environmental simulation 198
carbon dioxide 190
CDMICA *see* Cultural Diversity and Material Imagination in Canadian Architecture (CDMICA)
China 207
cleaning, data 138–9
ClearEdge 3D 157
ClientPro 259–60
Client Requirements Processing Model (CRPM) 259
climate change 190, 242
clustering approach, for outlier detection 169
clutter removal 181–2
color 70, 160
color coding, of model elements 103–4
community engagement 244–5
complexity, of historic buildings, as obstacle to retrofitting 192, 199
computer aided design (CAD) 1, 2, 3
concave polygon fitting 177a

conservation 136, 205–6, 208–10, 243–4
conservation, repair and maintenance (CRM) 16
contrast 70
convex polygon fitting 177
"cookie cutting" 24
coordination 104
Coria Cathedral 83
Craigmillar Castle 89, 90*f*
CRM *see* conservation, repair and maintenance (CRM)
CRPM *see* Client Requirements Processing Model (CRPM)
Cultural Diversity and Material Imagination in Canadian Architecture (CDMICA) 28
cultural landscapes 17

data acquisition 137–9, 137*f*, 195–6
DataCAD 194
data cleaning 138–9
data-driven registration 161
data formatting 160
data processing 138–9, 144, 168–71, 168*t*
data structuring 86–7
Dayr al-BaḤrī 39, 39*f*, 40*f*, 41
Declaration of San Antonio 34
"deep renovation" 244
DEM *see* digital elevation map (DEM)
density-based approach, for outlier detection 169
depth-based approach, for outlier detection 169
DHW *see* domestic hot water production (DHW)
digital divide 245–6
digital elevation map (DEM) 83
distance-based approach, for outlier detection 169
distribution-based approach, for outlier detection 169
districts, heritage 104–5, 134–5
documentation 104–6
domestic hot water production (DHW) 219
drones 73*f*, 74, 84
Dublin 136
Ducal Palace 224, 230, 233–9, 234*f*–238*f*, 234*t*
DURAARK 20, 24, 26, 28, 225, 226–7

EBIM *see* Existing Building Information Modeling (EBIM)
edge extraction 175
Egypt *see* Baron Empain; Bassili Pasha Villa; Dayr al-BaḤrī; Hatshepsut
Egyptian context 204–6, 210
Egyptian Society of Political Science, Statistics, and Legislation 47–9, 47*f*
EIR *see* Employer Information Requirements (EIR)
El-Hakim, Sabry 135
Employer Information Requirements (EIR): documents 257–8; impact of 258*f*; importance of 257; methodology 261; ontology 261–5, 263*f*, 265*f*; purpose of 260–1
energy efficiency 13
energy performance 218–19; *see also* environmental performance

energy retrofitting 190–1
energy simulation 193–7
engagement: community 244–5; forecasting 247; participatory sensing and 249; social media and 249–50; widening 247–8, 248*f*
English Heritage (agency) 220–1, 250
Enquiry by Design 246
environmental performance: complexity of buildings and 192; inconsistent simulations and 193; obstacles to retrofitting in 192–3; overview of 191; and shortage of data 193; uncertainty with buildings and 192; and value of historic buildings 192; *see also* Building Energy Modeling (BEM); energy performance; Green Building Information Modeling
environmental simulation 197–200
environmental sustainability 13, 76–7
Esslingen 184–5
estimation methodology, for classification models 165–6
European Structural Fund 244
Existing Building Information Modeling (EBIM) 6, 8–12
explicit data, implicit *vs.* 155–6

façade segmentation 228, 228*f*, 229*f*
FARO SCENE 138
feedback 251–3, 253*f*
FEM *see* Finite Element Method (FEM)
Finite Element Method (FEM) 83
fly-through films 70–1

galleting 93
gbXML *see* Green Building XML (gbXML)
GDL *see* geometric descriptive language (GDL)
Gehry Technologies Digital Project Designer 194
Generalized Hough Transform (GHT) 229, 231*f*; *see also* Hough transform
Geo-BIM 18
Geographic Information System (GIS) 3; in Bridge Information Modeling 116, 118*f*, 121–2, 123*f*, 124*f*; cookie cutters in 24; criticality of 18; in data collection 105; parametric objects and 23
geolocation 67
geometric descriptive language (GDL) 51, 115
geometric modeling 178–82, 180*f*
geometric reconstruction 178
geometry size fitting 178
georeferencing 162
Getty Conservation Institute 1
GHT *see* Generalized Hough Transform (GHT)
GIS *see* Geographic Information System (GIS)
global warming 190
Gone with the Wind (film) 34–5
Google Earth 117, 121, 122
Google SketchUp 117, 122, 139
Gothic architecture 45–7
Government Service Administration (GSA) 9–10
GPS *see* Geographic Information System (GIS)
Graphisoft ArchiCAD 226

Green Building Information Modeling: Building Energy Management in 208; retrofitting in 206–8
Green Building XML (gbXML) 178, 194, 197, 199, 225, 232f, 233, 238–9, 238f
Green Deal 244
greenhouse gases 190
ground-penetrating radar (GPR) 85–6
groups, of buildings 104–5
GSA *see* Government Service Administration (GSA)
Guimarães 224, 230, 233–9, 234f–238f, 234t

Harmondsworth 85
Harris, E. Vincent 220
Hatshepsut 39, 39f, 40f
HBIM *see* Heritage Building Information Modeling (HBIM)
Heliopolis 106–12, 107f–111f
heritage: authenticity and 33–41, 36f, 37f, 39f, 40f; intangible 17–18
Heritage BIM Execution Plan (HBEP) 257
Heritage Building Information Modeling (HBIM): 3D content model to 51–2; Building Information Management *vs.* 35–7, 43t; challenges with 136–7; in context 2; coordination 104; defined 50; energy simulation and 193–7; Existing Building Information Modeling and 8–12; explaining 102–3; focus of 16–19; generation stages 55f; history of 2–3; as information source 15; motivating use of, in communities 248; reasons for use of 103–4; as term 1
Heritage Cottage 9, 10–11
heritage districts 104–5, 134–5
Hijazi Architectural Objects Library (HAOL) 141–2, 144
holdout 166
hole-based window extraction 179, 180f
Hong Kong 207
Hough transform 164, 229, 231f

ICOMOS *see* International Council on Monuments and Sites (ICOMOS)
ICP *see* iterative closest point (ICP) method
ICT *see* information and communication technology (ICT)
IDEA 194
image survey campaign 53, 55–6, 56
IMAGINiT 197
implicit data, explicit *vs.* 155–6
information and communication technology (ICT) 1, 19
information loss 210–11
information resource, HBIM as 15
infrared cameras 85–6
inspections 13
intangible heritage 17–18
Integrated Project Delivery (IPD) 16
Intergraph Smart 3d for Plants 157
International Council on Monuments and Sites (ICOMOS) 13, 18, 243

Internet penetration 245–6
interoperability 199, 211
interpretation: primary 16–17; secondary 18
inventory 20–1
IPD *see* Integrated Project Delivery (IPD)
iterative closest point (ICP) method 161

James I 38
Jeddah 133–5, 134f
Jeddah Historical Building Information Modeling approach (JHBIM) 52, 139–49, 140f–151f
Jefferson, Richard 38
Jokkilehto, Jukka 34
Jones, Inigo 38
JP Morgan Bank Headquarters (London) 2

Karnak 41, 41f, 42f
Kubit PointSense Plant 157

landscapes, cultural 17
Lascaux Caves 2
laser scanners 61, 62f, 63–74, 66f, 68–70, 73f, 83, 90f, 137–8, 137f, 142–4, 195–6
Lassus, Jean-Baptiste-Antoine 45, 46
least squares hitting 177; *see also* total least squares
Le Corbusier 38
LEED 191
Leica CloudWorx 138, 157
Leica Cyclone 139
level of detail (LoD) 225
Levenberg-Marquardt algorithm 164
LiDAR data 24, 26–7, 83–5, 150f, 158
Linning, Chris 15
LoD *see* level of detail (LoD)
loss of information 210–11
Lowenthal, David 34
Lui Seng Chun 207

maintenance 13, 76
Manchester City Hall 220–2, *221*, *222*
Manshiet Nasser 246
masonry: access to 82; aerial photogrammetry with 84; biological activity in 78t; common defects in 78–9, 78t; contaminated 78t; costing with 81–2; cracks 78t; decision triggering for intervention with 79–80; erosion 78t; ground-penetrating radar with 85–6; hand-measured drawings for 79; laser scanners with 83, 84–5, 90f; measurement of 81–2; mortar joints in 90–1, 93, 94f, 95f; photogrammetry with 83–4, 84–5; pricing with 81–2; progressive survey techniques for 82–8; reality capture with 82–3; segmentation with 89, 91f, 91f–93f; structural movement of 78t; survey protocols for 79; thermographic cameras with 85–6; visual surveys of 79
MATLAB 251
mesh modeling 56, 57f, 58, 58f
minimum bounding rectangle 177
monitoring 104–6

monuments 106; *see also* Baron Empain
mortar joints 90–1, 93, 94*f*, 95*f*
MS Excel 117, 119
Mugamaa Complex 210
MultiView Stereo Software 85

Nara Document on Authenticity 34
Nasif Historical House 141, 141*f*, 147, 148*f*, 149, 150*f*
National Building Specifications (NBS) 260
National Museum of Welsh Life 26–7
National Organization of Urban Harmony (Egypt) 205
Navisworks 25, 104, 128, 226
Navisworks Manage 117, 119, 121
Nawwar, Sami 134–5
NBS *see* National Building Specifications (NBS)
Nefertari 102
Nemetschek Allplan Architecture 194
Nemetschek Graphisoft ArchiCAD 194
Nemetschek Vectorworks Architect 194
New Lanark 83
New Rules of Measurement (NRM) 81–2
New Zealand 243
noise, in mesh modeling 56, 57*f*, 58*f*
Notre-Dame de Paris 45–6
NRM *see* New Rules of Measurement (NRM)

Oasis World Heritage Site 26–7
occlusion labeling 181
occlusion reconstruction 182
occupation marks 35
ontology 261–2, 266*f*
opening detection 181–2
opening reconstruction 229–33, 230*f*–232*f*
orientation, in geometric feature recognition 179
outlier detection 169
outlier removal 169–70
outline generation 177

parametric data 155
parametric data compiling 156
parametric objects 23–4
participatory sensing 249
patch classification 174, 185–6, 186*t*
patch detection 174
patch intersection 181–2
photogrammetry 3, 51, 66, 83–4, 84–5, 106, 158*f*, 159*f*
pinning 93
Planning 4 Real 246
Plunkett, Walter 35
point cloud classification 165
point cloud data 24, 51, 69*f*, 156–7
point cloud manipulation 157–66, 158*f*, 159*f*, 163*f*
point cloud models 58–9, 58*f*, 59*f*
point cloud registration 139
point density, in geometric feature recognition 179, 180*f*
polygon fitting 177

Polyworks 138
Portugal *see* Ducal Palace
position, in geometric feature recognition 179
primary interpretation 16–17
processing, data 138–9, 144, 168–71
project management 13
Project Tango 250

Quality Function Development (QFD) 259, 260
Quebec City Declaration on the Preservation of the Spirit of Place 34

radar 85–6
Ramses II 102
Rani ki Vav 83
RANSAC 164, 171–2
Rapid Energy Modeling (REM) 200
reality capture 82–3
real-time data 21–2
Recording, Documentation & Information Management (RecorDIM) Initiative 1
rectangle, minimum building 177
region growing 162–3, 163*f*, 171, 175*f*
registration 161–2
REM *see* Rapid Energy Modeling (REM)
Remal Foundation 246
Remondino, Fabio 135
repair 13
restoration 45, 48
retrofitting 190–1, 206–8; *see also* environmental performance
Revit Family 104, 145–6
RhinoBIM (BETA) 194
Rhinoceros 139, 145
Rietveld, Gerrit 38
RiSCAN Pro 138
Rock Art 250
Rufford Abbey 61, 62*f*

scale 104–6
scanners, laser 61, 62*f*, 63–74, 66*f*, 68–70, 73*f*, 83, 84–5, 90*f*, 137–8, 137*f*, 142–4, 195–6
Scarlet O'Hara's dress 34–5
Schröderhuis 38
Scott, George Gilbert 35, 38
secondary interpretation 18
Second Life 251
segmentation 196; algorithms 162–5, 163*f*; automated 24; challenges with 162; façade 228, 228*f*, 229*f*; parameters 173–4, 174*f*; region growing 171; semantic 87–8; stone/mortar 89, 91*f*, 91*f*–93*f*
semantic feature recognition 178–9
semantic information 22–6
semantic labeling 87–8
semantic richness 209
semantic segmentation 87–8
sensor-driven registration 161

shapefiles 122
shape recognition 171–4, 171f–174f
simple split 166
size, in geometric feature recognition 179
Smithson, Robert 42
SMM *see* Standard Method of Measurement (SMM)
social media 249–50
Softtech Spirit 194
SOH *see* Sydney Opera House (SOH)
Solibri 104
Standard Method of Measurement (SMM) 81
Stonehenge 38
stratigraphic units 24
structural advanced analysis 125
St. Teilo's Church 26
subjectivity 80–1
surface recognition 171–4, 171f–174f
surveys 13
sustainability 13, 76–7, 242–4; *see also* Building Energy Modeling (BEM); energy performance; environmental performance; Green Building Information Modeling
Sutherland, Ivan 2
Sydney Opera House (SOH) 15, 22

Tekla BIMsight 194
Tekla Structures 117, 119, 122, 126, 128
tensor voting 169–70
terrestrial laser scanning (TLS) 83, 84–5, 137–8, 137f, 142–4, 195–6; *see also* laser scanners
thermographic imaging 85–6
TLS *see* terrestrial laser scanning (TLS)
topology, in geometric feature recognition 179
Toronto 135–6

total least squares 165; *see also* least squares hitting
Trimble RealWorks 157
Trimble SketchUp 8
TruView 109
Tschumi, Bernard 43n1

UAV *see* unmanned aerial vehicles (UAV)
UNESCO 3, 17, 38, 102, 133
unmanned aerial vehicles (UAV) 73f, 74, 84
use marks 35
utility 80–1

Valhalla project 25–7, 26f, 247–8
value: conservation and 243–4; of historic buildings, as obstacle to retrofitting 192
value map 52, 53f, 54f
values, embedded in heritage buildings 204–5
VEPs3D 251
Villa Savoye 38
Viollet-le-Duc, Eugène-Emmanuel: Egyptian Society of Political Science, Statistics, and Legislation and 47–9, 47f; Gothic architecture and 46–7; Notre Dame and 45–6; on truth in architecture 45
Visual Basic 118
visualisation 86–7
Visual Structure for Motion (VSfM) 85
Vlaardingen 184
voxelisation 160, 170–1

Walton Basin 20
Warsaw 37–8
Windsor Castle 42, 44n3

XML *see* Green Building XML (gbXML)